Transformations and Projections in
Computer Graphics

David Salomon

Transformations and Projections in Computer Graphics

 Springer

Professor David Salomon (emeritus)
Computer Science Department
California State University
Northridge, CA 91330-8281
USA
Email: david.salomon@csun.edu

Cover illustration: Adapted from figure 4.25, courtesy of Ari Salomon.

British Library Cataloguing in Publication Data
A catalogue record for this book is available from the British Library

Library of Congress Control Number: 2006923906

ISBN-10: 1-84628-392-2
ISBN-13: 978-1-84628-392-5

Printed on acid-free paper

Printed in the United States of America   (EB)

9 8 7 6 5 4 3 2 1

springer.com

Dedicated to Dick Termes, whose work and talent have contributed much to the quality of this book.

If there's a book you really want to read, but it hasn't been written yet, then you must write it.

—Toni Morrison

Preface

It is probably a coincidence that the three main terms discussed in this book, namely *transformations*, *projections*, and *perspective*, are ambiguous. Here is what the dictionary has to say about these terms.

Transformation

- (a) The act or an instance of transforming. (b) The state of being transformed.

- A marked change, as in appearance or character, usually for the better.

- Mathematical transformation. (a) Replacing a variable in an expression by its value. (b) Mapping a mathematical space onto another or onto itself.

- In geometry. Moving, rotating, reflecting, or otherwise systematically deforming a geometric figure (discussed in this book).

- In linguistics. (a) A rule to convert a syntactic form into another. (b) A sentence or sentential form derived by such a rule; a transform.

- In genetics. (a) The change undergone by a cell upon infection by a cancer-causing virus. (b) The alteration of a bacterial cell caused by the transfer of DNA from another bacterial cell, especially a pathogen.

Projection

- The act of projecting or the condition of being projected.

- (a) An object or part thereof that extends outward. (b) Spiky projections on top of a fence. (c) A projection of land along the coast.

- A prediction or an estimate of a future situation, based on current data or trends.

- (a) The process of projecting a recorded image onto a viewing surface. (b) An image so projected.

- In mathematics. The image of an n-dimensional geometric figure reproduced in $n-1$ or fewer dimensions. The most common case is for $n = 3$ (discussed in this book).

- In psychology. The attribution of one's own beliefs or suppositions to others (such as when a scientist projects his beliefs into the subjects of his research or into theories he develops).

Perspective

- (a) A view or scene. (b) A mental view or outlook.

- The appearance of objects in depth as perceived by normal binocular vision.

- (a) The relationship of aspects of a subject to each other and to a whole, "let's put this into perspective." (b) Subjective evaluation of relative significance, "in my perspective as an electrician, this wire is defective." (c) The ability to perceive things in their actual interrelations or comparative importance "in perspective, this flood is minor."

- The technique of representing three-dimensional objects and depth relationships on a two-dimensional surface (discussed in this book).

- (Adjective) Of, relating to, seen, or represented in perspective.

> This is why writing is such a liberating thing. You get to know what you didn't know you knew.
>
> —Richard Lederer

There is no question that computer graphics has become an important field that pervades our lives in many areas. Many advertisements on television and in magazines are graphical and are created on computers. The screens of computers, PDAs, cellular telephones, and similar devices interact graphically with the user. More and more full-length feature films are being created entirely by computers. Graphics software enables users to draw engineering plans, to create technical and artistic illustrations, and to develop fonts of text. (For a short history of computer graphics, see [hocg 06].)

Computer graphics is an immense discipline, encompassing many fields, but this book concentrates on the three key terms mentioned above. Following is a short discussion of each term.

The term "transformation" as discussed in this book refers to a geometric operation applied to all the points of an object. An object may be moved, rotated, or scaled (shrunk or stretched). It may be reflected about a plane (as in a mirror) or deformed in some way, as illustrated by Figures Intro.1 and 1.1. Several transformations may be combined and may completely change the position, orientation, and shape of the object. Many graphics operations are greatly simplified with the help of transformations. A forest can be created from a single tree by duplicating the tree several times and moving and transforming each copy differently. An object can be animated by moving it along a path in small steps while also rotating and scaling it slightly at each step. Transformations, both two-dimensional and three-dimensional, are discussed in Chapter 1.

Currently, virtually all our graphics output devices are two dimensional, but many graphics projects and objects are three-dimensional. Converting a three-dimensional graphics object or scene into two dimensions is a mathematical operation called *projection*. In general, a projection transforms an object from n dimensions to $n-1$ or fewer dimensions, but in computer graphics n is always 3. Because of the loss of dimensions, an object loses some of its details when projected. It is therefore important to study the various types of projections and always use the right one. Chapters 2 through 4 describe the three main classes of projections: parallel, perspective, and nonlinear.

◇ **Exercise Pre.1:** Discuss the impossible fork of Figure Pre.1.

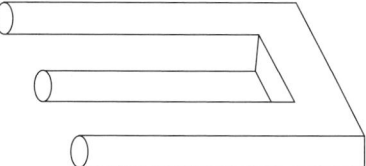

Figure Pre.1: An Impossible Fork.

Perspective (or more accurately, *linear perspective*) is the general name of several techniques that create the illusion of depth in a two-dimensional drawing. The rules of perspective determine where and how to place objects in a painting or drawing so that they appear to have depth and seem to be at the correct distance from the observer. A picture in perspective creates in the viewer's brain the same sensation as the original three-dimensional scene. The main tool employed by linear perspective is vanishing points. Perspective, including its history, its use in art, its applications to computer graphics, and its mathematical representation, is the topic of Chapter 3.

Following is a short description of the chapters and appendices of the book.

■ Chapter 1 introduces geometric transformations. Both two-dimensional and three-dimensional transformations are included, and it is shown that the latter are more plentiful and more complex than the former and are also more difficult to specify. A good example is rotations. In two dimensions there are only two directions, clockwise and counterclockwise, for a rotation and rotations are performed about a point. In three dimensions, rotations are about an axis and the terms clockwise and counterclockwise are ambiguous.

Fortunately, all the important two-dimensional transformations can be specified by a 3×3 transformation matrix, and this matrix is easy to extend to the three-dimensional case, where all the important transformations can be specified by means of a 4×4 matrix. Thus, the use of a transformation matrix is elegant and leads to a deep understanding of transformations.

Other topics discussed in this chapter are (1) the use of homogeneous coordinates, (2) combinations of transformations, such as a rotation followed by a reflection, and (3) transforming the coordinate system instead of the object.

■ The remainder of the book is devoted to projections, and Chapter 2 introduces parallel projections. These are used mostly in engineering drafting but can also be found in Eastern art. There are three classes of parallel projections: orthographic, axonometric, and oblique (although it is shown at the end of this chapter that the last two types are similar). An orthographic projection displays one side or one face of the object. The downside of this type is that three projections are needed in order to see the entire object. On the other hand, it is easy to compute dimensions of object details from measurements made on the projection.

Axonometric projections normally show three sides of the object. Thus, a single projection shows more of the object, but it is more difficult to compute dimensions of parts of the object because each face of the object may be shrunk by a different factor when drawn in the projection.

Oblique projections are similar to axonometric projections and employ certain projection angles in order to simplify the process of measuring and computing dimensions.

■   Perspective projections are the topic of Chapter 3. The chapter starts with an intuitive explanation of the important concept of vanishing points. It follows with a short history of perspective, its origins, and its applications to art. The short but important Section 3.3 is devoted to perspective projection in curved objects, a topic that is neglected by most texts on perspective. The bulk of the chapter develops the mathematics of perspective in a systematic way, approaching this topic from several points of view and illustrating it with examples. The chapter ends with a long presentation of stereoscopic images, an important application of perspective.

■   Chapter 4 treats the important (and alas neglected) topic of nonlinear projections. The most important nonlinear projections are the fisheye projection (Section 4.2), the panoramic projection (Section 4.4), and the many sphere projections (Section 4.14). In addition, this chapter includes material and examples on circle inversion (Section 4.3), six-point perspective (Section 4.8), panoramic cameras (Section 4.10), telescopic and microscopic projections (Sections 4.11 and 4.12), and anamorphosis (Section 4.13).

■   Appendix A, on vector products, and Appendix B, on quaternions, provide information on these mathematical topics that may be unfamiliar to some readers. Finally, Appendix C consists of color figures.

> The heart of mathematics consists of concrete examples and concrete problems.
> —Paul Halmos, *How to Write Mathematics* (1973).

I have collected and developed the material in this book over many years of studying and teaching. Some of it has been published in [Salomon 99] and in various class notes, but most of it is seeing the light of day for the first time in this book. I hope the readers will find the presentation clear and unambiguous and will immediately bring any errors, omissions, and misprints to my attention.

> I cannot tell my learned reader (whose eyebrows, I suspect, have by now traveled all the way to the back of his bald head), I cannot tell him how the knowledge came to me.
> —Vladimir Nabokov, *Lolita* (1955).

Readership of the Book

This book is aimed mostly at mathematically mature readers (i.e., those who can deal comfortably with mathematical abstractions), who are familiar with computers and computer graphics and are looking for a mathematically easy presentation of the transformations and projections used in computer graphics. The material presented here requires no previous knowledge of transformations, projections, or perspective. The key ideas are introduced slowly, are examined, whenever possible, from several points of view, and are illustrated by figures, examples, and (solved) exercises. The discussion must involve some mathematics, but it is nonrigorous and therefore easy to grasp. The mathematical background required is the basics of linear algebra, mostly vectors, vector operations, and matrices. The following features enhance the usefulness of the book:

■   The book has many figures. It is my belief that a book on aspects of graphics should have figures to illustrate the concepts under discussion. Drawings, paintings, and photographs are included. Most color figures have been printed in place in grayscale. All of them appear in color in Appendix C.

■   Many exercises are sprinkled throughout the text. These are important and should be worked out. The answers are also provided, but should be consulted only to verify the reader's own answer, or as a last resort.

> Learn from other people's mistakes. Life isn't long enough to make them all yourself.
> —Harry S. Truman

Books and Internet resources for transformations and projections.

Godel, Escher, Bach: An Eternal Golden Braid, by Douglas Hofstadter. Basic Books, 20th Anniversary edition, 1999. This classical volume discusses symmetries in art, literature, and science.

Transformation Geometry: An Introduction to Symmetry, by George E. Martin. Springer-Verlag, 1982. An excellent mathematical reference.

Symmetry Discovered: Concepts and Applications in Nature and Science, by Joe Rosen. Dover Press, 1975. An accessible introduction to the ideas of symmetry.

The New Ambidextrous Universe, by Martin Gardner. W. H. Freeman and Company, 1990. A beautifully written exploration of symmetry.

Symmetry, by Hermann Weyl. Princeton University Press, 1952. A classic illustrated introduction to symmetry.

The Renaissance and Rediscovery of Linear Perspective, by Samuel Y. Edgerton. Harper and Row, 1976 (especially chapters 9 and 18).

Secret Knowledge: Rediscovering the Lost Techniques of the Old Masters, by David Hockney. Viking, 2001.

The Science of Art, Optical Themes in Western Art From Brunelleschi to Seurat, by Martin Kemp. Yale University Press, 1990.

The Life of Brunelleschi, by Antonio Tuccio Manetti, edited by Howard Saalman. Penn State University, 1970.

Geometry: An Investigative Approach, and Laboratory Investigations in Geometry, by Phares G. O'daffer and Stanley R. Clemens. Addison-Wesley, 1976.

Reference [Wolfram 06] has information, examples of, and code to create many panoramic projections and map projections.

Reference [handprint 06] has a detailed discussion titled "Elements of Perspective."

Currently, the book's Web site is part of the author's Web site, which is located at `http://www.ecs.csun.edu/~dsalomon/`. Domain name `DavidSalomon.name` has been reserved and will always point to any future location of the Web site. The author's email address is `dsalomon@csun.edu`, but any email sent to email address ⟨*anyname*⟩`@DavidSalomon.name` will reach the author.

This book is dedicated to Dick Termes whose work and talent have contributed much to the quality of the book. The many images by Dick that are included in the book serve to illustrate important concepts. In addition, I would like to thank Ari Salomon for Figure 4.25 and Professor Shinji Araya, Fukuoka Institute of Technology, for Figure 4.32.

Lakeside, California   David Salomon

> The university as a step to anything but ordination seemed,
> to this man of fixed ideas, a preface without a volume.
> —Thomas Hardy, *Tess of the d'Urbervilles* (1891)

Contents

The contents, as in part I understand them, are to blame.

—William Shakespeare, *King Lear* (act I, scene II).

Introduction

The 1960s were the golden age of computer graphics. This was the time when many of the basic methods, algorithms, and techniques were developed, improved, and implemented. Two of the most important concepts that were identified and studied in those years were transformations and projections. Workers in the graphics field immediately recognized the importance of transformations. Once a graphical object is created, the use of transformations enables the designer to create copies of the object and modify them in significant ways. The necessity of projections was also realized early. Sophisticated graphics requires three-dimensional objects, but graphics output devices are two-dimensional. A three-dimensional object has to be projected on the flat output device in a way that will preserve its depth information. Thus, early researchers in computer graphics developed the mathematics of parallel and perspective projections and implemented these techniques. Nonlinear projections deform the projected image in various ways and are mostly used for artistic and ornamental purposes. These projections were also studied and implemented over the years by many people.

⋄ **Exercise Intro.1:** Most nonlinear projections are valued for their artistic and ornamental effects, but there is at least one type of nonlinear projection that has important applications. What is it?

Today, transformations and projections are important components of computer graphics and computer-aided design (CAD). Transformations save the designer work and time, while projections are necessary because three-dimensional output devices are still rare (but see [deeplight 06] for a new, revolutionary technique for three-dimensional displays) hence this book.

Figure Intro.1 shows the power of even the simplest two-dimensional transformations. It illustrates, from left to right, the following transformations: rotation, reflection, deformation (shearing), and scaling (see also Figure 1.1). It is not difficult to imagine the power of combining these transformations, but it is more difficult to imagine and visualize the power and flexibility of three-dimensional transformations.

The basic two-dimensional transformations are translation, rotation, reflection, scaling, and shearing. They are simple, but it is their combinations that make them powerful. It comes as a surprise to realize that these transformations can be specified

Figure Intro.1: Elementary Two-Dimensional Transformations.

by means of a single 3×3 matrix where only six of the nine elements are used. The same five basic transformations also exist in three dimensions, but have more degrees of freedom and therefore require more parameters to fully specify them. The general transformation matrix in three dimensions is 4×4, where 13 of the 16 elements control the transformations and 3 are used to specify the orientation of the projection plane in the case of perspective projections.

◇ **Exercise Intro.2:** What transformations are possible in one dimension?

In contrast with the five basic transformations, there are more than five types of projections. As Figure Intro.2 illustrates, we distinguish between linear and nonlinear projections. The former class consists of parallel and perspective projections, while the latter class includes many different types. Each type of projection has variants. Thus, parallel projections are classified into orthographic, axonometric, and oblique, while perspective projections include one-, two-, and three-point projections.

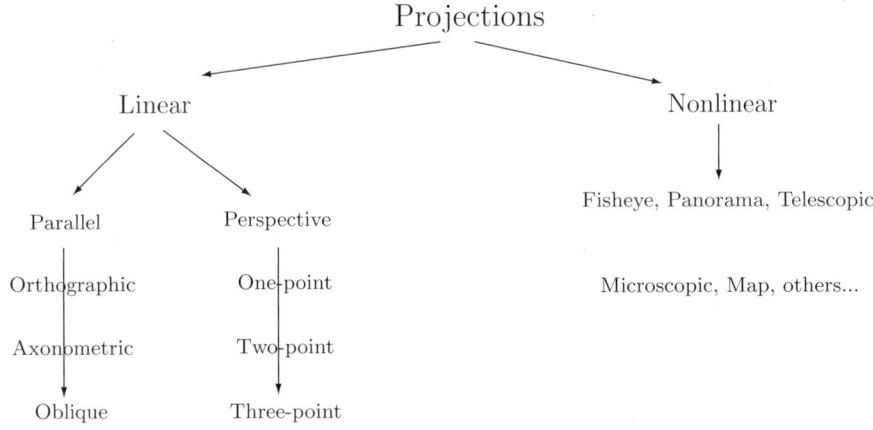

Figure Intro.2: Classification of Projections.

Nonlinear projections are all different and employ different approaches and ideas. Linear projections, on the other hand, are all based on the following simple rule of projection.

Rule. A three-dimensional object is projected on a two-dimensional plane called the projection plane. The object must be fully located on one side of the plane, and we imagine a viewer or an observer located on the other side. On that side, we select a point termed the *center of projection*, and it is the location of this point that determines the class of linear projection, parallel or perspective. A three-dimensional point **P** on the object is projected to a two-dimensional point **P*** on the projection plane by connecting **P** to the center of projection with a straight segment. Point **P*** is placed at the intersection of this segment with the projection plane. When the center of projection is at infinity, the result is a parallel projection. If the center of projection is at the observer, the projection is perspective.

dictionary definition of projection: the representation of a figure or solid on a plane as it would look from a particular direction.

Emma was not sorry to be pressed. She read, and was surprised. The style of the letter was much above her expectation. There were not merely no grammatical errors, but as a composition it would not have disgraced a gentleman; the language, though plain, was strong and unaffected, and the sentiments it conveyed very much to the credit of the writer. It was short, but expressed good sense, warm attachment, liberality, propriety, even delicacy of feeling.

—Jane Austen, *Emma* (1816)

1
Transformations

When working on computer graphics projects, we discover very quickly that transformations are an important part of the process of building an image. If an image has two identical (or even similar) parts, such as wheels, only one part need be constructed from scratch. The other ones can be obtained by copying the first and then moving, reflecting, and rotating it to bring it to the right shape, size, position, and orientation. Often, we want to zoom in on a small part of an image so more detail can be seen. Sometimes it is useful to zoom out, so a large image can be seen in its entirety on the screen, even though no details can then be discerned. Operations such as moving, rotating, reflecting, or scaling an image are called *geometric transformations* and are discussed in this chapter for two and three dimensions.

1.1 Introduction

Mathematically, a geometric transformation is a function f whose domain and range are points. We denote by $\mathbf{P}$ a general point before any transformation and by $\mathbf{P}^*$ the same point after a transformation. The notation $\mathbf{P}^* = f(\mathbf{P})$ implies that the transformed point $\mathbf{P}^*$ is obtained by applying f to $\mathbf{P}$. We call our transformations *geometric* because they have geometric interpretations. Thus, only certain functions f can be used. Years of study and practical experience have shown that in order for it to be meaningful as a geometric transformation, a function must satisfy two conditions: it has to be *onto* and *one-to-one*.

■ A general function f maps its domain D into its range R. If every point in R has a corresponding point in D, then the function maps its domain *onto* its range. An example is $f(x) = \lfloor x \rfloor$, which maps the real numbers onto the integers. Every integer has a real number (in fact, infinitely many real numbers) that map to it. Another example is $g(x) = 1/x$, a mapping from the real numbers into the real numbers. This mapping is not onto because no real number maps to zero. Requiring a transformation to be onto makes sense since it guarantees that there will not be any special points $\mathbf{P}^*$ that cannot be reached by the transformation.

■   An arbitrary function may map two distinct points x and y into the same point. Function $f(x)$ above maps the two distinct numbers 9.2 and 9.9 into the integer 9. A *one-to-one* function satisfies $x \neq y \rightarrow f(x) \neq f(y)$. Function $g(x)$ above is one-to-one. Requiring a transformation to be one-to-one makes sense because it implies that a given point $\mathbf{P}^*$ is the transformed image of one point only, thereby making it possible to reconstruct the inverse transformation.

Definition. A geometric transformation is a function that is both onto and one-to-one, and whose range and domain are points.

◇ **Exercise 1.1:** Do either of the two real functions $f_1(x, y) = (x^2, y)$ and $f_2(x, y) = (x^3, y)$ satisfy the definition above?

There are two ways to look at geometric transformations. We can interpret them as either moving the points to new locations or as moving the entire coordinate system while leaving the points alone. The latter interpretation is discussed in Section 1.5, but the reader should realize that whatever interpretation is used, the movement caused by a geometric transformation is *instantaneous*. We should not think of a point as moving along a path from its original location to a new location, but rather as being grabbed and immediately planted in its new location.

> The description of right lines and circles, upon which geometry is founded, belongs to mechanics. Geometry does not teach us to draw these lines, but requires them to be drawn.
>
> —Isaac Newton (1687)

Combining transformations is an important operation that is discussed in detail in Section 1.2.2. This paragraph intends to make it clear that such a combination (sometimes called a *product*) amounts to a *composition* of functions. If functions f and g represent two transformations, then the composition $g \circ f$ represents the product of the two transformations. Such a composition is often written as $\mathbf{P}^* = g(f(\mathbf{P}))$. It can be shown that combining transformations is associative (i.e., $g \circ (f \circ h) = (g \circ f) \circ h$). This fact, together with a few other basic properties of transformations, makes it possible to identify *groups* of transformations. A discussion of mathematical groups is beyond the scope of this book but can be found in many texts on linear algebra. A set of transformations constitutes a group if it includes the identity transformation, if it is closed, and if every transformation in the set has an inverse that is also included in the set.

An example of a group of transformations is the set of two-dimensional rotations about the origin through angles of 0° and 180°. This two-element set is a group since a zero-degree rotation is an identity transformation and since a 180° rotation is the inverse of itself.

⋄ **Exercise 1.2:** Is the operation of combining transformations commutative?

Another important example of a group of transformations is the set of *linear transformations* that map a point $\mathbf{P} = (x, y, z)$ to a point $\mathbf{P}^* = (x^*, y^*, z^*)$, where

$$
\begin{aligned}
x^* &= a_{11}x + a_{12}y + a_{13}z + a_{14}, \\
y^* &= a_{21}x + a_{22}y + a_{23}z + a_{24}, \\
z^* &= a_{31}x + a_{32}y + a_{33}z + a_{34}.
\end{aligned}
\tag{1.1}
$$

Each new coordinate depends on all three original coordinates, and the dependence is linear. Such transformations are called *affine* and are defined more rigorously on page 22.

A little thinking shows that the coefficients a_{i4} of Equation (1.1) represent quantities that are added to the transformed coordinates (x^*, y^*, z^*) regardless of the original coordinates, thereby simply *translating* $\mathbf{P}^*$ in space. This is why we start the detailed discussion here by temporarily ignoring these coefficients, which leads to the simple system of equations

$$
\begin{aligned}
x^* &= a_{11}x + a_{12}y + a_{13}z, \\
y^* &= a_{21}x + a_{22}y + a_{23}z, \\
z^* &= a_{31}x + a_{32}y + a_{33}z.
\end{aligned}
\tag{1.2}
$$

If the $3{\times}3$ coefficient matrix of this system of equations is nonsingular or, equivalently, if the determinant of the coefficient matrix is nonzero (see any text on linear algebra for a refresher on matrices and determinants), then the system is easy to invert and can be expressed in the form

$$
\begin{aligned}
x &= b_{11}x^* + b_{12}y^* + b_{13}z^*, \\
y &= b_{21}x^* + b_{22}y^* + b_{23}z^*, \\
z &= b_{31}x^* + b_{32}y^* + b_{33}z^*,
\end{aligned}
\tag{1.3}
$$

where the b_{ij}'s are expressed in terms of the a_{ij}'s. It is now easy to see that, for example, the two-dimensional line $Ax + By + C = 0$ is transformed by Equation (1.3) to the two-dimensional line

$$
(Ab_{11} + Bb_{21})x^* + (Ab_{12} + Bb_{22})y^* + C = 0.
$$

⋄ **Exercise 1.3:** Show that Equation (1.3) maps the general second-degree curve

$$
Ax^2 + Bxy + Cy^2 + Dx + Ey + F = 0
$$

to another second-degree curve.

In general, an affine transformation maps any curve of degree n to another curve of the same degree.

1.2 Two-Dimensional Transformations

In practice, a complete two-dimensional image is constructed on the screen object-by-object and it may be edited before it is deemed satisfactory. One aspect of editing is to *transform* objects. Typical transformations (Figures 1.1 and Intro.1) are moving or sliding (translation), reflecting or flipping (mirror image), zooming (scaling), rotating, and shearing (distorting).

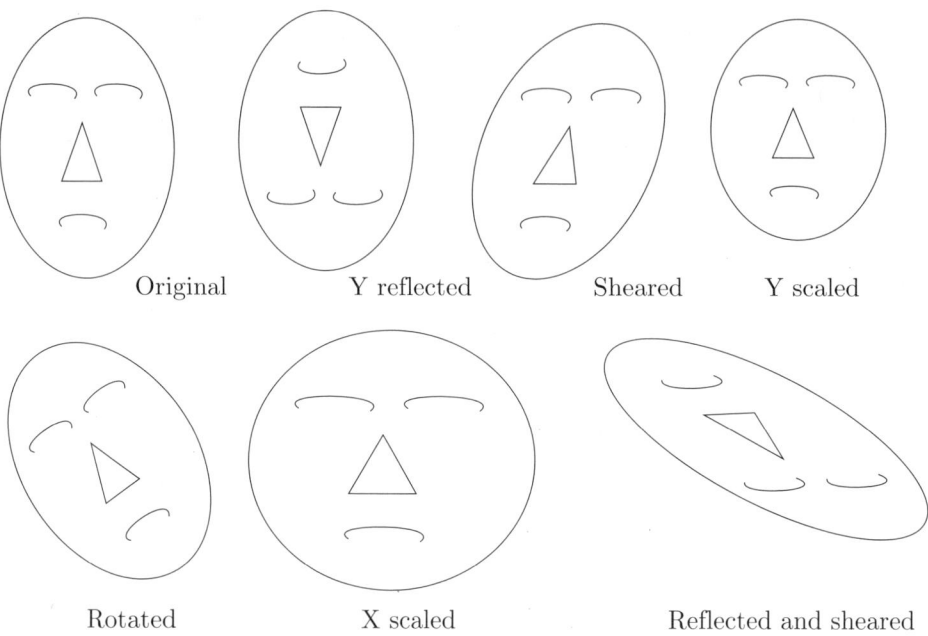

Original	Y reflected	Sheared	Y scaled

Rotated	X scaled	Reflected and sheared

Figure 1.1: Two-Dimensional Transformations.

The transformation can be applied to every pixel of the object. Alternatively, it can be applied only to some key points that fully define the object (such as the four corners of a rectangle), following which the transformed object is constructed from the transformed key points.

As soon as we use words like "image," we are already thinking of how one shape corresponds to the other—of how you might move one shape to bring it into coincidence with the other. Bilateral symmetry means that if you reflect the left half in a mirror, then you obtain the right half. Reflection is a mathematical concept, but it is not a shape, a number, or a formula. It is a *transformation*—that is, a rule for moving things around.

—Ian Stewart, *Nature's Numbers* (1995)

The same principle applies to a three-dimensional image. Such an image consists of one or more three-dimensional objects that can be transformed individually, following which the entire image should be projected on the two-dimensional screen (or other output device). We first take a look at the mathematics of two-dimensional transformations.

We use the notation $\mathbf{P} = (x, y)$ for a point and $\mathbf{P}^* = (x^*, y^*)$ for the transformed point. We are looking for a simple, fast transformation rule, so it is natural to try a linear transformation (i.e., a mathematical rule that does not use functions more complicated than x). The simplest linear transformation is $x^* = ax + cy$ and $y^* = bx + dy$, in which each of the new coordinates is a linear combination of the two old ones. This transformation can be written $\mathbf{P}^* = \mathbf{PT}$, where $\mathbf{T}$ is the 2×2 matrix $\left(\begin{smallmatrix} a & b \\ c & d \end{smallmatrix}\right)$. Thus, the transformation depends on just four parameters, which makes it easy to analyze and fully understand it.

To understand the effect of each of the four matrix elements, we start by setting $b = c = 0$. The transformation becomes $x^* = ax$, $y^* = dy$. Such a transformation is called *scaling*. If applied to all the points of an object, all the x dimensions are scaled by a factor of a and all the y dimensions are scaled by a factor of d. Note that a and d can also be less than 1, which causes shrinking of the object. If a or d (or both) equal -1, the transformation is a *reflection*. Any other negative values result in both scaling and reflection.

Note that scaling an object by factors of a and d changes its area by a factor of $a \times d$ and that this factor is also the value of the determinant of the scaling matrix $\left(\begin{smallmatrix} a & 0 \\ 0 & d \end{smallmatrix}\right)$. Here are examples of scaling and reflection. In $\mathbf{A}$, the y coordinates are scaled by a factor of 2. In $\mathbf{B}$, the x coordinates are reflected. In $\mathbf{C}$, the x dimensions are shrunk to 0.001 of their original values. In $\mathbf{D}$, the figure is shrunk to a vertical line.

$$\mathbf{A} = \begin{pmatrix} 1 & 0 \\ 0 & 2 \end{pmatrix}, \quad \mathbf{B} = \begin{pmatrix} -1 & 0 \\ 0 & 1 \end{pmatrix}, \quad \mathbf{C} = \begin{pmatrix} 0.001 & 0 \\ 0 & 1 \end{pmatrix}, \quad \mathbf{D} = \begin{pmatrix} 0 & 0 \\ 0 & 1 \end{pmatrix}.$$

⋄ **Exercise 1.4:** What scaling transformation changes a circle to an ellipse?

The next step is to set $a = 1$ and $d = 1$ (no scaling or reflection) and explore the effect of matrix elements b and c. The transformation becomes $x^* = x + cy$, $y^* = bx + y$. We first set $b = 1$ and $c = 0$ and look at how matrix $\left(\begin{smallmatrix} 1 & 1 \\ 0 & 1 \end{smallmatrix}\right)$ transforms the four points $(1, 0)$, $(3, 0)$, $(1, 1)$, and $(3, 1)$. They are transformed to $(1, 1)$, $(3, 3)$, $(1, 2)$, and $(3, 4)$. When we plot the original and the transformed points (Figure 1.2a), it becomes obvious that the original rectangle has been sheared vertically and was transformed into a parallelogram. A similar shearing effect results from matrix $\left(\begin{smallmatrix} 1 & 0 \\ 1 & 1 \end{smallmatrix}\right)$. The quantities b and c are therefore responsible for *shearing*. Figure 1.2b shows the connection between shearing and the operation of scissors. This is the reason for the name shearing.

⋄ **Exercise 1.5:** Apply the shearing transformation $\left(\begin{smallmatrix} 1 & -1 \\ 0 & 1 \end{smallmatrix}\right)$ to the four points $(1, 0)$, $(3, 0)$, $(1, 1)$, and $(3, 1)$. What are the transformed points? What geometrical figure do they represent?

The next important transformation is *rotation*. Figure 1.3 shows a point $\mathbf{P}$ rotated clockwise about the origin through an angle θ to become $\mathbf{P}^*$. Simple trigonometry yields

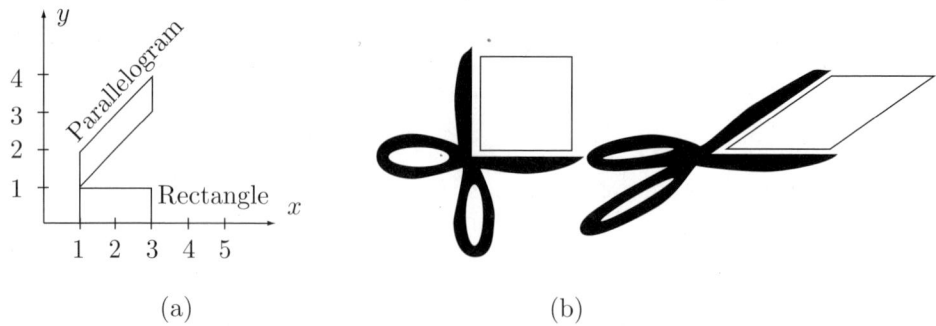

Figure 1.2: Scissors and Shearing.

$x = R\cos\alpha$ and $y = R\sin\alpha$. From this, we get the expressions for x^* and y^*

$$x^* = R\cos(\alpha - \theta) = R\cos\alpha\cos\theta + R\sin\alpha\sin\theta = x\cos\theta + y\sin\theta,$$
$$y^* = R\sin(\alpha - \theta) = -R\cos\alpha\sin\theta + R\sin\alpha\cos\theta = -x\sin\theta + y\cos\theta.$$

Hence, the clockwise rotation matrix in two dimensions is

$$\begin{pmatrix} \cos\theta & -\sin\theta \\ \sin\theta & \cos\theta \end{pmatrix}, \quad \text{which also equals the product} \quad \begin{pmatrix} \cos\theta & 0 \\ 0 & \cos\theta \end{pmatrix}\begin{pmatrix} 1 & -\tan\theta \\ \tan\theta & 1 \end{pmatrix}. \quad (1.4)$$

This shows that any rotation in two dimensions is a combination of scaling (and, perhaps, reflection) by a factor of $\cos\theta$ and shearing, an unexpected result (that's true for all angles where $\tan\theta$ is finite).

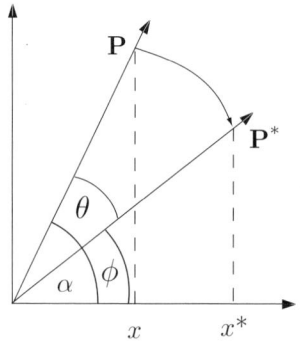

Figure 1.3: Clockwise Rotation.

◇ **Exercise 1.6:** Show how a 45° rotation can be achieved by scaling followed by shearing.

◇ **Exercise 1.7:** Discuss rotation in two dimensions using the polar coordinates (r, θ) of points instead of the Cartesian coordinates (x, y).

A rotation matrix has the following property: When any row is multiplied by itself, the result is 1, and when a row is multiplied by another row, the result is 0. The same is true for columns. Such a matrix is called *orthonormal*.

Matrix $\mathbf{T}_1$ below rotates counterclockwise. Matrix $\mathbf{T}_2$ reflects about the line $y = x$, and matrix $\mathbf{T}_3$ reflects about the line $y = -x$. Note the determinants of these matrices. In general, a determinant of $+1$ indicates pure rotation, whereas a determinant of -1 indicates pure reflection. (As a reminder, $\det \left(\begin{smallmatrix} a & b \\ c & d \end{smallmatrix} \right) = ad - bc$.)

$$\mathbf{T}_1 = \begin{pmatrix} \cos\theta & \sin\theta \\ -\sin\theta & \cos\theta \end{pmatrix}; \quad \mathbf{T}_2 = \begin{pmatrix} 0 & 1 \\ 1 & 0 \end{pmatrix}; \quad \mathbf{T}_3 = \begin{pmatrix} 0 & -1 \\ -1 & 0 \end{pmatrix}. \quad (1.5)$$

◇ **Exercise 1.8:** Show that a y-reflection (i.e., reflection about the x axis) followed by a reflection through the line $y = -x$ produces pure rotation.

◇ **Exercise 1.9:** Show that the transformation matrix

$$\begin{pmatrix} \dfrac{1 - t^2}{1 + t^2} & \dfrac{2t}{1 + t^2} \\ \dfrac{-2t}{1 + t^2} & \dfrac{1 - t^2}{1 + t^2} \end{pmatrix}$$

produces pure rotation.

◇ **Exercise 1.10:** For what values of A does the following matrix represent pure rotation and for what values does it represent pure reflection?

$$\begin{pmatrix} a/A & b/A \\ -b/A & a/A \end{pmatrix}.$$

■ **A 90° Rotation:** In the case of a 90° clockwise rotation, the rotation matrix is

$$\begin{pmatrix} \cos(90) & -\sin(90) \\ \sin(90) & \cos(90) \end{pmatrix} = \begin{pmatrix} 0 & -1 \\ 1 & 0 \end{pmatrix}. \quad (1.6)$$

A point $\mathbf{P} = (x, y)$ is therefore transformed to the point $(y, -x)$. For a counterclockwise 90° rotation, (x, y) is transformed to $(-y, x)$. This is called the *negate and exchange* rule.

Representations rotated not always by one hundred and eighty degrees, but sometimes by ninety or forty-five, completely subvert habitual perceptions of space; the outline of Europe, for instance, a shape familiar to anyone who has been even only to junior school, when swung around ninety degrees to the right, with the west at the top, begins to look like Denmark.

—Georges Perec, *Life, A User's Manual* (1976)

The Golden Ratio

Start with a straight segment of length l and divide it into two parts a and b such that $a + b = l$ and $l/a = a/b$.

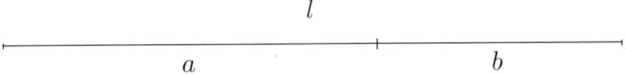

The ratio a/b is a constant called the *Golden Ratio* and is denoted ϕ. It is one of the important mathematical constants, like π and e, and was already known to the ancient Greeks. It is believed that geometric figures can be made more pleasing to the human eye if they involve this ratio. One example is the golden rectangle, whose sides are x and $x\phi$ long. Many classical buildings and paintings involve this ratio. [Huntley 70] is a lively introduction to the Golden Ratio. It illustrates properties such as

$$\phi = \sqrt{1 + \sqrt{1 + \sqrt{1 + \sqrt{1 + \cdots}}}} \quad \text{and} \quad \phi = 1 + \cfrac{1}{1 + \frac{1}{1 + \frac{1}{\cdots}}}.$$

The value of ϕ is easy to calculate. The basic ratio $l/a = a/b = \phi$ implies $(a+b)/a = a/b = \phi$, which, in turn, means $1 + b/a = \phi$ or $1 + 1/\phi = \phi$, an equation that can be written $\phi^2 - \phi - 1 = 0$. This equation is easy to solve, yielding $\phi = (1+\sqrt{5})/2 \approx 1.618\ldots$.

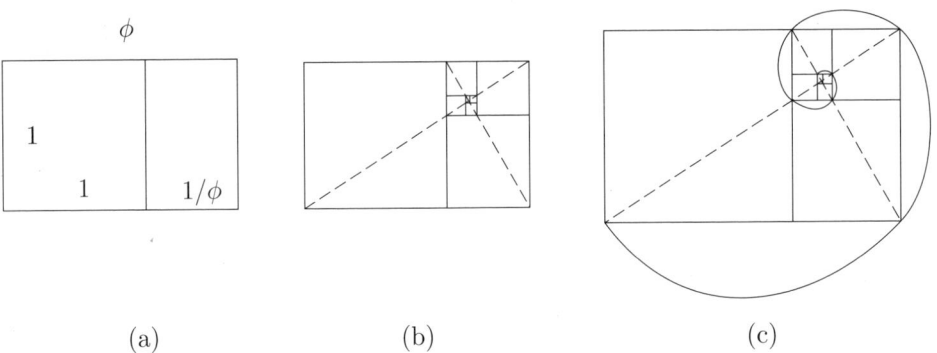

(a)   (b)   (c)

Figure 1.4: The Golden Ratio.

The equation $\phi = 1 + 1/\phi$ illustrates another unusual property of ϕ. Imagine the golden rectangle with sides $1 \times \phi$ (Figure 1.4a). Such a rectangle can be divided into a 1×1 square and a smaller golden rectangle of dimensions $1 \times 1/\phi$. The smaller rectangle can now be divided into a $1/\phi \times 1/\phi$ square and an even smaller golden rectangle (Figure 1.4b). When this process continues, the rectangles converge to a point. Figure 1.4c shows how a logarithmic spiral can be drawn through corresponding corners of the rectangles.

1.2.1 Homogeneous Coordinates

Unfortunately, our simple 2×2 transformation matrix cannot generate all the basic transformations that are needed in practice! In particular, it cannot generate *translation*. This is easy to see by arguing that any object containing the origin will, after any of the transformations above, still contain the origin [i.e., the result of $(0,0)\mathbf{T}$ is $(0,0)$ for any matrix $\mathbf{T}$].

Translations can be expressed by $x^* = x + m$, $y^* = y + n$, and one way to implement them is to generalize our transformations to $\mathbf{P}^* = \mathbf{PT} + (m, n)$, where $\mathbf{T}$ is the familiar 2×2 transformation matrix. A more elegant approach, however, is to stay with the compact notation $\mathbf{P}^* = \mathbf{PT}$ and to extend $\mathbf{T}$ to the 3×3 matrix

$$\mathbf{T} = \begin{pmatrix} a & b & 0 \\ c & d & 0 \\ m & n & 1 \end{pmatrix}. \tag{1.7}$$

This approach is called *homogeneous coordinates* and is commonly used in projective geometry. It makes it possible to unify all the two-dimensional transformations within one 3×3 matrix with six parameters. The problem is that a two-dimensional point (a pair) cannot be multiplied by a 3×3 matrix. This is solved by representing our points in homogeneous coordinates, which is done by extending the point (x, y) to the triplet $(x, y, 1)$. The rules for using homogeneous coordinates are the following:

1. To transform a point (x, y) to homogeneous coordinates, simply add a third component of 1. Hence, $(x, y) \Rightarrow (x, y, 1)$.

2. To transform the triplet (a, b, c) from homogeneous coordinates back into a pair (x, y), divide by the third component. Hence, $(a, b, c) \Rightarrow (a/c, b/c)$.

This means that a point (x, y) has an infinite number of representations in homogeneous coordinates. Any triplet (ax, ay, a) where a is nonzero is a valid representation of the point. This suggests a way to intuitively understand homogeneous coordinates. We can consider the triplet (ax, ay, a) a point in three-dimensional space. When a varies from 0 to ∞, the point travels along a straight ray from the origin to infinity. The direction of the ray is determined by x and y but not by a. Therefore, each two-dimensional point (x, y) corresponds to a ray in three-dimensional space. To find the "real" location of the point, we look at the $z = 1$ plane. All points on this plane have coordinates $(x, y, 1)$, so we only have to strip off the "1" in order to see where the point is located. Section 1.4 shows that homogeneous coordinates can also be applied to three-dimensional points.

⋄ **Exercise 1.11:** Write the transformation matrix that performs (1) a y-reflection, (2) a translation by -1 in the x and y directions, and (3) a 180° counterclockwise rotation about the origin. Apply this compound transformation to the four corners $(1, 1)$, $(1, -1)$, $(-1, 1)$, and $(-1, -1)$ of a square centered on the origin. What are the transformed corners?

Matrix (1.7) is the general transformation matrix in two dimensions. It produces the most general linear transformation, $x^* = ax + cy + m$, $y^* = bx + dy + n$, and it shows that this transformation depends on just six numbers.

We can gain a deeper understanding of homogeneous coordinates when we add two more parameters to matrix (1.7), writing it as

$$\begin{pmatrix} a & b & p \\ c & d & q \\ m & n & 1 \end{pmatrix}.$$

(1.8)

A general point (x, y) is now transformed to

$$(x, y, 1) \begin{pmatrix} a & b & p \\ c & d & q \\ m & n & 1 \end{pmatrix} = (ax + cy + m, bx + dy + n, px + qy + 1).$$

Applying rule 2 shows that the transformed point (x^*, y^*) is given by

$$x^* = \frac{ax + cy + m}{px + qy + 1}, \quad y^* = \frac{bx + dy + n}{px + qy + 1}.$$

To understand what this means, we apply this result to the four points $(2, 1)$, $(6, 1)$, $(2, 5)$, and $(6, 5)$ that constitute the four corners of a square (Figure 1.5a). Using the simple transformation

$$\begin{pmatrix} 1 & 0 & 1 \\ 0 & 1 & 1 \\ 0 & 0 & 1 \end{pmatrix}$$

(i.e., no scaling, rotation, shearing, or translation and $p = q = 1$), the points are transformed to

$$\mathbf{P}_1 = (2, 1) \rightarrow (2, 1, 4) \rightarrow (1/2, 1/4),$$
$$\mathbf{P}_2 = (6, 1) \rightarrow (6, 1, 8) \rightarrow (3/4, 1/8),$$
$$\mathbf{P}_3 = (2, 5) \rightarrow (2, 5, 8) \rightarrow (1/4, 5/8),$$
$$\mathbf{P}_4 = (6, 5) \rightarrow (6, 5, 12) \rightarrow (1/2, 5/12).$$

The transformed points (Figure 1.5b) also seem to form a square, but one that's viewed from a different direction and seen in perspective. This suggests that our transformation (using just p and q, without scaling, reflection, rotation, or shearing) has moved the square from its original position in the xy plane to another plane. Such transformations are called *projections* and are useful when dealing with objects in three-dimensional space.

1.2.2 Combining Transformations

Matrix notation is useful when working with transformations since it makes it easy to combine transformations. To combine transformations $\mathbf{A}$, $\mathbf{B}$, and $\mathbf{C}$, we write the three transformation matrices and multiply them. An example is an x-reflection, followed by a y-scaling, followed by a 45° rotation

$$\begin{pmatrix} -1 & 0 \\ 0 & 1 \end{pmatrix} \begin{pmatrix} 1 & 0 \\ 0 & 2 \end{pmatrix} \begin{pmatrix} 0.707 & -0.707 \\ 0.707 & 0.707 \end{pmatrix} = \begin{pmatrix} -0.707 & 0.707 \\ 1.414 & 1.414 \end{pmatrix}.$$

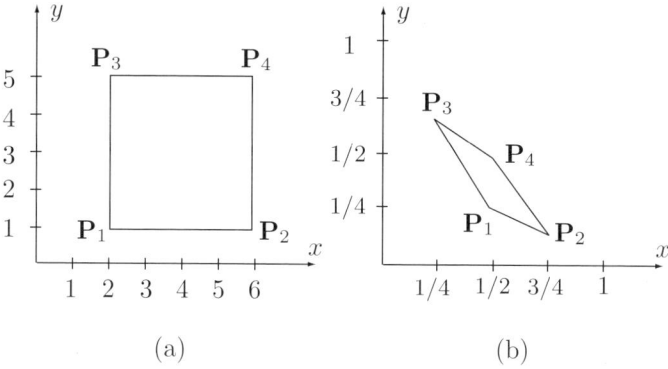

Figure 1.5: A Two-Dimensional Projection of a Square.

In general, matrix multiplication is noncommutative, reflecting the fact that geometric transformations are also noncommutative. It is easy to convince yourself that, for example, a rotation about the origin followed by a translation is not the same as a translation followed by a rotation about the origin.

Note that all the transformations discussed earlier are performed about the origin. Figure 1.6a shows an object rotated $40°$ clockwise. It is easy to see that the center of rotation is the origin. If, for example, we want to rotate an object about a point **P**, we have to translate both the object and the point such that **P** goes to the origin (Figure 1.6b), then rotate the object, and finally translate back (Figure 1.6c). Similarly, to reflect an object through an arbitrary line, we have to (1) translate the line (and the object) until it passes through the origin, (2) rotate the line (and the object) until it coincides with one of the coordinate axes, (3) reflect through that axis, (4) rotate back, and (5) translate back.

(Transformations are usually done about the origin. See Exercise 3.10 for an example on how this affects scaling in three dimensions.)

◇ **Exercise 1.12:** Derive the rotation matrix for a two-dimensional rotation about a point (x_0, y_0) using just trigonometry (i.e., without using translation).

Example: Reflection through the line y = x + 1. This line has a slope of 1 (i.e., it makes an angle of $45°$ with the x axis) and it intercepts the y axis at $y = 1$. We first translate down one unit, then rotate clockwise by $45°$, then reflect through the x axis, rotate back, and translate back. The result is (α stands for both $\sin 45°$ and $\cos 45°$)

$$\mathbf{T} = \begin{pmatrix} 1 & 0 & 0 \\ 0 & 1 & 0 \\ 0 & -1 & 1 \end{pmatrix} \begin{pmatrix} \alpha & -\alpha & 0 \\ \alpha & \alpha & 0 \\ 0 & 0 & 1 \end{pmatrix} \begin{pmatrix} 1 & 0 & 0 \\ 0 & -1 & 0 \\ 0 & 0 & 1 \end{pmatrix} \begin{pmatrix} \alpha & \alpha & 0 \\ -\alpha & \alpha & 0 \\ 0 & 0 & 1 \end{pmatrix} \begin{pmatrix} 1 & 0 & 0 \\ 0 & 1 & 0 \\ 0 & 1 & 1 \end{pmatrix}$$

$$= \begin{pmatrix} 0 & 2\alpha^2 & 1 \\ 2\alpha^2 & 0 & 0 \\ -2\alpha^2 & 1 & 1 \end{pmatrix} = \begin{pmatrix} 0 & 1 & 0 \\ 1 & 0 & 0 \\ -1 & 1 & 1 \end{pmatrix}$$

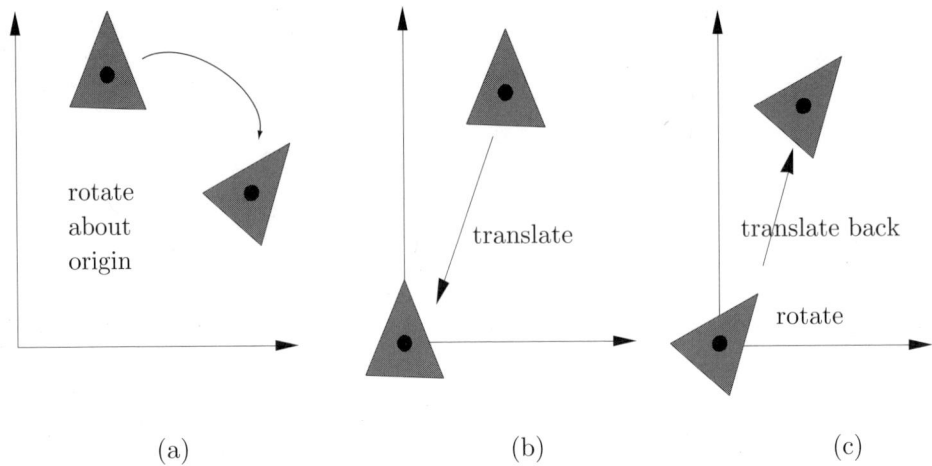

Figure 1.6: Rotation About a Point.

(because $2\alpha^2 = \sin^2 45° + \cos^2 45° = 1$). Note that $\det \mathbf{T} = -1$ (i.e., pure reflection).

◇ **Exercise 1.13:** Show that the result in the example is correct.

Example: Reflection about an arbitrary line. Given the line $y = ax + b$, it is possible to reflect a point about this line by transforming the line to the x axis, reflecting about that axis, and transforming the line back. Since a is the slope (i.e., the tangent of the angle α between the line and the x axis) and b is the y intercept, the individual transformations needed are (1) a translation of $-b$ units in the y direction, (2) a clockwise rotation of α degrees about the origin, (3) a reflection about the x axis, (4) a counterclockwise rotation, and (5) a reverse translation. The combined transformation matrix is therefore

$$
\mathbf{T}_{\text{reflect}} = \begin{pmatrix} 1 & 0 & 0 \\ 0 & 1 & 0 \\ 0 & -b & 1 \end{pmatrix} \begin{pmatrix} \cos\alpha & -\sin\alpha & 0 \\ \sin\alpha & \cos\alpha & 0 \\ 0 & 0 & 1 \end{pmatrix} \begin{pmatrix} 1 & 0 & 0 \\ 0 & -1 & 0 \\ 0 & 0 & 1 \end{pmatrix}
$$

$$
\times \begin{pmatrix} \cos\alpha & \sin\alpha & 0 \\ -\sin\alpha & \cos\alpha & 0 \\ 0 & 0 & 1 \end{pmatrix} \begin{pmatrix} 1 & 0 & 0 \\ 0 & 1 & 0 \\ 0 & b & 1 \end{pmatrix}
$$

$$
= \begin{pmatrix} \cos(2\alpha) & \sin(2\alpha) & 0 \\ \sin(2\alpha) & -\cos(2\alpha) & 0 \\ -b\sin(2\alpha) & 2b\cos^2\alpha & 1 \end{pmatrix}. \tag{1.9}
$$

The determinant of this transformation matrix equals -1, as should be for pure reflection. For the two special cases $\alpha = b = 0$ and $\alpha = 45°$ and $b = 0$, Equation (1.9) becomes

$$\begin{pmatrix} 1 & 0 & 0 \\ 0 & -1 & 0 \\ 0 & 0 & 1 \end{pmatrix} \quad \text{and} \quad \begin{pmatrix} 0 & 1 & 0 \\ 1 & 0 & 0 \\ 0 & 0 & 1 \end{pmatrix}, \quad \text{respectively.}$$

One feature that makes Equation (1.9) less than general is the way the sine and cosine are obtained from the tangent of a known angle. Given that the slope a equals $\tan\alpha$, we can calculate

$$a = \tan\alpha = \frac{\sin\alpha}{\cos\alpha} = \frac{\sin\alpha}{\sqrt{1 - \sin^2\alpha}},$$

which yields $\sin^2\alpha = a^2/(1 + a^2)$ or

$$\sin\alpha = \pm\frac{a}{\sqrt{1 + a^2}} \quad \text{and} \quad \cos\alpha = \pm\frac{1}{\sqrt{1 + a^2}}.$$

The signs depend on the angle (or rather the quadrant in which the angle happens to be) and cannot be determined in a general way.

⋄ **Exercise 1.14:** Calculate the numerical value of matrix $\mathbf{T}_{\text{reflect}}$ for the case $\alpha = 30°$ and $b = 1$.

⋄ **Exercise 1.15:** Digital images displayed on a screen or printed on paper consist of pixels. Even smooth curves are made of pixels. Thus, there is a need for efficient algorithms to compute the best pixels for a given curve or geometric figure. The circle has a high degree of symmetry, which is why it is possible to determine the best pixels for a given circle by computing the pixels for one octant and duplicating and transforming each pixel seven times to complete the remaining seven octants. The question is, is it possible to improve such an algorithm even more by doing half an octant and duplicating each pixel 15 times?

Another feature that makes Equation (1.9) less than general is the use of the explicit representation $y = ax + b$. This representation is limited because it cannot express vertical lines (for which a would be infinite). When reflecting a point about an arbitrary line, it is better to use the more general implicit representation of a straight line $ax + by + c = 0$, where a or b but not both can be zero. The slope of this line is $-a/b$, and substituting $b = 0$ yields a vertical line.

Given such a line, we start with a point $\mathbf{P} = (x, y)$ and its reflection $\mathbf{P}^* = (x^*, y^*)$ about the line. It is clear that the segment $\mathbf{PP}^*$ must be perpendicular to the line, so its equation must be $bx - ay + d = 0$. Since both $\mathbf{P}$ and $\mathbf{P}^*$ are on such a line, they satisfy $bx - ay + d = 0$ and $bx^* - ay^* + d = 0$. Subtracting these two expressions yields

$$b(x - x^*) = a(y - y^*). \tag{1.10}$$

We assume that $\mathbf{P}^*$ is the reflection of $\mathbf{P}$ about the line $ax + by + c = 0$, so the midpoint of segment $\mathbf{PP}^*$, which is the point $\big((x + x^*)/2, (y + y^*)/2\big)$, must be on this line and must therefore satisfy

$$a\frac{x + x^*}{2} + b\frac{y + y^*}{2} + c = 0. \tag{1.11}$$

Equations (1.10) and (1.11) can easily be solved for x^* and y^*. The solutions are

$$\mathbf{P}^* = (x^*, y^*) = \left(x - \frac{2a(ax + by + c)}{a^2 + b^2}, y - \frac{2b(ax + by + c)}{a^2 + b^2} \right)$$

$$= \left(\frac{(b^2 - a^2)x - 2aby - 2ac}{a^2 + b^2}, \frac{-2abx + (a^2 - b^2)y - 2bc}{a^2 + b^2} \right). \tag{1.12}$$

Equation (1.12) is easy to verify intuitively for vertical and for horizontal lines. When b is zero, the line becomes the vertical line $x = -c/a$ and Equation (1.12) reduces to

$$\mathbf{P}^* = (x^*, y^*) = \left(x - \frac{2a(ax + c)}{a^2}, y \right) = \left(-x - \frac{2c}{a}, y \right).$$

When $a = 0$, the line is the horizontal $y = -c/b$, and the same equation reduces to

$$\mathbf{P}^* = (x^*, y^*) = \left(x, y - \frac{2b(by + c)}{b^2} \right) = \left(x, -y - \frac{2c}{b} \right).$$

The transformation matrix for reflection about an arbitrary line $ax + by + c = 0$ is directly obtained from Equation (1.12)

$$\mathbf{T} = \begin{pmatrix} b^2 - a^2 & -2ab & 0 \\ -2ab & a^2 - b^2 & 0 \\ -2ac & -2bc & \frac{1}{a^2 + b^2} \end{pmatrix}. \tag{1.13}$$

Its determinant is

$$\det \mathbf{T} = \frac{(b^2 - a^2)(a^2 - b^2) - 4a^2b^2}{a^2 + b^2} = -\frac{a^4 + 2a^2b^2 + b^4}{a^2 + b^2} = -(a^2 + b^2),$$

which equals -1 (pure reflection) for lines expressed in the standard form (defined as the case where $a^2 + b^2 = 1$).

⋄ **Exercise 1.16:** Use Equation (1.12) to obtain the transformation rule for reflection about a line that passes through the origin.

We turn now to the product of two reflections about the two arbitrary lines L_1: $ax + by + c = 0$ and L_2: $dx + ey + f = 0$ (Figure 1.7a). This product can be calculated from Equation (1.13) as the matrix product

$$\begin{pmatrix} b^2 - a^2 & -2ab & 0 \\ -2ab & a^2 - b^2 & 0 \\ -2ac & -2bc & \frac{1}{a^2 + b^2} \end{pmatrix} \begin{pmatrix} e^2 - d^2 & -2de & 0 \\ -2de & d^2 - e^2 & 0 \\ -2df & -2ef & \frac{1}{d^2 + e^2} \end{pmatrix},$$

but this product is complex and hard to interpret geometrically. In order to simplify it, we assume, without loss of generality, that both lines pass through the origin and that the first is also horizontal (Figure 1.7b). The first assumption means that the lines intersect at the origin and that $c = f = 0$. The second assumption means that the first line is identical to the x axis, so $a = 0$ and $b = 1$. Also, $f = 0$ implies $dx + ey = 0$ or $y = -(d/e)x$. The quantity $-d/e$ is the slope (i.e., $\tan \theta$) of the second line, so we conclude that

$$-\frac{d}{e} = -\tan \theta = -\frac{\sin \theta}{\cos \theta}, \quad \text{implying } d^2 + e^2 = 1.$$

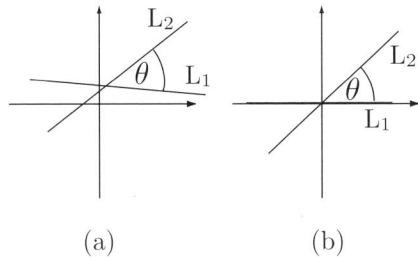

Figure 1.7: Reflections About Two Intersecting Lines.

Under these assumptions, the matrix product above becomes

$$
\begin{pmatrix} 1 & 0 & 0 \\ 0 & -1 & 0 \\ 0 & 0 & 1 \end{pmatrix}
\begin{pmatrix} e^2 - d^2 & -2de & 0 \\ -2de & d^2 - e^2 & 0 \\ 0 & 0 & 1 \end{pmatrix}
$$

$$
= \begin{pmatrix} e^2 - d^2 & -2de & 0 \\ 2de & e^2 - d^2 & 0 \\ 0 & 0 & 1 \end{pmatrix}
$$

$$
= \begin{pmatrix} \cos(2\theta) & -\sin(2\theta) & 0 \\ \sin(2\theta) & \cos(2\theta) & 0 \\ 0 & 0 & 1 \end{pmatrix}, \tag{1.14}
$$

leading to the important conclusion that the product of two reflections about arbitrary lines is a rotation through an angle 2θ about the intersection point of the lines, where θ is the angle between the lines. It can be shown that the opposite is also true; any rotation is the product of two reflections about two intersecting lines.

The discussion above assumes that both lines pass through the origin. In the special case where $\theta = 0$, such lines would be identical, so reflecting a point $\mathbf{P}$ about them would move it back to itself. However, for $\theta = 0$, matrix (1.14) reduces to the identity matrix, so it is valid even for identical lines.

In the special case where the lines are parallel, their intersection point is at infinity and a rotation about a center at infinity is a translation.

⬦ **Exercise 1.17:** Given the two parallel lines $y = 0$ and $y = c$, calculate the double reflection of a point (x, y) about them.

⬦ **Exercise 1.18:** Consider the shearing transformation $\mathbf{T}_a$ of Equation (1.15), followed by the 90° rotation $\mathbf{T}_b$. What is the combined transformation, and what kind of transformation is it?

$$\mathbf{T}_a = \begin{pmatrix} 0 & 1 & 0 \\ 2 & 0 & 0 \\ 0 & 0 & 1 \end{pmatrix}, \qquad \mathbf{T}_b = \begin{pmatrix} \cos 90° & -\sin 90° & 0 \\ \sin 90° & \cos 90° & 0 \\ 0 & 0 & 1 \end{pmatrix}. \qquad (1.15)$$

⬦ **Exercise 1.19:** Given the two rotations

$$\mathbf{T}_1 = \begin{pmatrix} \cos\theta_1 & -\sin\theta_1 & 0 \\ \sin\theta_1 & \cos\theta_1 & 0 \\ 0 & 0 & 1 \end{pmatrix} \quad \text{and} \quad \mathbf{T}_2 = \begin{pmatrix} \cos\theta_2 & -\sin\theta_2 & 0 \\ \sin\theta_2 & \cos\theta_2 & 0 \\ 0 & 0 & 1 \end{pmatrix},$$

calculate the combined transformation $\mathbf{T}_1 \mathbf{T}_2$. Is it identical to a rotation through $(\theta_1 + \theta_2)$?

⬦ **Exercise 1.20:** Given the two shearing transformations

$$\mathbf{T}_1 = \begin{pmatrix} 1 & b & 0 \\ 0 & 1 & 0 \\ 0 & 0 & 1 \end{pmatrix} \quad \text{and} \quad \mathbf{T}_2 = \begin{pmatrix} 1 & 0 & 0 \\ c & 1 & 0 \\ 0 & 0 & 1 \end{pmatrix},$$

calculate the combined transformation $\mathbf{T}_1 \mathbf{T}_2$. Is it identical to a shearing by factors b and c?

⬦ **Exercise 1.21:** Prove that three successive shearings about the x, y, and x axes is equivalent to a rotation about the origin.

⬦ **Exercise 1.22:** Matrix $\begin{pmatrix} a & 0 \\ 0 & d \end{pmatrix}$ scales an object by factors a and d along the x and y axes, respectively. If we want to scale the object by the same factors, but in the i and j directions (see Figure 1.8, where i and j are perpendicular and form an angle θ with the x and y axes, respectively), we need to (1) rotate the object θ degrees clockwise, (2) scale along the x and y axes using matrix $\begin{pmatrix} a & 0 \\ 0 & d \end{pmatrix}$, and (3) rotate back. Write the three transformation matrices and their product. Discuss the case $a = d$ (uniform scaling).

⬦ **Exercise 1.23:** We can perform an exercise with shearing, similar to Exercise 1.22. Matrix $\begin{pmatrix} 1 & b \\ c & 1 \end{pmatrix}$ shears an object by factors c and b along the x and y axes, respectively. Calculate the matrix that shears the object by the same factors, but in the i and j directions (Figure 1.8).

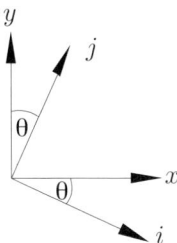

Figure 1.8: Scaling Along Rotated Axes.

$\diamond$ **Exercise 1.24:** Discuss scaling relative to a point (x_0, y_0), and show that the result is identical to the product of a translation followed by scaling, followed by a reverse translation.

Using Equation (Ans.2) in the Answers to Exercises, it is easy to explore the effect of two consecutive scaling transformations, with scaling factors of k_1 and k_2 and about points $\mathbf{P}_1 = (x_1, y_1)$ and $\mathbf{P}_2 = (x_2, y_2)$, respectively. We simply multiply the two transformation matrices

$$
\begin{pmatrix} k_1 & 0 & 0 \\ 0 & k_1 & 0 \\ x_1(1-k_1) & y_1(1-k_1) & 1 \end{pmatrix} \begin{pmatrix} k_2 & 0 & 0 \\ 0 & k_2 & 0 \\ x_2(1-k_2) & y_2(1-k_2) & 1 \end{pmatrix}
$$

$$
= \begin{pmatrix} k_1 k_2 & 0 & 0 \\ 0 & k_1 k_2 & 0 \\ k_2(1-k_1)x_1 + (1-k_2)x_2 & k_2(1-k_1)y_1 + (1-k_2)y_2 & 1 \end{pmatrix}. \tag{1.16}
$$

The result is similar to Equation (Ans.2) except for the bottom row. It seems that the product of two scalings is a third scaling with a factor $k_1 k_2$, but about what point? To write Equation (1.16) in the form of Equation (Ans.2), we write

$$
k_2(1-k_1)x_1 + (1-k_2)x_2 = x_c(1-k_1 k_2),
$$
$$
k_2(1-k_1)y_1 + (1-k_2)y_2 = y_c(1-k_1 k_2),
$$

and solve for (x_c, y_c), obtaining

$$
x_c = \frac{k_2(1-k_1)x_1 + (1-k_2)x_2}{1 - k_1 k_2},
$$
$$
y_c = \frac{k_2(1-k_1)y_1 + (1-k_2)y_2}{1 - k_1 k_2}.
$$

The center of the double scaling is therefore point

$$
\mathbf{P}_c = \frac{k_2(1-k_1)}{1 - k_1 k_2}\mathbf{P}_1 + \frac{1-k_2}{1 - k_1 k_2}\mathbf{P}_2 = a\mathbf{P}_1 + b\mathbf{P}_2.
$$

Notice that $a + b = 1$, which is why $\mathbf{P}_c$ is a point on the straight segment connecting $\mathbf{P}_1$ and $\mathbf{P}_2$ (see also Equation (Ans.7)).

In the special case $\mathbf{P}_1 = \mathbf{P}_2$, it is easy to see that the center of the double scaling is $\mathbf{P}_c = \mathbf{P}_1 = \mathbf{P}_2$.

⋄ **Exercise 1.25:** What is the result of two consecutive scalings with the same scaling factors but about two different points?

⋄ **Exercise 1.26:** Show that all the points with coordinates (t^2, t), where $0 \le t \le 1$, after being transformed by

$$\begin{pmatrix} -1 & 0 & 1 \\ 0 & 2 & 0 \\ 1 & 0 & 1 \end{pmatrix},$$

lie on the perimeter of the unit circle $x^2 + y^2 = 1$. (Hint: See Figure 1.9.)

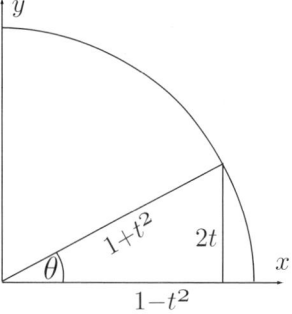

Figure 1.9: A Unit Circle.

It is easy to see that the transformations discussed here can change lengths and angles. Scaling changes the lengths of objects. Rotation and shearing change angles. One feature that's preserved, though, is parallel lines. A pair of parallel lines will remain parallel after any scaling, reflection, rotation, shearing, and translation. A transformation that preserves parallelism (and also maps finite points to finite points) is called *affine*.

1.2.3 Fast Rotations

Rotation requires the calculation of the transcendental sine and cosine functions, which is time-consuming. If many rotations are needed, it is preferable to precalculate the trigonometric functions for many angles and store them in a table. This section shows how to do this using integers only, a method that results in much faster rotations than using floating-point numbers.

The method is illustrated for the first quadrant (rotation angles of 0° to 90°) in increments of 1°. Notice that rotations in other quadrants can be achieved by a first-quadrant rotation followed by a reflection. The following *Mathematica* code generates

91 sine values, from $\sin 0° = 0$ to $\sin 90° = 1$, multiplies each by $2^{14} = 16{,}384$, rounds them, and stores them in a table as 16-bit integers ranging from 0 to 16,384.

```
d2r=Pi/180;
Table[Round[N[16384*Sin[i*d2r]]], {i,0,90}]
```

θ	$\sin\theta$	θ	$\sin\theta$	θ	$\sin\theta$	θ	$\sin\theta$	θ	$\sin\theta$
0	0	1	286	2	572	3	857	4	1143
5	1428	6	1713	7	1997	8	2280	9	2563
10	2845	11	3126	12	3406	13	3686	14	3964
15	4240	16	4516	17	4790	18	5063	19	5334
20	5604	21	5872	22	6138	23	6402	24	6664
25	6924	26	7182	27	7438	28	7692	29	7943
30	8192	31	8438	32	8682	33	8923	34	9162
35	9397	36	9630	37	9860	38	10087	39	10311
40	10531	41	10749	42	10963	43	11174	44	11381
45	11585	46	11786	47	11982	48	12176	49	12365
50	12551	51	12733	52	12911	53	13085	54	13255
55	13421	56	13583	57	13741	58	13894	59	14044
60	14189	61	14330	62	14466	63	14598	64	14726
65	14849	66	14968	67	15082	68	15191	69	15296
70	15396	71	15491	72	15582	73	15668	74	15749
75	15826	76	15897	77	15964	78	16026	79	16083
80	16135	81	16182	82	16225	83	16262	84	16294
85	16322	86	16344	87	16362	88	16374	89	16382
90	16384								

Table 1.10: Sine Values as 16-Bit Integers.

The 91 values are listed in Table 1.10, but notice that they are only approximations of the true sine values. (Even floating-point sine values are, in general, just approximations, but normally better than our integers.) This means that the use of this table for many successive rotations of a point may place it farther and farther away from its true position. When we perform many successive rotations of an object that consists of many points, placing points away from where they should be generally results in a deformation of the object.

We assume that the points are represented by coordinates that are 16-bit integers. Calculating the rotated coordinates (x^*, y^*) of a point (x, y) can now be done, for example, by

$$x^* = \text{rshift}(x \times \text{Table}(90 - \theta), 14) - \text{rshift}(y \times \text{Table}(\theta), 14),$$
$$y^* = \text{rshift}(x \times \text{Table}(\theta), 14) + \text{rshift}(y \times \text{Table}(90 - \theta), 14).$$

Notice how the required cosine values are obtained from the end of the table. This method works because the table has 91 entries. Multiplying a 16-bit integer coordinate by a 16-bit integer sine value yields a 32-bit product. The right shift effectively divides the product by $2^{14} = 16,384$, a necessary operation because our integer sine values have been premultiplied by this scale factor.

◇ **Exercise 1.27:** Use this method to calculate the results of rotating point $(1, 2)$ by $60°$ and by $80°$. In each case, compare the results with those obtained when built-in sine and cosine functions are used.

1.2.4 CORDIC Rotations

We routinely use calculators to compute values of common functions, but have you ever wondered how a calculator determines the value of, say, $\tan 72.81°$ so fast? Many calculators use CORDIC (COordinate Rotation, DIgital Computer), a general method for computing many elementary functions. CORDIC was originally proposed by [Volder 59] and was extended by [Walther 71]. The original references are hard to find but are included in [Swartzlander 90]. Here, we show how CORDIC can be used to implement fast rotations.

It is sufficient to consider a rotation about the origin where the rotation angle θ is in the interval $[0, 90°)$ (the first quadrant). The special case $\theta = 90°$ can be implemented by the negate and exchange rule [Equation (1.6)]. Rotations in other quadrants can be achieved by a first-quadrant rotation, followed by a reflection.

The rotation is expressed by [see Equation (1.4)]

$$(x^*, y^*) = (x, y) \begin{pmatrix} \cos\theta & -\sin\theta \\ \sin\theta & \cos\theta \end{pmatrix}. \tag{1.17}$$

Because θ is less than $90°$, we know that $\cos\theta$ is nonzero, so we can factor out $\cos\theta$, yielding

$$(x^*, y^*) = \cos\theta\,(x, y) \begin{pmatrix} 1 & -\tan\theta \\ \tan\theta & 1 \end{pmatrix}.$$

We now express θ as the sum $\sum_{i=0}^{m} \theta_i$, where angles θ_i are defined by the relation $\tan\theta_i = 2^{-i}$ or $\theta_i \overset{\text{def}}{=} \arctan(2^{-i})$. The first 16 θ_i, for $i = 0, 1, \ldots, 15$, are listed in Table 1.11.

In order to express any angle θ as the sum of these particular θ_i, some θ_i will have to be subtracted. Consider, for example, $\theta = 58°$. We start with $\theta_0 = 45°$. Since $\theta_0 < \theta$, we add θ_1. The sum $\theta_0 + \theta_1 = 45 + 26.5651 = 71.5651$ is greater than θ, so we subtract θ_2. The new sum, 57.5289, is less than θ, so we add θ_3, and so on.

◇ **Exercise 1.28:** We want to be able to express any angle θ in the range $[0°, 90°)$ by adding and subtracting a number of consecutive θ_i, from θ_0 to some θ_m, without skipping any θ_i in between. Is that possible?

It is easy to write a program that decides which of the θ_i's should be added and which should be subtracted. Thus, we end up with

$$\theta = \sum_{i=0}^{m} d_i\theta_i = \sum_{i=0}^{m} d_i \arctan(2^{-i}), \quad \text{where} \quad d_i = \pm 1.$$

i	θ_i (degrees)	θ_i (radians)	K_i
0	45.	0.785398	0.70710678118654746
1	26.5651	0.463648	0.63245553203367577
2	14.0362	0.244979	0.61357199107789628
3	7.12502	0.124355	0.60883391251775243
4	3.57633	0.0624188	0.60764825625616825
5	1.78991	0.0312398	0.60735177014129604
6	0.895174	0.0156237	0.60727764409352614
7	0.447614	0.00781234	0.60725911229889284
8	0.223811	0.00390623	0.60725447933256249
9	0.111906	0.00195312	0.60725332108987529
10	0.0559529	0.000976562	0.60725303152913446
11	0.0279765	0.000488281	0.60725295913894495
12	0.0139882	0.000244141	0.60725294104139727
13	0.00699411	0.00012207	0.60725293651701029
14	0.00349706	0.0000610352	0.60725293538591352
15	0.00174853	0.0000305176	0.60725293510313938

Table 1.11: The First 16 θ_i's and Scale Factors.

Once the number m of necessary d_i's and their values have been determined, we rotate (x, y) to (x^*, y^*) in a loop where each iteration rotates a point (x_i, y_i) through an angle $d_i\theta_i$ to a point (x_{i+1}, y_{i+1}). A general iteration can be expressed in the form

$$
\begin{aligned}
(x_{i+1}, y_{i+1}) &= \cos(d_i\theta_i)\, (x_i, y_i) \begin{pmatrix} 1 & -\tan\theta_i \\ \tan\theta_i & 1 \end{pmatrix} \\
&= \cos(d_i\theta_i)\, (x_i, y_i) \begin{pmatrix} 1 & -d_i 2^{-i} \\ d_i 2^{-i} & 1 \end{pmatrix} \\
&= \cos(d_i\theta_i)\, (x_i + y_i d_i 2^{-i}, y_i - x_i d_i 2^{-i}).
\end{aligned}
\tag{1.18}
$$

We interpret the result (x_{i+1}, y_{i+1}) of an iteration as the vector from the origin to point (x_{i+1}, y_{i+1}). Equation (1.18) shows that this vector is the product of two terms. The second term, $(x_i + y_i d_i 2^{-i}, y_i - x_i d_i 2^{-i})$, determines the direction of the vector, while the first term, $\cos(d_i\theta_i)$, affects only the magnitude of the vector. The second term is easy to calculate since it just involves shifts. We know that d_i is just a sign and that a product of the form $x_i 2^{-i}$ can be computed by shifting x_i i positions to the right. The problem is to calculate the first term, $\cos(d_i\theta_i)$, and to multiply the two terms.

This is why CORDIC proceeds by first performing all the iterations

$$
(x_{i+1}, y_{i+1}) \leftarrow (x_i + y_i d_i 2^{-i}, y_i - x_i d_i 2^{-i})
$$

using just right shifts and additions/subtractions; the cosine terms are ignored. The result is a vector that points in the right direction but is too long (Figure 1.13). To bring this vector to its correct size, it should be multiplied by the scale factor

$$K_m = \prod_{i=0}^{m} \cos \theta_i.$$

(Notice that $\cos(d_i \theta_i) = \cos \theta_i$ since cosine is an even function.) This is discouraging because it suggests that m multiplications are needed just to calculate the scale factor K_m. However, the first 16 scale factors are listed in Table 1.11 and even a quick glance shows that they converge to the number 0.60725.... Reference [Vachss 87] shows that K_m can be obtained simply by using the m most significant bits of this number and ignoring the rest.

Using the identity $\sin^2 \theta + \cos^2 \theta = 1$ and the definition $\tan \theta_i = 2^{-i}$, we get

$$\cos \theta_i = \frac{1}{\sqrt{1 + \tan^2 \theta_i}} = \frac{1}{\sqrt{1 + 2^{-2i}}},$$

which is why the scale factors of Table 1.11 were so easily calculated to a high precision by the code

`N[Table[Product[(2^(-2i)+1)^(-1/2),{i,0,n}],{n,0,16}],17]//TableForm.`

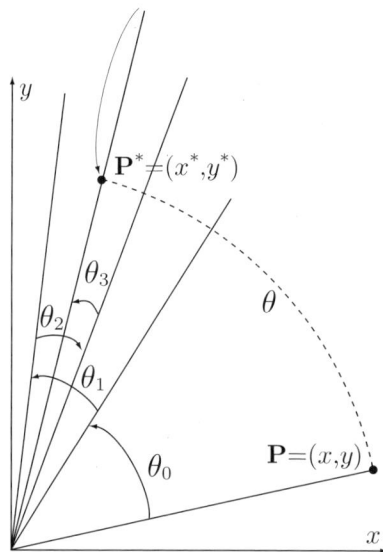

Figure 1.12: CORDIC Rotation.

$\diamond$ **Exercise 1.29:** Suggest another way to calculate K_m.

Any practical CORDIC implementation (see [Jarvis 90] for a C program) should have the following two features.

1. CORDIC employs only shifts and additions/subtractions, so any implementation should use fixed-point, instead of floating-point, arithmetic. This is fast since shifting and adding fixed-point numbers can be done with integer operations. Notice that all the numbers involved in the computations are less than unity, except perhaps the original coordinates (x, y). A software package for graphics employing this method should therefore use normalized coordinates (fixed-point numbers in the interval $[0, 1]$) throughout and perform all the calculations on these small numbers. Each iteration results in a pair (x_{i+1}, y_{i+1}) that's slightly larger than its predecessor, but the last iteration results in a pair that can be larger than (x, y) by a factor of at most $1/0.60725\ldots = 1.64676\ldots$. This pair is then scaled down when multiplied by K_m. The final step is to scale the final coordinates up.

All this suggests a 32-bit fixed-point format where the leftmost bit is reserved, as usual, for the sign, the next two bits are the integer part, and the remaining 29 bits are the fractional part (29 bits being equivalent to 9 decimal digits). The largest number that can be represented by this format is $11.11\ldots1_2 = 3.999\ldots$ and the smallest one is $110\ldots0_2 = -4$. It's a good idea to reserve two bits for the integer part because (1) even though all the numbers involved are 1 or smaller, some intermediate results may be greater than 1 and (2) this convention makes it possible to represent the important constants π, e, and ϕ (the Golden Ratio).

2. Earlier, we said, "It is easy to write a program that decides which of the θ_i's should be added and which should be subtracted." The practical way to do this is to initialize a variable z to θ and try to drive z to zero during the iterations. In iteration i the program should calculate both $z + \theta_i$ and $z - \theta_i$, select the value that's closer to zero, use it to decide whether to add or subtract θ_i, and then update z. If $z - \theta_i$ is closer to zero, then θ_i should be added; otherwise, θ_i should be subtracted. An example is $\theta = 58°$. We initialize z to 58. In iteration 0, it is clear that $58 - 45 = 13$ is closer to zero than $58 + 45$. The program therefore adds θ_0 and updates z to 13. In iteration 1, the program finds that $13 - 26.5651 = -13.5651$ is closer to zero than $13 + 26.5651$, so it adds θ_1 and updates z to -13.5651. In iteration 2, the program discovers that $-13.5651 + 14.0362 = 0.4711$ is closer to zero than $-13.5651 - 14.0362$, so it subtracts θ_2 and updates z to 0.4711.

Finally, we realize that there is really no need to compare $z+\theta_i$ and $z-\theta_i$ in iteration i. We simply start by selecting $d_0 = +1$ and update z by subtracting $z \leftarrow z - \theta_0$, $z \leftarrow z - \theta_1$, etc., until we get a negative value in z. We then change d_i to -1 (the new sign of z) and update z by $z \leftarrow z - d_i\theta_i$ (which now amounts to adding θ_i to z). This is summarized by the *Mathematica* code of Figure 1.13. (But note that the `Sign` function of *Mathematica* returns $+1$, 0, or -1, while we need a result of $+1$ or -1. The code as shown is simple but not completely general.)

> Compared to other approaches, CORDIC is a clear winner when a hardware multiplier is unavailable (e.g. in a microcontroller) or when you want to save the gates required to implement one (e.g. in an FPGA). On the other hand, when a hardware multiplier is available (e.g. in a DSP microprocessor), table-lookup methods and good old-fashioned power series are generally faster than CORDIC.
> —Grant R. Griffin, www.dspguru.com/info/faqs/cordic.htm

```
t=Table[ArcTan[2.^{-i}], {i,0,15}]; (* arctans in radians *)
d=1; x=2.1; y=0.34; z=46. Degree;
Do[{Print[i,", ",x,", ",y,", ",z,", ",d],
 xn=x+y d 2^{-i}, yn=y-x d 2^{-i},
 zn=z-d t[[i+1]], d=Sign[zn], x=xn, y=yn, z=zn}, {i,0,14}]
Print[0.60725x,", ",0.60725y]
```

Figure 1.13: *Mathematica* Code for CORDIC Rotations.

⋄ **Exercise 1.30:** Instead of using the complex CORDIC method, wouldn't it be simpler to perform a rotation by a direct use of Equation (1.17)? After all, this only requires the calculation of one sine and one cosine values.

1.2.5 Similarities

A *similarity* is a transformation that scales all distances by a fixed factor. It is easy to show that a similarity is produced by the special transformation matrix

$$\begin{pmatrix} a & c & 0 \\ -c & a & 0 \\ m & n & 1 \end{pmatrix}.$$

To show this, we observe that translations preserve distances, so we can ignore the translation part of the matrix above and restrict ourselves to the matrix $\begin{pmatrix} a & c \\ -c & a \end{pmatrix}$. It transforms a point $\mathbf{P} = (x, y)$ to the point $\mathbf{P}^* = (x^*, y^*) = (ax - cy, cx + ay)$. Given the two transformations $\mathbf{P}_1 \to \mathbf{P}_1^*$ and $\mathbf{P}_2 \to \mathbf{P}_2^*$, it is straightforward to illustrate the relation

$$
\begin{aligned}
\texttt{distance}^2(\mathbf{P}_1^*\mathbf{P}_2^*) &= \left((\Delta x^*)^2 + (\Delta y^*)^2\right) \\
&= [(ax_2 - cy_2) - (ax_1 - cy_1)]^2 + [(cx_2 + ay_2) - (cx_1 + ay_1)]^2 \\
&= (a\Delta x - c\Delta y)^2 + (c\Delta x + a\Delta y)^2 \\
&= a^2 \Delta x^2 - 2a\Delta x c\Delta y + c^2 \Delta y^2 + c^2 \Delta x^2 + 2c\Delta x a\Delta y + a^2 \Delta y^2 \\
&= (a^2 + c^2)(\Delta x^2 + \Delta y^2) \\
&= (a^2 + c^2)\texttt{distance}^2(\mathbf{P}_1\mathbf{P}_2),
\end{aligned}
$$

implying that $\texttt{distance}(\mathbf{P}_1^*\mathbf{P}_2^*) = \sqrt{a^2 + c^2}\,\texttt{distance}(\mathbf{P}_1\mathbf{P}_2)$. Thus, all distances are scaled by a factor of $\sqrt{a^2 + c^2}$.

In general, a similarity is a transformation of the form $\mathbf{P}^* = (x^*, y^*) = (ax - cy + m, \pm(cx + ay) + n)$, where the ratio of expansion (or shrinking) is $k = \sqrt{a^2 + c^2}$. If k is positive, the similarity is called *direct*; if k is negative, the similarity is *opposite*.

⋄ **Exercise 1.31:** Discuss the case $k = 0$.

Using the ratio k, we can write a similarity (ignoring the translation part) as the product

$$\begin{pmatrix} a & c & 0 \\ -c & a & 0 \\ 0 & 0 & 1 \end{pmatrix} \begin{pmatrix} k & 0 & 0 \\ 0 & k & 0 \\ 0 & 0 & 1 \end{pmatrix} \begin{pmatrix} a/k & c/k & 0 \\ -c/k & a/k & 0 \\ 0 & 0 & 1 \end{pmatrix},$$

which shows that a similarity is a combination of a scaling/reflection (by a factor k) and a rotation. (The definition of k implies that $(a/k)^2 + (c/k)^2 = 1$, so we can consider c/k and a/k the sine and cosine of the rotation angle, respectively.)

1.2.6 A 180° Rotation

Another interesting example of combining transformations is a 180° rotation about a fixed point $\mathbf{P} = (P_x, P_y)$. This combination is called a *halfturn*. It is performed, as usual, by translating $\mathbf{P}$ to the origin, rotating about the origin, and translating back. The transformation matrix is (notice that $\cos(180°) = -1$)

$$\mathbf{T} = \begin{pmatrix} 1 & 0 & 0 \\ 0 & 1 & 0 \\ -P_x & -P_y & 1 \end{pmatrix} \begin{pmatrix} -1 & 0 & 0 \\ 0 & -1 & 0 \\ 0 & 0 & 1 \end{pmatrix} \begin{pmatrix} 1 & 0 & 0 \\ 0 & 1 & 0 \\ P_x & P_y & 1 \end{pmatrix} = \begin{pmatrix} -1 & 0 & 0 \\ 0 & -1 & 0 \\ 2P_x & 2P_y & 1 \end{pmatrix}.$$

A general point (x, y) is therefore transformed by a halfturn to

$$(x, y, 1) \begin{pmatrix} -1 & 0 & 0 \\ 0 & -1 & 0 \\ 2P_x & 2P_y & 1 \end{pmatrix} = (-x + 2P_x, -y + 2P_y, 1) \qquad (1.19)$$

(Figure 1.14a), but it's more interesting to explore the effect of two consecutive halfturns, about points $\mathbf{P}$ and $\mathbf{Q}$. The second halfturn transforms point $(-x + 2P_x, -y + 2P_y, 1)$ to

$$(-x + 2P_x, -y + 2P_y, 1) \begin{pmatrix} -1 & 0 & 0 \\ 0 & -1 & 0 \\ 2Q_x & 2Q_y & 1 \end{pmatrix} = (x - 2P_x + 2Q_x, y - 2P_y + 2Q_y, 1). \quad (1.20)$$

If $\mathbf{P} = \mathbf{Q}$, then the result of the second halfturn is (x, y), showing how two identical 180° rotations return a point to its original location. If $\mathbf{P}$ and $\mathbf{Q}$ are different, the result is a *translation* of the original point (x, y) by factors $-2P_x + 2Q_x$ and $-2P_y + 2Q_y$ (Figure 1.14b).

$\diamond$ **Exercise 1.32:** What is the result of three consecutive halfturns about the distinct points $\mathbf{P}$, $\mathbf{Q}$, and $\mathbf{R}$?

> Things turn out best for the people who make the best out of the way things turn out.
>
> —Art Linkletter

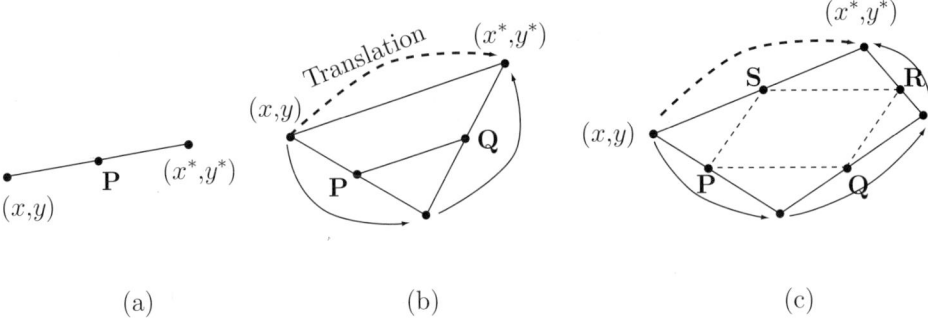

(a)   (b)   (c)

Figure 1.14: Halfturns.

1.2.7 Glide Reflections

This transformation is a special combination of three reflections. Imagine the two vertical parallel lines $x = L$ and $x = M$ and the horizontal line $y = N$ (Figure 1.15a). Reflecting a point $\mathbf{P} = (x, y)$ about the line $x = L$ is done by translating the line to the y axis, reflecting about that axis, and translating back. The transformation matrix is

$$\begin{pmatrix} 1 & 0 & 0 \\ 0 & 1 & 0 \\ -L & 0 & 1 \end{pmatrix} \begin{pmatrix} -1 & 0 & 0 \\ 0 & 1 & 0 \\ 0 & 0 & 1 \end{pmatrix} \begin{pmatrix} 1 & 0 & 0 \\ 0 & 1 & 0 \\ L & 0 & 1 \end{pmatrix} = \begin{pmatrix} -1 & 0 & 0 \\ 0 & 1 & 0 \\ 2L & 0 & 1 \end{pmatrix},$$

and the transformed point is

$$(x, y, 1) \begin{pmatrix} -1 & 0 & 0 \\ 0 & 1 & 0 \\ 2L & 0 & 1 \end{pmatrix} = (-x + 2L, y, 1).$$

Reflecting this point about the line $x = M$ results in

$$(-x + 2L, y, 1) \begin{pmatrix} -1 & 0 & 0 \\ 0 & 1 & 0 \\ 2M & 0 & 1 \end{pmatrix} = (x - 2L + 2M, y, 1)$$

(a translation), and reflecting this about the horizontal line $y = N$ yields

$$(x - 2L + 2M, y, 1) \begin{pmatrix} 1 & 0 & 0 \\ 0 & -1 & 0 \\ 0 & 2N & 1 \end{pmatrix} = (x - 2L + 2M, -y + 2N, 1).$$

This particular glide reflection is therefore a translation in x and a reflection in y. A general glide reflection is the product of three reflections, the first two about parallel lines L and M and the third about a line N perpendicular to them (Figure 1.15b).

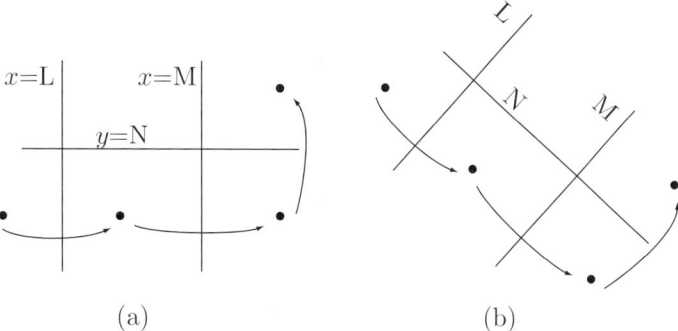

Figure 1.15: Glide Reflection.

1.2.8 Improper Rotations

A rotation followed by a reflection about one of the coordinate axes is called an *improper rotation*. The transformation matrices for the two possible improper rotations in two dimensions (Figure 1.16) are

$$\begin{pmatrix} \cos\theta & -\sin\theta \\ \sin\theta & \cos\theta \end{pmatrix} \begin{pmatrix} 1 & 0 \\ 0 & -1 \end{pmatrix} = \begin{pmatrix} \cos\theta & \sin\theta \\ \sin\theta & -\cos\theta \end{pmatrix},$$

$$\begin{pmatrix} \cos\theta & -\sin\theta \\ \sin\theta & \cos\theta \end{pmatrix} \begin{pmatrix} -1 & 0 \\ 0 & 1 \end{pmatrix} = \begin{pmatrix} -\cos\theta & -\sin\theta \\ -\sin\theta & \cos\theta \end{pmatrix},$$

and the transformation rules therefore are

$$x^* = x\cos\theta + y\sin\theta, \quad y^* = x\sin\theta - y\cos\theta,$$
$$x^* = -x\cos\theta - y\sin\theta, \quad y^* = -x\sin\theta + y\cos\theta.$$

Notice that the determinant of an improper rotation matrix equals -1, like that of a pure reflection.

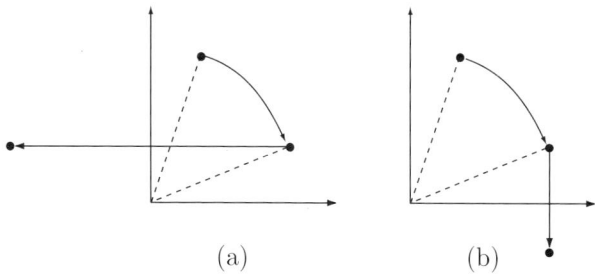

Figure 1.16: Improper Rotations.

An improper rotation differs from a rotation in one important aspect. When we rotate an object through a small angle and repeat this transformation, the object seems to move smoothly along a circle. Each time we repeat an improper rotation, however, the object "jumps" from one side of the coordinate plane to the other. The total effect is very different from that of a smooth circular movement.

1.2.9 Decomposing Transformations

Sometimes, a certain transformation A may be equivalent to the combined effects of several different transformations B, C, and D. We say that A can be *decomposed* into B, C, and D. Mathematically, this is equivalent to saying that the original transformation matrix $\mathbf{T}_A$ equals the product $\mathbf{T}_B\mathbf{T}_C\mathbf{T}_D$. We have already seen that a rotation in two dimensions can be decomposed into a scaling followed by a shearing; here are other examples.

It may come as a surprise that the general two-dimensional transformation matrix, Equation (1.7), can be written as a product of shearing, scaling, rotation, and translation as follows:

$$
\begin{bmatrix} a & b & 0 \\ c & d & 0 \\ m & n & 1 \end{bmatrix} =
$$

$$
\begin{bmatrix} 1 & 0 & 0 \\ (ac+bd)/A^2 & 1 & 0 \\ 0 & 0 & 1 \end{bmatrix}
\begin{bmatrix} A & 0 & 0 \\ 0 & (ad-bc)/A & 0 \\ 0 & 0 & 1 \end{bmatrix}
\begin{bmatrix} a/A & b/A & 0 \\ -b/A & a/A & 0 \\ 0 & 0 & 1 \end{bmatrix}
\begin{bmatrix} 1 & 0 & 0 \\ 0 & 1 & 0 \\ m & n & 1 \end{bmatrix},
$$

$$(1.21)$$

where $A = \sqrt{a^2 + b^2}$. The third matrix produces rotation since $(a/A)^2 + (b/A)^2 = 1$.

Even something as simple as shearing in one direction can be written as the product of a unit shearing and two scalings:

$$
\begin{pmatrix} 1 & 0 & 0 \\ c & 1 & 0 \\ 0 & 0 & 1 \end{pmatrix} =
\begin{pmatrix} 1/c & 0 & 0 \\ 0 & 1 & 0 \\ 0 & 0 & 1 \end{pmatrix}
\begin{pmatrix} 1 & 0 & 0 \\ 1 & 1 & 0 \\ 0 & 0 & 1 \end{pmatrix}
\begin{pmatrix} c & 0 & 0 \\ 0 & 1 & 0 \\ 0 & 0 & 1 \end{pmatrix}.
$$

Even the simple transformation of a unit shearing can be decomposed into a product that involves a scaling and two rotations. Note that the Golden Ratio ϕ is involved,

$$
\begin{pmatrix} 1 & 0 & 0 \\ 1 & 1 & 0 \\ 0 & 0 & 1 \end{pmatrix} =
\begin{pmatrix} \cos\alpha & -\sin\alpha & 0 \\ \sin\alpha & \cos\alpha & 0 \\ 0 & 0 & 1 \end{pmatrix}
\begin{pmatrix} \phi & 0 & 0 \\ 0 & 1/\phi & 0 \\ 0 & 0 & 1 \end{pmatrix}
\begin{pmatrix} \cos\beta & \sin\beta & 0 \\ -\sin\beta & \cos\beta & 0 \\ 0 & 0 & 1 \end{pmatrix},
$$

where $\alpha = \tan^{-1}\phi \approx 58.28°$ and $\beta = \tan^{-1}(1/\phi) \approx 31.72°$.

(This is indeed a surprising result. It means that a clockwise rotation of $58.28°$, followed by a scaling of ϕ in the x direction and $1/\phi$ in the y direction, followed by a counterclockwise rotation of $31.72°$, is equivalent to a unit shear in the x direction. This is illustrated by Figure 1.17.)

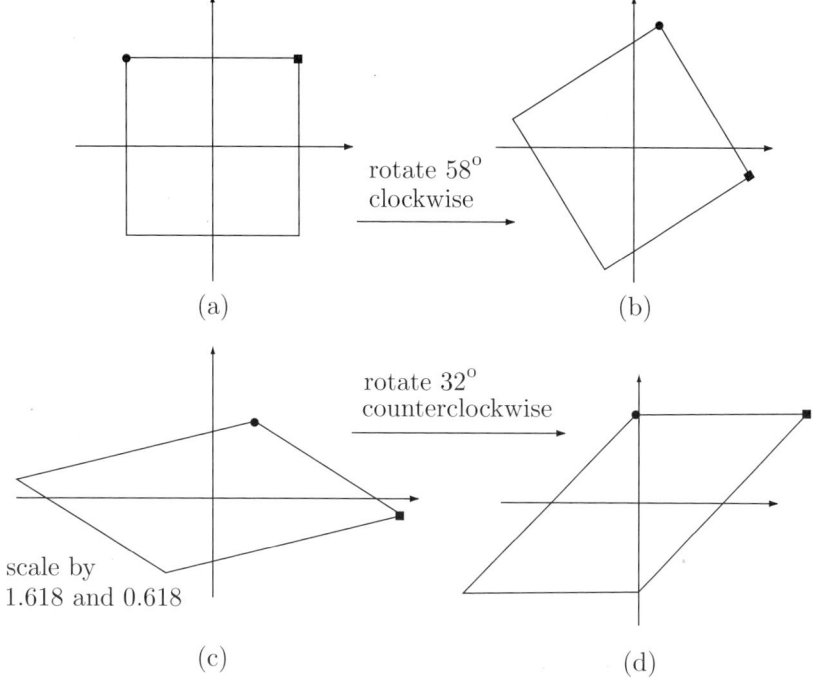

rotate 58°
clockwise

(a)

(b)

rotate 32°
counterclockwise

scale by
1.618 and 0.618

(c)

(d)

Figure 1.17: Shearing Decomposed into Rotation and Scaling.

> Geometry has two great treasures: one the Theorem of Pythagoras; the other,
> the division of a line into extreme and mean ratio. The first we may compare
> to a measure of gold; the second we may name a precious jewel.
>
> —Johannes Kepler

⋄ **Exercise 1.33:** Given the transformation

$$x^* = 3x - 2y + 1, \quad y^* = 4x + 5y - 6,$$

calculate the transformation matrix and decompose it into a product of four matrices
as shown in Equation (1.21).

1.2.10 Reconstructing Transformations

Given a sequence of two-dimensional transformations, we normally write the 3×3 matrix
for each and then multiply the matrices. The result is another 3×3 matrix which is used
to transform all the points of an object. An interesting question is: Given the points
of an object before and after a transformation, can we reconstruct the transformation
matrix from them?

The answer is yes! The general two-dimensional transformation matrix depends
on six numbers, so all we need are six equations involving transformed points. Since

each point consists of two numbers, three points are enough to reconstruct the transformation matrix. Given three points both before ($\mathbf{P}_1$, $\mathbf{P}_2$, $\mathbf{P}_3$) and after ($\mathbf{P}_1^*$, $\mathbf{P}_2^*$, $\mathbf{P}_3^*$) a transformation, we can write the three equations $\mathbf{P}_1^* = \mathbf{P}_1\mathbf{T}$, $\mathbf{P}_2^* = \mathbf{P}_2\mathbf{T}$, and $\mathbf{P}_3^* = \mathbf{P}_3\mathbf{T}$ and solve for the six elements of $\mathbf{T}$.

Example. The three points $(1, 1)$, $(1, 0)$, and $(0, 1)$ are transformed to $(3, 4)$, $(2, -1)$, and $(0, 2)$, respectively. We write the general transformation $(x^*, y^*) = (ax + cy + m, bx + dy + n)$ for the three sets

$$(3, 4) = (a + c + m, b + d + n),$$
$$(2, -1) = (a + m, b + n),$$
$$(0, 2) = (c + m, d + n),$$

and this is easily solved to yield $a = 3$, $b = 2$, $c = 1$, $d = 5$, $m = -1$, and $n = -3$. The transformation matrix is therefore

$$\mathbf{T} = \begin{pmatrix} 3 & 2 & 0 \\ 1 & 5 & 0 \\ -1 & -3 & 1 \end{pmatrix}.$$

◇ **Exercise 1.34: Inverse transformations.** From $\mathbf{P}^* = \mathbf{PT}$, we get $\mathbf{P}^*\mathbf{T}^{-1} = \mathbf{PTT}^{-1}$ or $\mathbf{P} = \mathbf{P}^*\mathbf{T}^{-1}$. We can therefore reconstruct an original point $\mathbf{P}$ from the transformed one, $\mathbf{P}^*$, if we know the inverse of the transformation matrix $\mathbf{T}$. In general, the inverse of the 3×3 matrix

$$\mathbf{T} = \begin{pmatrix} a & b & 0 \\ c & d & 0 \\ m & n & 1 \end{pmatrix}$$

is

$$\mathbf{T}^{-1} = \frac{1}{ad - bc} \begin{pmatrix} d & -b & 0 \\ -c & a & 0 \\ cn - dm & bm - an & 1 \end{pmatrix}. \tag{1.22}$$

Calculate the inverses of the transformation matrices for scaling, shearing, rotation, and translation, and discuss their properties.

◇ **Exercise 1.35:** Given that the four points

$$\mathbf{P}_1 = (0, 0), \quad \mathbf{P}_2 = (0, 1), \quad \mathbf{P}_3 = (1, 1), \quad \text{and } \mathbf{P}_4 = (1, 0)$$

are transformed to

$$\mathbf{P}_1^* = (0, 0), \quad \mathbf{P}_2^* = (2, 3), \quad \mathbf{P}_3^* = (8, 4), \quad \text{and } \mathbf{P}_4^* = (6, 1),$$

reconstruct the transformation matrix.

1.2.11 A Note

All the expressions derived so far for transformations are based on the basic relation $\mathbf{P}^* = \mathbf{PT}$. Some authors prefer the equivalent relation $\mathbf{P}^* = \mathbf{TP}$, which changes the mathematics somewhat. If we want the coordinates of the transformed point to be the same as before (i.e., $x^* = ax + cy + m$, $y^* = bx + dy + n$), we have to write the relation $\mathbf{P}^* = \mathbf{TP}$ in the form

$$\begin{pmatrix} x^* \\ y^* \\ 1 \end{pmatrix} = \begin{pmatrix} a & c & m \\ b & d & n \\ 0 & 0 & 1 \end{pmatrix} \begin{pmatrix} x \\ y \\ 1 \end{pmatrix}.$$

The first difference is that both $\mathbf{P}$ and $\mathbf{P}^*$ are columns instead of rows. This is because of the rules of matrix multiplication. The second difference is that the new transformation matrix $\mathbf{T}$ is the transpose of the original one. Hence, rotation, for example, is achieved by the matrices

$$\begin{pmatrix} \cos\theta & \sin\theta & 0 \\ -\sin\theta & \cos\theta & 0 \\ 0 & 0 & 1 \end{pmatrix}$$

for a clockwise rotation, and

$$\begin{pmatrix} \cos\theta & -\sin\theta & 0 \\ \sin\theta & \cos\theta & 0 \\ 0 & 0 & 1 \end{pmatrix}$$

for a counterclockwise rotation.

Similarly, translation is done by $\begin{pmatrix} 1 & 0 & m \\ 0 & 1 & n \\ 0 & 0 & 1 \end{pmatrix}$ instead of $\begin{pmatrix} 1 & 0 & 0 \\ 0 & 1 & 0 \\ m & n & 1 \end{pmatrix}$.

1.2.12 Summary

The general two-dimensional affine transformation is given by $x^* = ax + cy + m$, $y^* = bx + dy + n$. This section shows the values or constraints that should be assigned to the four coefficients a, b, c, and d in order to obtain certain types of transformations (we ignore translations).

■   A general affine transformation is obtained when $ad - bc \neq 0$. For $ad - bc = +1$, the transformation is rotation, and for $ad - bc = -1$, it is reflection.

■   The case $ad - bc = 0$ corresponds to a singular transformation.

■   The identity transformation is obtained when $a = d = 1$ and $b = c = 0$.

■   An isometry is obtained by $a^2 + b^2 = c^2 + d^2 = 1$ and $ac + bd = 0$. An isometry is a transformation that preserves distances. If $\mathbf{P}$ and $\mathbf{Q}$ are two points on an object, then the distance d between them is preserved, meaning that the distance d between $\mathbf{P}^*$ and $\mathbf{Q}^*$ is the same. Rotations, reflections, and translations are isometries.

■ A similarity is obtained for $a^2 + b^2 = c^2 + d^2$ and $ac + bd = 0$. A similarity is a transformation that preserves the ratios of lengths. A typical similarity is scaling, but it may be combined with rotation, reflection, and translation.

■ An *equiareal* transformation (preserving areas) is obtained when $|ad - bc| = 1$.

■ A shearing in the x direction is caused by $a = d = 1$ and $b = 0$. Similarly, a shearing in the y direction corresponds to $a = d = 1$ and $c = 0$.

■ A uniform scaling is $a = d > 0$ and $b = c = 0$. (The identity is a special case of scaling.)

■ A uniform reflection is $a = d < 0$ and $b = c = 0$.

■ A rotation is the result of $a = d = \cos\theta$ and $b = -c = \sin\theta$.

1.3 Three-Dimensional Coordinate Systems

We now turn to transformations in three dimensions. In most cases, the mathematics of linear transformations is easy to extend from two dimensions to three, but the discussion here demonstrates that certain transformations, most notably rotations, are more complex in three dimensions because there are more directions about which to rotate and because the simple terms clockwise and counterclockwise no longer apply. We start with a short discussion of coordinate systems in three dimensions.

In two dimensions, there is only one Cartesian coordinate system, with two perpendicular axes labeled x and y (actually, the axes don't have to be perpendicular, but this is irrelevant for our discussion of transformations). A coordinate system in three dimensions consists similarly of three perpendicular axes labeled x, y, and z, but there are two such systems, a left-handed and a right-handed (Figure 1.18a), and they are different. A right-handed coordinate system is constructed by the following rule. Align your right thumb with the positive x axis and your right index finger with the positive y axis. Your right middle finger will then point in the direction of positive z. The rule for a left-handed system uses the left hand in a similar manner. It is also possible to define a left-handed coordinate system as the mirror image (reflection) of a right-handed one. Notice that one coordinate system cannot be transformed into the other by translating or rotating it.

The difference between left-handed and right-handed coordinate systems becomes important when a three-dimensional object is projected on a two-dimensional screen (Chapter 3). We assume that the screen is positioned at the xy plane with its origin (i.e., its bottom left corner) at the origin of the three-dimensional system. We also assume that the object to be projected is located on the positive side of the z axis and the viewer is located on the negative side, looking at the projection of the image on the screen. Figure 1.18b shows that in a left-handed three-dimensional coordinate system, the directions of the positive x and y axes on the screen coincide with those of the three-dimensional x and y axes. In a right-handed system (Figure 1.18c), though, the two-dimensional x axis (on the screen) and the three-dimensional x axis point in opposite directions.

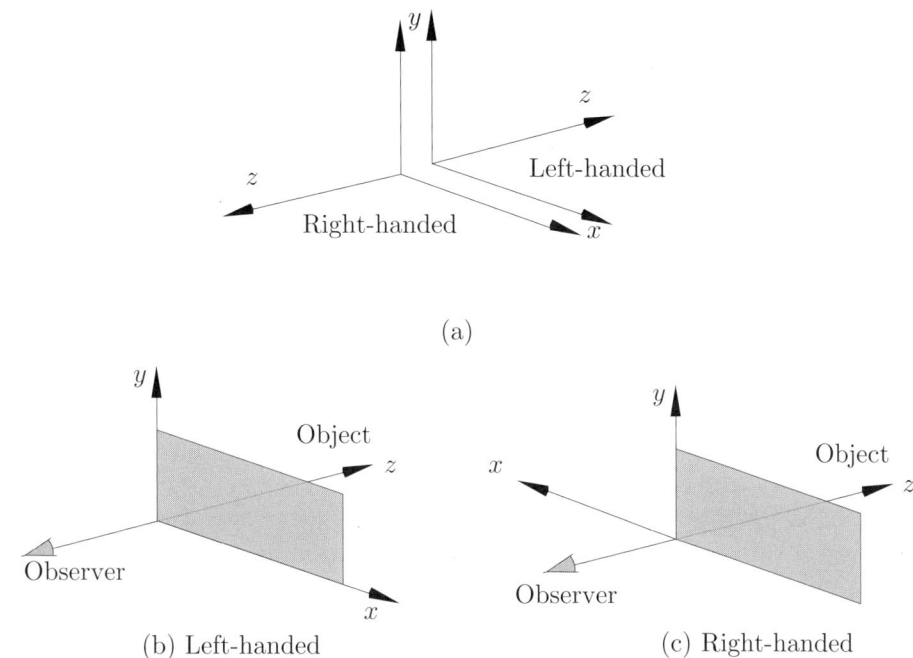

(a)

(b) Left-handed   (c) Right-handed

Figure 1.18: Three-Dimensional Coordinate Systems.

Principle: Express co-ordinate ideas in similar form.

This principle, that of parallel construction, requires that expressions of similar content and function should be outwardly similar. The likeness of form enables the reader to recognize more readily the likeness of content and function. Familiar instances from the Bible are the Ten Commandments, the Beatitudes, and the petitions of the Lord's Prayer.

—W. Strunk Jr. and E. B. White, *The Elements of Style*

1.4 Three-Dimensional Transformations

We develop three-dimensional transformations by extending the methods used in two-dimensional transformations, especially the concept of homogeneous coordinates. A three-dimensional point $\mathbf{P} = (x, y, z, 1)$ is transformed to a point $\mathbf{P}^* = (x^*, y^*, z^*, 1)$ by multiplying it by a 4×4 matrix

$$\mathbf{T} = \begin{pmatrix} a & b & c & p \\ d & e & f & q \\ h & i & j & r \\ l & m & n & s \end{pmatrix}. \tag{1.23}$$

The last column of $\mathbf{T}$ is not $(0,0,0,1)^T$ because it is used for projections. (See the discussion of n-point perspective on page 110.) As a result, the product $\mathbf{PT}$ is the 4-tuple (X, Y, Z, H), where H equals $xp + yq + zr + s$ and is generally not 1. The three coordinates (x^*, y^*, z^*) of $\mathbf{P}^*$ are obtained by dividing (X, Y, Z) by H. Hence, $(x^*, y^*, z^*) = (X/H, Y/H, Z/H)$.

The top left 3×3 part of $\mathbf{T}$ is responsible for scaling and reflection (a, e, and j), shearing (b, c, f and d, h, i), and rotation (all nine elements). The three quantities l, m, and n are responsible for translation, and the only new parameters are those in the last column (p, q, r, s).

To understand the meaning of s, we examine the matrix $\mathbf{T} = \begin{pmatrix} 1 & & & \\ & 1 & & \\ & & 1 & \\ & & & s \end{pmatrix}$. Multiplying $\mathbf{P}$ by $\mathbf{T}$ transforms $(x, y, z, 1)$ into (x, y, z, s), so the new point has coordinates $(x/s, y/s, z/s)$. The parameter s is therefore responsible for global scaling (by a factor of $1/s$). Its effect is identical to transforming by $\begin{pmatrix} 1/s & & & \\ & 1/s & & \\ & & 1/s & \\ & & & 1 \end{pmatrix}$.

Translation in three dimensions is a direct extension of the two-dimensional case. A point can be translated in the direction of any of the coordinate axes.

Scaling in three dimensions is simple. An object can be scaled about the origin along any of the three coordinate axes. To scale about another point $\mathbf{P}_0$, a sequence of three transformations is needed. The point should be translated to the origin, the scaling performed, and the point translated back. Notice that scaling an object is done by scaling all its points. Scaling a point does not change its dimensions (since a point has no dimensions) but simply moves it to another location.

Shearing in three dimensions is difficult to visualize. It is controlled by the six off-diagonal matrix elements b, c, f, d, h, and i, which is why many variations are possible. Perhaps the best way to become familiar with three-dimensional shearing is to experiment with the effect of varying each of the six parameters. Figure 1.19 shows a few possible shearings of a rectangular box.

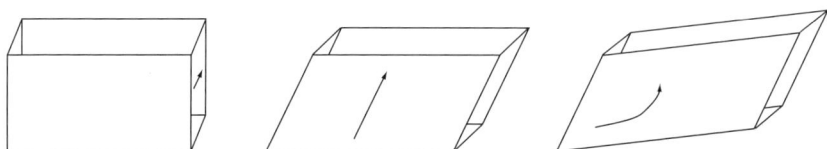

Figure 1.19: Shearing in Three Dimensions.

Shearing: A transformation in which all points along a given line L remain fixed while other points are shifted parallel to L by a distance proportional to their perpendicular distance from L. Shearing a plane figure does not change its area. This can also be generalized to three dimensions, where planes are translated instead of lines.
—Eric W. Weisstein, http://mathworld.wolfram.com/Shear.html

1.4.1 Reflection

It is easy to reflect a point (x, y, z) about any of the three coordinate planes xy, xz, or yz. All that's needed is to change the sign of one of the point's coordinates. In this section, we discuss and explain the general case where an arbitrary plane and a point are given and we want to reflect the point about the plane. We proceed in three steps as follows: (1) We discuss planes and their equations, (2) show how to determine the distance of a point from a given plane, and (3) explain how to compute the reflection of a point about a plane.

The (implicit) equation of a straight line is $Ax + By + C = 0$, where A or B but not both can be zero. The equation of a flat plane is the direct extension $Ax + By + Cz + D = 0$, where A, B, and C cannot all be zero. Four equations are needed to calculate the four unknown coefficients A, B, C, and D. On the other hand, we know that any three independent (i.e., noncollinear) points $\mathbf{P}_i = (x_i, y_i, z_i)$, $i = 1, 2, 3$ define a plane. Thus, we can write a set of four equations, three of which are based on three given points and the fourth one expressing the condition that a general point (x, y, z) lies on the plane

$$
0 = \begin{vmatrix} x & y & z & 1 \\ x_1 & y_1 & z_1 & 1 \\ x_2 & y_2 & z_2 & 1 \\ x_3 & y_3 & z_3 & 1 \end{vmatrix}
$$

$$
= x \begin{vmatrix} y_1 & z_1 & 1 \\ y_2 & z_2 & 1 \\ y_3 & z_3 & 1 \end{vmatrix} - y \begin{vmatrix} x_1 & z_1 & 1 \\ x_2 & z_2 & 1 \\ x_3 & z_3 & 1 \end{vmatrix} + z \begin{vmatrix} x_1 & y_1 & 1 \\ x_2 & y_2 & 1 \\ x_3 & y_3 & 1 \end{vmatrix} - \begin{vmatrix} x_1 & y_1 & z_1 \\ x_2 & y_2 & z_2 \\ x_3 & y_3 & z_3 \end{vmatrix}.
$$

We cannot solve this system of equations because x, y, and z can have any values, but we don't need to solve it! We just have to guarantee that this system has a solution. In general, a system of linear algebraic equations has a solution if and only if its determinant is zero. The expression below assumes this and also expands the determinant by its top row:

$$
0 = \begin{vmatrix} x & y & z & 1 \\ x_1 & y_1 & z_1 & 1 \\ x_2 & y_2 & z_2 & 1 \\ x_3 & y_3 & z_3 & 1 \end{vmatrix}
$$

$$
= x \begin{vmatrix} y_1 & z_1 & 1 \\ y_2 & z_2 & 1 \\ y_3 & z_3 & 1 \end{vmatrix} - y \begin{vmatrix} x_1 & z_1 & 1 \\ x_2 & z_2 & 1 \\ x_3 & z_3 & 1 \end{vmatrix} + z \begin{vmatrix} x_1 & y_1 & 1 \\ x_2 & y_2 & 1 \\ x_3 & y_3 & 1 \end{vmatrix} - \begin{vmatrix} x_1 & y_1 & z_1 \\ x_2 & y_2 & z_2 \\ x_3 & y_3 & z_3 \end{vmatrix}.
$$

This is of the form $Ax + By + Cz + D = 0$, so we conclude that

$$
A = \begin{vmatrix} y_1 & z_1 & 1 \\ y_2 & z_2 & 1 \\ y_3 & z_3 & 1 \end{vmatrix} \quad B = - \begin{vmatrix} x_1 & z_1 & 1 \\ x_2 & z_2 & 1 \\ x_3 & z_3 & 1 \end{vmatrix} \quad C = \begin{vmatrix} x_1 & y_1 & 1 \\ x_2 & y_2 & 1 \\ x_3 & y_3 & 1 \end{vmatrix} \quad D = - \begin{vmatrix} x_1 & y_1 & z_1 \\ x_2 & y_2 & z_2 \\ x_3 & y_3 & z_3 \end{vmatrix}.
$$

$$(1.24)$$

◇ **Exercise 1.36:** Calculate the expression of the plane containing the z axis and passing through the point $(1, 1, 0)$.

◇ **Exercise 1.37:** In the plane equation $Ax + By + Cz + D = 0$ if $D = 0$, then the plane passes through the origin. Assuming $D \neq 0$, we can write the same equation as $x/a + y/b + z/c = 1$, where $a = -D/A$, $b = -D/B$, and $c = -D/C$. What is the geometrical interpretation of a, b, and c?

> We operate with nothing but things which do not exist, with lines, planes, bodies, atoms, divisible time, divisible space—how should explanation even be possible when we first make everything into an image, into our own image!
>
> —Friedrich Nietzsche

In some practical applications, the normal to the plane and one point on the plane are known. It is easy to derive the plane equation in such a case.

We assume that $\mathbf{N}$ is the (known) normal vector to the plane, $\mathbf{P}_1$ is a known point on the plane, and $\mathbf{P}$ is an arbitrary point in the plane. The vector $\mathbf{P} - \mathbf{P}_1$ is perpendicular to $\mathbf{N}$, so their dot product $\mathbf{N} \bullet (\mathbf{P} - \mathbf{P}_1)$ equals zero. Since the dot product is associative, we can write $\mathbf{N} \bullet \mathbf{P} = \mathbf{N} \bullet \mathbf{P}_1$. The dot product $\mathbf{N} \bullet \mathbf{P}_1$ is just a number, to be denoted by s, so we get

$$\mathbf{N} \bullet \mathbf{P} = s \quad \text{or} \quad N_x x + N_y y + N_z z - s = 0. \tag{1.25}$$

Equation (1.25) can now be written as $Ax + By + Cz + D = 0$, where $A = N_x$, $B = N_y$, $C = N_z$, and $D = -s = -\mathbf{N} \bullet \mathbf{P}_1$. The three unknowns A, B, and C are the components of the normal vector, and D can be calculated from any known point $\mathbf{P}_1$ on the plane. The expression $\mathbf{N} \bullet \mathbf{P} = s$ is a useful equation of the plane and is used in many applications.

◇ **Exercise 1.38:** Given $\mathbf{N}(u, w) = (1, 1, 1)$ and $\mathbf{P}_1 = (1, 1, 1)$, calculate the plane equation.

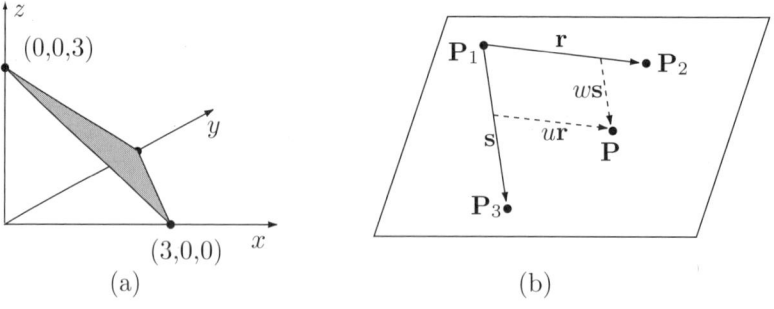

Figure 1.20: (a). A Plane. (b) Three Points on a Plane.

Note that the direction in which the normal is pointing is irrelevant for the plane equation. Substituting $(-A, -B, -C)$ for (A, B, C) would also change the sign of D, resulting in the same equation. However, the direction of the normal is important when a surface is to be shaded. We want the normal, in such a case, to point *outside* the surface. Often, this has to be done manually since the computer has no concept of the shape of the object in question and the meaning of the terms "inside" and "outside." However, in cases where a plane is defined by three points, the direction of the normal can be specified by arranging the three points (in the data structure in memory) in a certain order.

It is also easy to derive the equation of a plane when three points on the plane, $\mathbf{P}_1$, $\mathbf{P}_2$, and $\mathbf{P}_3$, are known. In order for the points to define a plane, they should not be collinear. We consider the vectors $\mathbf{r} = \mathbf{P}_2 - \mathbf{P}_1$ and $\mathbf{s} = \mathbf{P}_3 - \mathbf{P}_1$ a local coordinate system on the plane. Any point $\mathbf{P}$ on the plane can be expressed as a linear combination $\mathbf{P} = u\mathbf{r} + w\mathbf{s}$, where u and w are real numbers. Since $\mathbf{r}$ and $\mathbf{s}$ are local coordinates on the plane, the position of point $\mathbf{P}$ relative to the origin is expressed as (Figure 1.20b)

$$\mathbf{P}(u, w) = \mathbf{P}_1 + u\mathbf{r} + w\mathbf{s}, \quad -\infty < u, w < \infty. \tag{1.26}$$

$\diamond$ **Exercise 1.39:** Given the three points $\mathbf{P}_1 = (3, 0, 0)$, $\mathbf{P}_2 = (0, 3, 0)$, and $\mathbf{P}_3 = (0, 0, 3)$, write the equation of the plane defined by them.

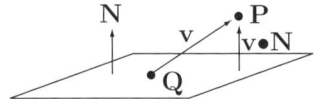

Figure 1.21: Distance of a Point from a Plane.

The next step is to determine the distance between a point and a plane. Given the point $\mathbf{P} = (x, y, z)$ and the plane $Ax + By + Cz + D = 0$, we select an arbitrary point $\mathbf{Q} = (x_0, y_0, z_0)$ on the plane. Since $\mathbf{Q}$ is on the plane, it satisfies $Ax_0 + By_0 + Cz_0 + D = 0$ or $-Ax_0 - By_0 - Cz_0 = D$. We construct the vector $\mathbf{v}$ from $\mathbf{Q}$ to $\mathbf{P}$ as the difference $\mathbf{v} = \mathbf{P} - \mathbf{Q} = (x - x_0, y - y_0, z - z_0)$. Figure 1.21 shows that the required distance (the size of the vector from the plane to $\mathbf{P}$ that's perpendicular to the plane) is the component $\mathbf{v}_N$ of $\mathbf{v}$ in the direction of the normal $\mathbf{N} = (A, B, C)$. This component is given by

$$
\begin{aligned}
\mathbf{v}_N &= \frac{|\mathbf{v} \bullet \mathbf{N}|}{|\mathbf{N}|} = \frac{|A(x - x_0) + B(y - y_0) + C(z - z_0)|}{\sqrt{A^2 + B^2 + C^2}} \\
&= \frac{|Ax + By + Cz - Ax_0 - By_0 - Cz_0|}{\sqrt{A^2 + B^2 + C^2}} \\
&= \frac{|Ax + By + Cz + D|}{\sqrt{A^2 + B^2 + C^2}}.
\end{aligned}
\tag{1.27}
$$

If we omit the absolute value, then the distance becomes a signed quantity. We can think of the plane as if it divides all of space into two parts, one in the direction of $\mathbf{N}$ and the other on the other side of the plane. The distance is positive if $\mathbf{P}$ is located in that part of space pointed to by the normal (which is the case in Figure 1.21), and it is negative in the opposite case.

◇ **Exercise 1.40:** What's the distance of a plane from the origin?

Now that we can figure out the distance between a point and a plane, the last step is to reflect a point about a given plane. We start with a point $\mathbf{P} = (x, y, z)$ and a plane $Ax + By + Cz + D = 0$. We denote the normal unit vector by $\mathbf{N} = (A, B, C)/\sqrt{A^2 + B^2 + C^2}$ and the (signed) distance between $\mathbf{P}$ and the plane by d. To get from $\mathbf{P}$ to the plane, we have to travel a distance d in the direction of $\mathbf{N}$. To arrive at the reflection point $\mathbf{P}^*$, we should travel another d units in the same direction. Thus, the reflection $\mathbf{P}^*$ of $\mathbf{P}$ is given by

$$\mathbf{P}^* = \mathbf{P} - 2d\mathbf{N} = \mathbf{P} - \frac{2(Ax + By + Cz + D)}{A^2 + B^2 + C^2}(A, B, C). \qquad (1.28)$$

◇ **Exercise 1.41:** Why $\mathbf{P} - 2d\mathbf{N}$ and not $\mathbf{P} + 2d\mathbf{N}$?

Most neurotics have been mindful of their five W's since grammar school: why, why, why, why, why.

—Terri Guillemets

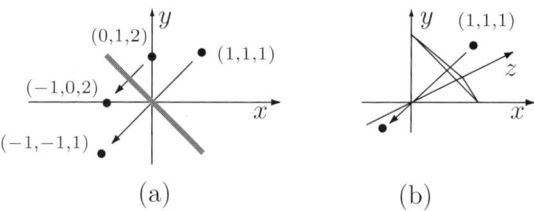

(a)   (b)

Figure 1.22: Reflection in Three Dimensions: Examples.

Examples. We select (Figure 1.22a) the plane $x + y = 0$ and the point $\mathbf{P} = (1, 1, 1)$. Equation (1.28) becomes

$$\mathbf{P}^* = (1, 1, 1) - \frac{2(1+1)}{1+1+0}(1, 1, 0) = (-1, -1, 1).$$

Similarly, point $\mathbf{P} = (0, 1, 2)$ is reflected to

$$\mathbf{P}^* = (0, 1, 2) - \frac{2(0+1)}{1+1+0}(1, 1, 0) = (-1, 0, 2).$$

We now select (Figure 1.22b) the plane $x + y + z - 1 = 0$ and the point $\mathbf{P} = (1, 1, 1)$. Equation (1.28) becomes

$$\mathbf{P}^* = (1, 1, 1) - \frac{2(1 + 1 + 1 - 1)}{1 + 1 + 1}(1, 1, 1) = -\frac{1}{3}(1, 1, 1).$$

Similarly, point $\mathbf{P} = (0, 0, 0)$ is reflected to

$$\mathbf{P}^* = (0, 0, 0) - \frac{2(0 + 0 + 0 - 1)}{1 + 1 + 1}(1, 1, 1) = \frac{2}{3}(1, 1, 1).$$

The special case of a reflection about one of the coordinate planes is also obtained from Equation (1.28). The equation of the xy plane, for example, is $z = 0$, where Equation (1.28) yields

$$\mathbf{P}^* = (x, y, z) - \frac{2(0 + 0 + z + 0)}{0^2 + 0^2 + 1^2}(0, 0, 1) = (x, y, -z).$$

1.4.2 Rotation

Rotation in three dimensions is difficult to visualize and is often confusing. One approach is to write three rotation matrices that rotate about the three coordinate axes:

$$\begin{pmatrix} \cos\theta & -\sin\theta & 0 & 0 \\ \sin\theta & \cos\theta & 0 & 0 \\ 0 & 0 & 1 & 0 \\ 0 & 0 & 0 & 1 \end{pmatrix}, \begin{pmatrix} \cos\theta & 0 & -\sin\theta & 0 \\ 0 & 1 & 0 & 0 \\ \sin\theta & 0 & \cos\theta & 0 \\ 0 & 0 & 0 & 1 \end{pmatrix}, \begin{pmatrix} 1 & 0 & 0 & 0 \\ 0 & \cos\theta & -\sin\theta & 0 \\ 0 & \sin\theta & \cos\theta & 0 \\ 0 & 0 & 0 & 1 \end{pmatrix}.$$

$$(1.29)$$

Let's look at the first of these matrices. Its third row and third column are $(0, 0, 1, 0)$, which is why multiplying a point $(x, y, z, 1)$ by this matrix leaves its z coordinate unchanged. The sines and cosines in the first two rows and two columns mix up the x and y coordinates in a way similar to a two-dimensional rotation [Equation (1.4)]. Thus, this transformation matrix causes a rotation about the z axis. The two other matrices rotate about the y and x axes.

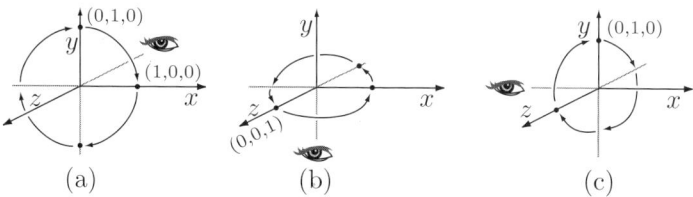

Figure 1.23: Rotating About the Coordinate Axes.

> Okay, so I assume going into this tutorial that you know how to perform matrix multiplication. I don't care to explain it, and it's available all over the Internet. However, once you know how to perform that operation, you should be good to go for this tutorial.
>
> (Found on the Internet)

It is therefore easy to identify the axis of rotation for each of the three rotation matrices of Equation (1.29), but what about their direction of rotation? To figure out the directions, we select $\theta = 90°$ and substitute $\sin\theta = 1$ and $\cos\theta = 0$. Simple tests in a right-handed coordinate system show that the first matrix of Equation (1.29) (rotation about the z axis) rotates point $(1, 0, 0)$ to $(0, -1, 0)$ and point $(0, 1, 0)$ to $(1, 0, 0)$. Thus, when we observe this 90° rotation looking in the direction of positive z, the rotation is counterclockwise (Figure 1.23a). The second matrix, however, behaves differently. It rotates point $(1, 0, 0)$ to $(0, 0, -1)$ and point $(0, 0, 1)$ to $(1, 0, 0)$. When we observe this 90° rotation about the y axis looking in the direction of positive y, the rotation is clockwise (Figure 1.23b). The third matrix (rotation about the x axis) rotates point $(0, 1, 0)$ to $(0, 0, -1)$ and point $(0, 0, 1)$ to $(0, 1, 0)$. When we observe this 90° rotation looking in the direction of positive x, the rotation is counterclockwise (Figure 1.23c).

We therefore decide (somewhat arbitrarily) to switch the signs (positive and negative) of the sine functions in the matrices that rotate about the z and x axes. The result,

$$
\begin{pmatrix} \cos\theta & \sin\theta & 0 & 0 \\ -\sin\theta & \cos\theta & 0 & 0 \\ 0 & 0 & 1 & 0 \\ 0 & 0 & 0 & 1 \end{pmatrix}, \quad
\begin{pmatrix} \cos\theta & 0 & -\sin\theta & 0 \\ 0 & 1 & 0 & 0 \\ \sin\theta & 0 & \cos\theta & 0 \\ 0 & 0 & 0 & 1 \end{pmatrix}, \quad
\begin{pmatrix} 1 & 0 & 0 & 0 \\ 0 & \cos\theta & \sin\theta & 0 \\ 0 & -\sin\theta & \cos\theta & 0 \\ 0 & 0 & 0 & 1 \end{pmatrix},
$$

(1.30)

is a set of three rotation matrices that rotate a point about the three coordinate axes in such a way that if we look in the positive direction of that axis, the rotation is clockwise.

(Surprisingly, it turns out that there is an elegant way to specify the direction of rotation that's generated by the rotation matrices of Equation (1.29), and this is described below.)

The rotation matrices of Equations (1.29) and (1.30) are simple but not very useful because in practice we rarely know how to break a general rotation into three rotations about the coordinate axes. There are some cases, however, where rotations about the coordinate axes are common. One such case is discussed in Section 2.2; two more are presented here.

Case 1: Rotations about the coordinate axes are common in the motion of a submarine or an airplane. These vehicles have three degrees of freedom and have three natural, mutually perpendicular axes of rotation that are called *roll*, *pitch*, and *yaw* (Figure 1.24). Roll is a rotation about the direction of motion of the vehicle. An airplane rolls when it banks by dipping one wing and lifting the other. Pitch is an up or down rotation about an axis that goes through the wings. An airplane uses its elevators for this. Yaw is a left–right rotation about a vertical axis, accomplished by the rudder. These terms originated with sailors because a ship can yaw and also has limited roll and pitch capabilities.

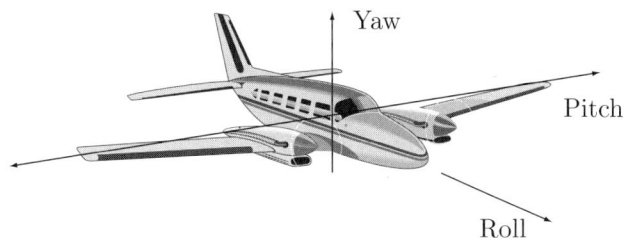

Figure 1.24: Roll, Pitch, and Yaw.

Case 2: Another example of an application where rotations about the three coordinate axes are common is L-systems. This is a system of formal notation developed by the biologist Aristid Lindenmayer (hence the "L") in 1968 as a tool to describe the morphology of plants [Lindenmayer 68]. In the 1970s, this notation was adopted by computer scientists and used to define formal languages. Since 1984, it has also been used to describe and draw many types of fractals. Today, L-systems are used to generate tilings, geometric art, and even music.

The main idea of L-systems is to define a complex object by (1) defining an initial simple object, called the *axiom*, and (2) writing rules that show how to replace parts of the axiom. The rules are written in terms of *turtle moves*, a concept originally introduced in the LOGO programming language [Abelson and DiSessa 82]. L-systems, however, specify the structure of three-dimensional objects, so the turtle must move in three dimensions and can rotate about its three main axes. For more information on L-systems, see [Prusinkiewicz 89].

It has already been mentioned that rotation in three dimensions is more complex than in two dimensions. One reason for this is that rotation in two dimensions is about a point, whereas rotation in three dimensions is about an axis (any axis, not just one of the three coordinate axes). Another reason is that the direction of rotation in two dimensions can be only clockwise or counterclockwise, but the direction of rotation in three dimensions is more complex to specify. The rotation is about an axis, but its direction, clockwise or counterclockwise, about this axis depends on how we look at the axis. Thus, a general rule is needed to specify the direction of a three-dimensional rotation unambiguously. We state such a rule for the rotation matrices of Equation (1.29).

The direction of a three-dimensional rotation generated by the matrices of (1.29) in a right-handed coordinate system is determined by the following rule: Write down the sequence "x, y, z" and erase the symbol that corresponds to the axis of rotation. The two remaining symbols are denoted by l and r. Draw the coordinate axes such that the positive direction of l will be up and the positive direction of r will be to the right. (This is not a necessary requirement, but it conforms to Figure 1.25.) The rotation will then be from positive r to positive l to negative r to negative l (Figure 1.25 and see also Exercise 3.12).

Example: A rotation about the z axis produced by the leftmost matrix of (1.29). After erasing z, the two symbols left are x and y. We draw the coordinate axes such that positive x is up and positive y is to the right. The matrix produces

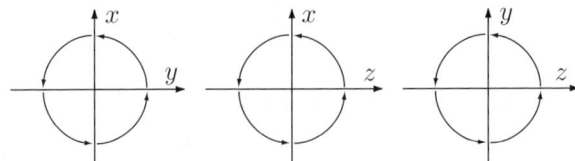

Figure 1.25: Direction of Three-Dimensional Rotations.

counterclockwise rotation. To achieve clockwise rotation, either use a negative angle or the inverse of the rotation matrix. Inverting our rotation matrices is especially easy and requires only that we change the signs of the sine functions.

Example. Consider the following compound transformation: (1) a translation by l, m, and n units along the three coordinate axes, (2) a rotation of θ degrees about the x axis, (3) a rotation of ϕ degrees about the y axis, and (4) the reverse translation. The four transformation matrices are

$$
\mathbf{T}_r = \begin{pmatrix} 1 & 0 & 0 & 0 \\ 0 & 1 & 0 & 0 \\ 0 & 0 & 1 & 0 \\ l & m & n & 1 \end{pmatrix}, \quad
\mathbf{T}_{rr} = \begin{pmatrix} 1 & 0 & 0 & 0 \\ 0 & 1 & 0 & 0 \\ 0 & 0 & 1 & 0 \\ -l & -m & -n & 1 \end{pmatrix},
$$

$$
\mathbf{R}_x = \begin{pmatrix} 1 & 0 & 0 & 0 \\ 0 & \cos\theta & \sin\theta & 0 \\ 0 & -\sin\theta & \cos\theta & 0 \\ 0 & 0 & 0 & 1 \end{pmatrix}, \quad
\mathbf{R}_y = \begin{pmatrix} \cos\phi & 0 & -\sin\phi & 0 \\ 0 & 1 & 0 & 0 \\ \sin\phi & 0 & \cos\phi & 0 \\ 0 & 0 & 0 & 1 \end{pmatrix}.
$$

Their product equals the 4×4 matrix

$$
\mathbf{T} = \mathbf{T}_r \mathbf{R}_x \mathbf{R}_y \mathbf{T}_{rr}
$$

$$
= \begin{pmatrix}
\cos\phi & 0 & -\sin\phi & 0 \\
\sin\phi\sin\theta & \cos\theta & \cos\phi\sin\theta & 0 \\
\cos\theta\sin\phi & -\sin\theta & \cos\phi\cos\theta & 0 \\
\begin{matrix} -l+l\cos\phi \\ +m\cos(\phi-\theta)/2 \\ -m\cos(\phi+\theta)/2 \\ +n\sin(\phi-\theta)/2 \\ +n\sin(\phi+\theta)/2 \end{matrix} &
\begin{matrix} -m \\ +m\cos\theta \\ -n\sin\theta \end{matrix} &
\begin{matrix} [-2n+n\cos(\phi-\theta) \\ +n\cos(\phi+\theta) \\ -2l\sin\phi \\ -m\sin(\phi-\theta) \\ +m\sin(\phi+\theta)]/2 \end{matrix} & 1
\end{pmatrix}.
$$

Substituting the values $\theta = 30°$, $\phi = 45°$, and $l = m = n = -1$, we get the 4×4 matrix

$$
\mathbf{T} = \begin{pmatrix}
0.7071 & 0 & -0.7071 & 0 \\
0.3540 & 0.866 & 0.3540 & 0 \\
0.6124 & -0.50 & 0.6124 & 0 \\
-0.673 & 0.634 & 0.7410 & 1
\end{pmatrix}.
$$

A point at $(1, 2, 3)$, for example, is transformed by $\mathbf{T}$ to the point

$$(1, 2, 3, 1)\mathbf{T} = (2.5793, 0.866, 2.5791, 1).$$

◇ **Exercise 1.42:** Do the same operations for the compound transformation $\mathbf{T}_r \mathbf{R}_x \mathbf{T}_{rr}$.

1.4.3 General Rotations

In practice, we normally don't know how to express an arbitrary rotation as a product of rotations about the coordinate axes, so we have to derive the important transformation of general rotation explicitly. The problem is easy to state. A point $\mathbf{P}$ is to be rotated through an angle θ about a specified axis. We first have to realize that there is a difference between an axis and a vector. A vector is fully specified by three numbers. It has direction and magnitude, but no specific location in space. An axis has both direction and location (it starts at a certain point), but its magnitude is normally irrelevant. A full specification of an axis requires a start point and a vector, a total of six numbers. (However, because the magnitude of the vector is irrelevant, it can be represented by two numbers only.) In order to simplify our derivation, we assume that our axis of rotation starts at the origin. If it starts at point $\mathbf{P}_0$, we have to precede the rotation by a translation of $\mathbf{P}_0$ to the origin and follow the rotation by the inverse translation.

We therefore denote by $\mathbf{u}$ a unit vector located on an axis that starts at the origin. We can now fully specify a general rotation in three dimensions by four numbers—the rotation angle θ and the three components of $\mathbf{u}$. The rotated point $\mathbf{P}$ ends up at $\mathbf{P}^*$. We connect $\mathbf{P}$ to the origin and call the resulting vector $\mathbf{r}$. Rotating point $\mathbf{P}$ to $\mathbf{P}^*$ is identical to rotating vector $\mathbf{r}$ to $\mathbf{r}^*$.

Figure 1.26a shows that the component OC of $\mathbf{r}$ along $\mathbf{u}$ is left unchanged, but the component CP is rotated to CP*. The distance OC is seen from the diagram to be $(\mathbf{r} \bullet \mathbf{u})$, so the vector $\vec{\text{OC}}$ can be written $(\mathbf{r} \bullet \mathbf{u})\mathbf{u}$. From $\mathbf{r} = \vec{\text{OC}} + \vec{\text{CP}}$, we get $\vec{\text{CP}} = \mathbf{r} - (\mathbf{r} \bullet \mathbf{u})\mathbf{u}$ or, in terms of magnitudes, $|\vec{\text{CP}}| = |\mathbf{r} - (\mathbf{r} \bullet \mathbf{u})\mathbf{u}|$. It can also be seen from the diagram that $|\vec{\text{CP}}| = |\mathbf{r}| \sin \phi$. Since $\mathbf{u}$ is a unit vector, we can write $|\mathbf{u} \times \mathbf{r}| = |\mathbf{r}| \sin \phi$. We thus obtain $|\vec{\text{CP}}| = |\mathbf{r} - (\mathbf{r} \bullet \mathbf{u})\mathbf{u}| = |\mathbf{u} \times \mathbf{r}|$.

Figure 1.26b shows the situation when looking from the origin in the positive $\mathbf{u}$ direction. (The diagram shows the tail of $\mathbf{u}$.) Note that the vector $\vec{\text{CQ}}$ is perpendicular to both $\mathbf{u}$ and $\mathbf{r}$, so it is in the direction of $\mathbf{u} \times \mathbf{r}$.

The next step is to resolve CP* into its components. From Figure 1.26b, we get

$$\vec{\text{CP}}^* = \cos \theta [\mathbf{r} - (\mathbf{r} \bullet \mathbf{u})\mathbf{u}] + \sin \theta [\mathbf{r} - (\mathbf{r} \bullet \mathbf{u})\mathbf{u}] = \cos \theta [\mathbf{r} - (\mathbf{r} \bullet \mathbf{u})\mathbf{u}] + \sin \theta (\mathbf{u} \times \mathbf{r}),$$

which can be used to express $\mathbf{r}^*$:

$$\mathbf{r}^* = \vec{\text{OC}} + \vec{\text{CP}}^* = (\mathbf{r} \bullet \mathbf{u})\mathbf{u} + \cos \theta [\mathbf{r} - (\mathbf{r} \bullet \mathbf{u})\mathbf{u}] + \sin \theta (\mathbf{u} \times \mathbf{r}). \qquad (1.31)$$

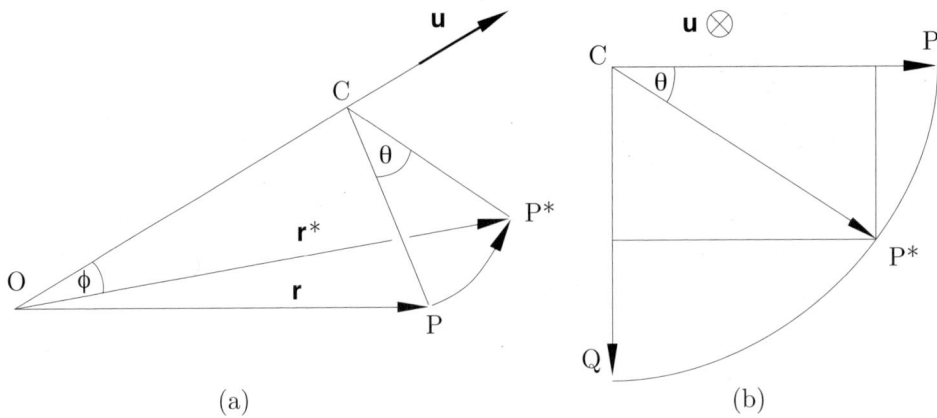

Figure 1.26: A General Rotation.

Using Equations (A.3) and (A.5) (page 222), we can rewrite this as $\mathbf{r}^* = (\mathbf{u}\mathbf{u}^T)\mathbf{r} + \cos\theta\mathbf{r} - \cos\theta(\mathbf{u}\mathbf{u}^T)\mathbf{r} + \sin\theta\mathbf{U}\mathbf{r}$, where

$$\mathbf{U} = \begin{pmatrix} 0 & -u_z & u_y \\ u_z & 0 & -u_x \\ -u_y & u_x & 0 \end{pmatrix}.$$

The result can now be summarized as $\mathbf{r}^* = \mathbf{M}\mathbf{r}$, where

$$\mathbf{M} = \mathbf{u}\mathbf{u}^T + \cos\theta(\mathbf{I} - \mathbf{u}\mathbf{u}^T) + \sin\theta\mathbf{U} \tag{1.32}$$
$$= \begin{bmatrix} u_x^2 + \cos\theta(1 - u_x^2) & u_x u_y(1 - \cos\theta) - u_z\sin\theta & u_x u_z(1 - \cos\theta) + u_y\sin\theta \\ u_x u_y(1 - \cos\theta) + u_z\sin\theta & u_y^2 + \cos\theta(1 - u_y^2) & u_y u_z(1 - \cos\theta) - u_x\sin\theta \\ u_x u_z(1 - \cos\theta) - u_y\sin\theta & u_y u_z(1 - \cos\theta) + u_x\sin\theta & u_z^2 + \cos\theta(1 - u_z^2) \end{bmatrix}.$$

Direction cosines. If $\mathbf{v} = (v_x, v_y, v_z)$ is a three-dimensional vector, its *direction cosines* are defined as

$$N_1 = \frac{v_x}{|\mathbf{v}|}, \quad N_2 = \frac{v_y}{|\mathbf{v}|}, \quad N_3 = \frac{v_z}{|\mathbf{v}|}.$$

These are the cosines of the angles between the direction of $\mathbf{v}$ and the three coordinate axes. It is easy to verify that $N_1^2 + N_2^2 + N_3^2 = 1$. If $\mathbf{u} = (u_x, u_y, u_z)$ is a unit vector, then $|\mathbf{u}| = 1$ and u_x, u_y, and u_z are the direction cosines of $\mathbf{u}$.

It can be shown that a rotation through an angle $-\theta$ is performed by the transpose $\mathbf{M}^T$. Consider the two successive and opposite rotations $\mathbf{r}^* = \mathbf{M}\mathbf{r}$ and $\mathbf{r}' = \mathbf{M}^T\mathbf{r}^*$. On the one hand, they can be expressed as the product $\mathbf{r}' = \mathbf{M}^T\mathbf{r}^* = \mathbf{M}^T\mathbf{M}\mathbf{r}$. On the other hand, they rotate in opposite directions, so they return all points to their original positions; therefore $\mathbf{r}'$ must be equal to $\mathbf{r}$. We end up with $\mathbf{r} = \mathbf{M}^T\mathbf{M}\mathbf{r}$ or $\mathbf{M}\mathbf{M}^T = \mathbf{I}$, where $\mathbf{I}$ is the identity matrix. The transpose $\mathbf{M}^T$ therefore equals the inverse, $\mathbf{M}^{-1}$, of $\mathbf{M}$, which shows that a rotation matrix $\mathbf{M}$ is orthogonal.

Example. Consider a rotation about the z axis. The rotation axis is $\mathbf{u} = (0, 0, 1)$, resulting in

$$\mathbf{u}\mathbf{u}^T = \begin{pmatrix} 0 & 0 & 0 \\ 0 & 0 & 0 \\ 0 & 0 & 1 \end{pmatrix} \text{ and } \mathbf{U} = \begin{pmatrix} 0 & -1 & 0 \\ 1 & 0 & 0 \\ 0 & 0 & 0 \end{pmatrix}, \text{ and hence } \mathbf{M} = \begin{pmatrix} \cos\theta & -\sin\theta & 0 \\ \sin\theta & \cos\theta & 0 \\ 0 & 0 & 1 \end{pmatrix},$$

which is the familiar rotation matrix about the z axis. It is identical to the z-rotation matrix of Equation (1.29), so we conclude that it rotates counterclockwise when viewed from the direction of positive z.

The general rotation matrix of Equation (1.32) can also be constructed as the product of five simple rotations about various coordinate axes. Given a unit vector $\mathbf{u} = (u_x, u_y, u_z)$, consider the following rotations.

1. Rotate $\mathbf{u}$ about the z axis into the xz plane, so its y coordinate becomes zero. This is done by a rotation matrix of the form

$$\mathbf{A} = \begin{bmatrix} \cos\psi & -\sin\psi & 0 \\ \sin\psi & \cos\psi & 0 \\ 0 & 0 & 1 \end{bmatrix},$$

and the angle ψ of rotation can be computed from the requirement that the y component of vector $\mathbf{v} = \mathbf{u}\mathbf{A}$ be zero. This component is $-u_x \sin\psi + u_y \cos\psi$, which implies $\cos\psi = u_x / \sqrt{u_x^2 + u_y^2}$ and $\sin\psi = u_y / \sqrt{u_x^2 + u_y^2}$. Notice that rotating $\mathbf{u}$ does not affect its magnitude, so $\mathbf{v}$ is also a unit vector. In addition, since the rotation is about the z axis, the z component of $\mathbf{u}$ does not change, so $v_z = u_z$.

2. Rotate vector $\mathbf{v}$ about the y axis until it coincides with the z axis. This is accomplished by the matrix

$$\mathbf{B} = \begin{bmatrix} \cos\phi & 0 & \sin\phi \\ 0 & 1 & 0 \\ -\sin\phi & 0 & \cos\phi \end{bmatrix}.$$

The angle ϕ of rotation is computed from the dot product $\cos\phi = \mathbf{v} \cdot (0, 0, 1) = v_z = u_z$, implying that $\sin\phi = \sqrt{1 - u_z^2}$. Since $\mathbf{v}$ is a unit vector, it is rotated by $\mathbf{B}$ to vector $(0, 0, 1)$.

3. Rotate $(0, 0, 1)$ about the z axis through an angle θ. This is done by matrix

$$\mathbf{C} = \begin{bmatrix} \cos\theta & -\sin\theta & 0 \\ \sin\theta & \cos\theta & 0 \\ 0 & 0 & 1 \end{bmatrix}.$$

This is a trivial rotation that does not change $(0, 0, 1)$.

4. Rotate the result of step 3 by $\mathbf{B}^{-1}$ (which equals $\mathbf{B}^T$).

5. Rotate the result of step 4 by $\mathbf{A}^{-1}$ (which equals $\mathbf{A}^T$).

When these five steps are performed on a point (x, y, z), the effect is to rotate the point through an angle θ about $\mathbf{u}$. In practice, the five steps are combined by multiplying the five matrices above, as shown in the listing of Figure 1.27. The result is identical to Equation (1.32).

```
tm=Sqrt[x^2+y^2];
a={{x/tm,-y/tm,0},{y/tm,x/tm,0},{0,0,1}};
b={{z,0,Sqrt[1-z^2]},{0,1,0},{-Sqrt[1-z^2],0,z}};
c={{Cos[t],-Sin[t],0},{Sin[t],Cos[t],0},{0,0,1}};
FullSimplify[a.b.c.Transpose[b].Transpose[a] /. x^2+y^2->1-z^2]
```

Figure 1.27: *Mathematica* Code for a General Rotation.

1.4.4 Givens Rotations

The general rotation matrix, Equation (1.32), can be constructed for any general rotation in three dimensions. Given such a matrix $\mathbf{A}$, it is possible to reduce it to a product of rotation matrices that cause the same rotation by performing a sequence of rotations about the coordinate axes. This process [Givens 58] is based on the QR decomposition of matrices, a subject discussed in any text on matrices, and it results in a set of *Givens rotations*. Each Givens rotation matrix $\mathbf{T}_{i,j}$ is identified by two indexes, i and j, where $i > j$. The matrix is an identity matrix except for the two diagonal elements (i, i) and (j, j) that are cosines of some angle and for the two off-diagonal elements (i, j) and (j, i) that are the $\pm \sin$ of the same angle. Specifically, $\mathbf{T}_{i,j}[i, i] = \mathbf{T}_{i,j}[j, j] = c$ and $\mathbf{T}_{i,j}[j, i] = -\mathbf{T}_{i,j}[i, j] = s$, where $c = \mathbf{A}[j, j]/D$, $s = \mathbf{A}[i, j]/D$, and $D = \sqrt{\mathbf{A}[j, j]^2 + \mathbf{A}[i, j]^2}$. The special construction of $\mathbf{T}_{i,j}$ implies that the matrix product $\mathbf{T}_{i,j}\mathbf{A}$ transforms $\mathbf{A}$ to a matrix whose (i, j)th element is zero.

Once a general rotation matrix $\mathbf{A}$ is given, its Givens rotations can be found by preparing the Givens rotation matrices $\mathbf{T}_{i,j}$ that zero those elements of $\mathbf{A}$ located below the main diagonal, column by column, from the bottom up. Figure 1.28 is a listing of *Matlab* code that does that for the rotation matrix that rotates point $(1, 1, 1)$ to the x axis.

The three rotation matrices produced by this computation are listed in Figure 1.29, where they are used to rotate point $(1, 1, 1)$ to the x axis. Matrix T1 rotates $(1, 1, 1)$ $45°$ about the y axis to $(1.4142, 1, 0)$, which is rotated by T2 $35.26°$ about the z axis to $(1.7321, 0, 0)$, which is trivially rotated by T3 $15°$ about the x axis to itself.

1.4.5 Quaternions

Appendix B is a general introduction to quaternions and should be reviewed before reading ahead. Quaternions can elegantly express arbitrary rotations in three dimensions. Those familiar with complex numbers may have noticed that a rotation in two

```
n=3;
A=[.5774,-.5774,-.5774; .5774,.7886,-.2115; .5774,-.2115,.7886]
% Rotation from 1,1,1 to x-axis
Q=eye(n);
for j=1:n-1,
  for i=n:-1:j+1,
    T=eye(n);
    D=sqrt(A(j,j)^2+A(i,j)^2);
    cos=A(j,j)/D; sin=A(i,j)/D;
    T(j,j)=cos; T(j,i)=sin; T(i,j)=-sin; T(i,i)=cos; T
    A=T*A;
Q=Q*T';
  end;
end;
Q
A
```

Figure 1.28: Computing Three Givens Matrices.

```
T1=[0.7071,0,0.7071; 0,1,0; -0.7071,0,0.7071];
T2=[0.8165,0.5774,0; -0.5774,0.8165,0; 0,0,1];
T3=[1,0,0; 0,0.9660,0.2587; 0,-0.2587,0.9660];
p=[1;1;1];
a=T1*p
b=T2*a
c=T3*b
```

Figure 1.29: Rotating Point $(1,1,1)$ to the x Axis.

J. Wallace Givens, Jr. (1910–1993) pioneered the use of plane rotations in the early days of automatic matrix computations. Givens graduated from Lynchburg College in 1928, and he completed his Ph.D. at Princeton University in 1936. After spending three years at the Institute for Advanced Study in Princeton as an assistant of Oswald Veblen, Givens accepted an appointment at Cornell University, but later moved to Northwestern University. In addition to his academic career, Givens was the director of the Applied Mathematics Division at Argonne National Laboratory and, like his counterpart Alston Scott Householder at Oak Ridge National Laboratory, Givens served as an early president of SIAM. He published his work on the rotations in 1958.

—Carl D. Meyer

dimensions is similar to multiplying two complex numbers because the product

$$(a, b) \begin{pmatrix} c & d \\ -d & c \end{pmatrix} = (ac - bd, ad + bc)$$

is identical to the product $(a + ib)(c + id)$. Quaternions extend this similarity to three dimensions as follows. To rotate a point $\mathbf{P}$ by an angle θ about a direction $\mathbf{v}$, we first prepare the quaternion $\mathbf{q} = [\cos(\theta/2), \sin(\theta/2)\mathbf{u}]$, where $\mathbf{u} = \mathbf{v}/|\mathbf{v}|$ is a unit vector in the direction of $\mathbf{v}$. The rotation can then be expressed as the triple product $\mathbf{q} \cdot [0, \mathbf{P}] \cdot \mathbf{q}^{-1}$. Note that our $\mathbf{q}$ is a unit quaternion since $\sin^2(\theta/2) + \cos^2(\theta/2) = 1$.

⋄ **Exercise 1.43:** Prove that the triple product $\mathbf{q} \cdot [0, \mathbf{P}] \cdot \mathbf{q}^{-1}$ really performs a rotation of $\mathbf{P}$ about $\mathbf{v}$ (or $\mathbf{u}$). [Hint: Perform the multiplications and show that they produce Equation (1.31).]

As an example of quaternion rotation, consider a $90°$ rotation of point $\mathbf{P} = (0, 1, 1)$ about the y axis. The quaternion required is $\mathbf{q} = [\cos 45°, \sin 45°(0, 1, 0)]$. It is a unit quaternion, so its inverse is $\mathbf{q}^{-1} = [\cos 45°, -\sin 45°(0, 1, 0)]$. The rotated point is thus

$$\mathbf{q}[0, \mathbf{P}]\mathbf{q}^{-1}$$
$$= [-\sin 45°, (\sin 45°, \cos 45°, \cos 45°)] [0, (0, 1, 1)] [\cos 45°, -\sin 45°(0, 1, 0)]$$
$$= [0, (1, 1, 0)].$$

The quaternion resulting from the triple product always has a zero scalar. We ignore the scalar and find that the point has been moved, by the rotation, from the $x = 0$ plane to the $z = 0$ plane.

Figure 1.30 illustrates this particular rotation about the y axis and also makes it easy to understand the rule for the direction of the quaternion rotation $\mathbf{q}[0, \mathbf{P}]\mathbf{q}^{-1}$. The rule is: Let $\mathbf{q} = [s, \mathbf{v}]$ be a rotation quaternion in a right-handed three-dimensional coordinate system. To an observer looking in the direction of $\mathbf{v}$, the triple product $\mathbf{q}[0, \mathbf{P}]\mathbf{q}^{-1}$ rotates point $\mathbf{P}$ clockwise. For a negative rotation angle, the rotation is counterclockwise. In a left-handed coordinate system (Figure 1.30b), the direction of rotation is the opposite.

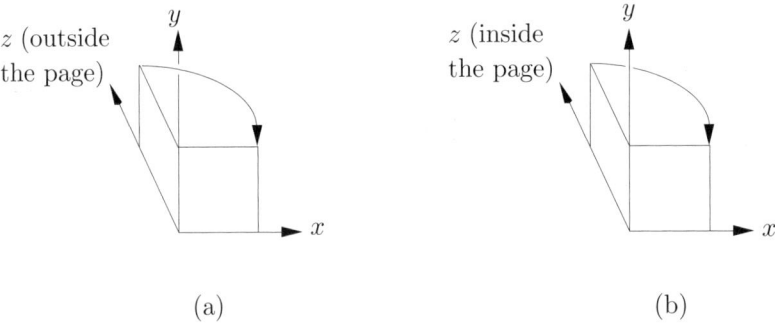

(a)   (b)

Figure 1.30: Rotation in a Right-Handed (a) and in a Left-Handed (b) Coordinate System.

1.4.6 Concatenating Rotations

Sometimes, we have to perform two consecutive rotations on an object. This turns out to be easy and numerically stable with a quaternion representation.

If $\mathbf{q}_1$ and $\mathbf{q}_2$ are unit quaternions representing the two rotations, then associativity of quaternion multiplication implies that the combined rotation of $\mathbf{q}_1$ followed by $\mathbf{q}_2$ is represented by the quaternion $\mathbf{q}_2 \cdot \mathbf{q}_1$. The proof is

$$\mathbf{q}_2 \cdot (\mathbf{q}_1 \cdot \mathbf{P} \cdot \mathbf{q}_1^{-1}) \cdot \mathbf{q}_2^{-1} = (\mathbf{q}_2 \cdot \mathbf{q}_1) \cdot \mathbf{P} \cdot (\mathbf{q}_1^{-1} \cdot \mathbf{q}_2^{-1}) = (\mathbf{q}_2 \cdot \mathbf{q}_1) \cdot \mathbf{P} \cdot (\mathbf{q}_2 \cdot \mathbf{q}_1)^{-1}.$$

Quaternion multiplication involves fewer operations than matrix multiplication, so combining rotations by means of quaternions is faster. Performing fewer multiplications also implies better numerical accuracy.

In general, we use 4×4 transformation matrices to express three-dimensional transformations, so we would like to be able to express the rotation $\mathbf{P}^* = \mathbf{q}[0, \mathbf{P}]\mathbf{q}^{-1}$ as $\mathbf{P}^* = \mathbf{PM}$, where $\mathbf{M}$ is a 4×4 matrix. Given the two quaternions $\mathbf{q}_1 = w_1 + x_1\mathbf{i} + y_1\mathbf{j} + z_1\mathbf{k} = (w_1, x_1, y_1, z_1)$ and $\mathbf{q}_2 = w_2 + x_2\mathbf{i} + y_2\mathbf{j} + z_2\mathbf{k} = (w_2, x_2, y_2, z_2)$, their product is

$$\mathbf{q}_1 \cdot \mathbf{q}_2 = (w_1w_2 - x_1x_2 - y_1y_2 - z_1z_2) + (w_1x_2 + x_1w_2 + y_1z_2 - z_1y_2)\mathbf{i}$$
$$+ (w_1y_2 - x_1z_2 + y_1w_2 + z_1x_2)\mathbf{j} + (w_1z_2 + x_1y_2 - y_1x_2 + z_1w_2)\mathbf{k}.$$

The first step is to realize that each term in this product depends linearly on the coefficients of $\mathbf{q}_1$. This product can therefore be expressed as

$$\mathbf{q}_1 \cdot \mathbf{q}_2 = \mathbf{q}_2 \cdot \mathbf{L}(\mathbf{q}_1) = (x_2, y_2, z_2, w_2) \begin{pmatrix} w_1 & z_1 & -y_1 & -x_1 \\ -z_1 & w_1 & x_1 & -y_1 \\ y_1 & -x_1 & w_1 & -z_1 \\ x_1 & y_1 & z_1 & w_1 \end{pmatrix}.$$

When $\mathbf{L}(\mathbf{q}_1)$ multiplies the row vector $\mathbf{q}_2$, the result is a row vector representation for $\mathbf{q}_1 \cdot \mathbf{q}_2$. Each term also depends linearly on the coefficients of $\mathbf{q}_2$, so the same product can also be expressed as

$$\mathbf{q}_1 \cdot \mathbf{q}_2 = \mathbf{q}_1 \cdot \mathbf{R}(\mathbf{q}_2) = (x_1, y_1, z_1, w_1) \begin{pmatrix} w_2 & -z_2 & y_2 & -x_2 \\ z_2 & w_2 & -x_2 & -y_2 \\ -y_2 & x_2 & w_2 & -z_2 \\ x_2 & y_2 & z_2 & w_2 \end{pmatrix}.$$

When $\mathbf{R}(\mathbf{q}_2)$ multiplies the row vector $\mathbf{q}_1$, the result is also a row vector representation for $\mathbf{q}_1 \cdot \mathbf{q}_2$.

We can now write the triple product $\mathbf{q} \cdot [0, \mathbf{P}] \cdot \mathbf{q}^{-1}$ in terms of the matrices $\mathbf{L}(\mathbf{q})$ and $\mathbf{R}(\mathbf{q})$:

$$\mathbf{q}[0, \mathbf{P}]\mathbf{q}^{-1} = \mathbf{q}([0, \mathbf{P}] \cdot \mathbf{q}^{-1}) = \mathbf{q}([0, \mathbf{P}]\mathbf{R}(\mathbf{q}^{-1}))$$
$$= ([0, \mathbf{P}]\mathbf{R}(\mathbf{q}^{-1}))\mathbf{L}(\mathbf{q}) = [0, \mathbf{P}](\mathbf{R}(\mathbf{q}^{-1})\mathbf{L}(\mathbf{q}))$$
$$= [0, \mathbf{P}]\mathbf{M},$$

where matrix $\mathbf{M}$ is

$$\mathbf{M} = \mathbf{R}(\mathbf{q}^{-1}) \cdot \mathbf{L}(\mathbf{q})$$

$$= \begin{pmatrix} w & z & -y & x \\ -z & w & x & y \\ y & -x & w & z \\ -x & -y & -z & w \end{pmatrix} \begin{pmatrix} w & z & -y & -x \\ -z & w & x & -y \\ y & -x & w & -z \\ x & y & z & w \end{pmatrix}$$

$$= \begin{pmatrix} w^2 + x^2 - y^2 - z^2 & 2xy + 2wz & 2xz - 2wy & 0 \\ 2xy - 2wz & w^2 - x^2 + y^2 - z^2 & 2yz + 2wx & 0 \\ 2xz + 2wy & 2yz - 2wx & w^2 - x^2 - y^2 + z^2 & 0 \\ 0 & 0 & 0 & w^2 + x^2 + y^2 + z^2 \end{pmatrix}.$$

Since we have unit quaternions, they satisfy $w^2 + x^2 + y^2 + z^2 = 1$, so we can write the final result

$$\mathbf{M} = \begin{pmatrix} 1 - 2y^2 - 2z^2 & 2xy + 2wz & 2xz - 2wy & 0 \\ 2xy - 2wz & 1 - 2x^2 - 2z^2 & 2yz - 2wx & 0 \\ 2xz + 2wy & 2yz - 2wx & 1 - 2x^2 - 2y^2 & 0 \\ 0 & 0 & 0 & 1 \end{pmatrix}. \tag{1.33}$$

In a left-handed coordinate system, the same rotation is expressed by the triple product $\mathbf{q}^{-1}[0, \mathbf{P}]\mathbf{q}$ or, equivalently, by $\mathbf{P}^* = \mathbf{P} \cdot \mathbf{M}^T$, where $\mathbf{M}^T$ is the transpose of $\mathbf{M}$.

1.5 Transforming the Coordinate System

Our discussion so far has assumed that points are transformed in a static coordinate system. It is also possible (and sometimes useful) to transform the coordinate system instead of the points. To understand the main idea, let's consider the simple example of translation. Suppose that a two-dimensional point $\mathbf{P}$ is transformed to a point $\mathbf{P}^*$ by translating it m and n units in the x and y directions, respectively. How can the transformation be reversed? We consider two ways.

1. Suppose that the original transformation was $\mathbf{P}^* = \mathbf{PT}$, where

$$\mathbf{T} = \begin{pmatrix} 1 & 0 & 0 \\ 0 & 1 & 0 \\ m & n & 1 \end{pmatrix}.$$

It is clear that the transformation matrix

$$\mathbf{S} = \begin{pmatrix} 1 & 0 & 0 \\ 0 & 1 & 0 \\ -m & -n & 1 \end{pmatrix}$$

will transform $\mathbf{P}^*$ back to $\mathbf{P}$. However, it is trivial to show, by using Equation (1.22), that $\mathbf{S}$ is the inverse of $\mathbf{T}$.

2. The transformation can be reversed by translating the coordinate system in the reverse directions (i.e., by $-m$ and $-n$ units) by using an (unknown) transformation matrix $\mathbf{M}$.

Since the two methods produce the same result, we conclude that $\mathbf{M} = \mathbf{S} = \mathbf{T}^{-1}$. Transforming the coordinate axes is therefore done by a matrix that's the inverse of transforming a point. This is true for any affine transformations, not just translation.

> The magic transformation from the minute to the vast has not been so cunningly effected but that the rich adornment still counteracts the impression of space and loftiness.
>
> —Nathaniel Hawthorne, *The Marble Faun* (1860)

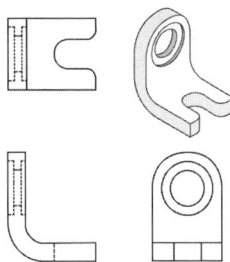

2
Parallel Projections

There are several variants of parallel projections, but they are all based on the following principle: Select a direction **v** and construct a ray that starts at a general point **P** on the object and goes in the direction **v**. The point **P*** where this ray intercepts the projection plane becomes the projection of **P**. The process is repeated for all the points on the object, creating a set of parallel rays, which is why this class of projections is called parallel. Figure 2.1 illustrates the principle of parallel projections. In Figure 2.1a the rays are perpendicular to the projection plane and in Figure 2.1b they strike at a different angle. This is why the latter method is called *oblique projection* (Section 2.3).

Figure 2.1c shows a different interpretation of parallel projections. Because the rays are parallel, we can imagine that they originate at a *center of projection* located at infinity. This interpretation unifies parallel and perspective projections and is in accordance with the general rule of projections (page 2) which distinguishes between parallel and perspective projections by the location of the center of projection.

The three types of parallel projections are orthographic, axonometric, and oblique.

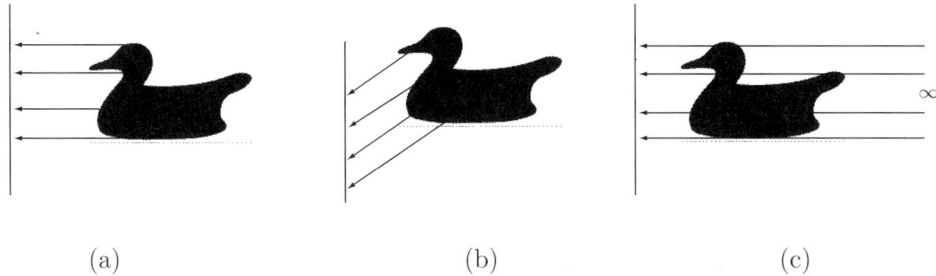

(a)   (b)   (c)

Figure 2.1: Parallel Projections.

I will sette as I doe often in woorke use, a paire of paralleles, or [twin] lines of one lengthe, thus =, bicause noe 2. thynges, can be moare equalle.

—Robert Recorde, 1557

2.1 Orthographic Projections

The term orthographic (or orthography) is derived from the Greek $o\rho\theta o$ (correct) and $\gamma\rho\alpha\varphi o\zeta$ (that writes). This term is used in several areas, such as orthographic projection of a sphere (page 206) and the orthography of a language. The latter is the set of rules that specify correct writing in a language. An example of an orthographic rule in English is i comes before e (as in "view") except after a c (as in "ceiling").

The family of orthographic projections is the simplest of the three types of parallel projections. The principle is to imagine a box around the object to be projected and to project the object "flat" on each of the six sides of the box (Figure 2.2a). If the object is simple and familiar, three projections, on three orthogonal sides, may be enough (Figure 2.2b). If the object is complex or is unfamiliar, a perspective projection may be needed in addition to the three or six parallel projections. For even more complex objects, sectional views may be necessary. Such a view is obtained by passing an imaginary plane through the object and drawing a projection of the plane.

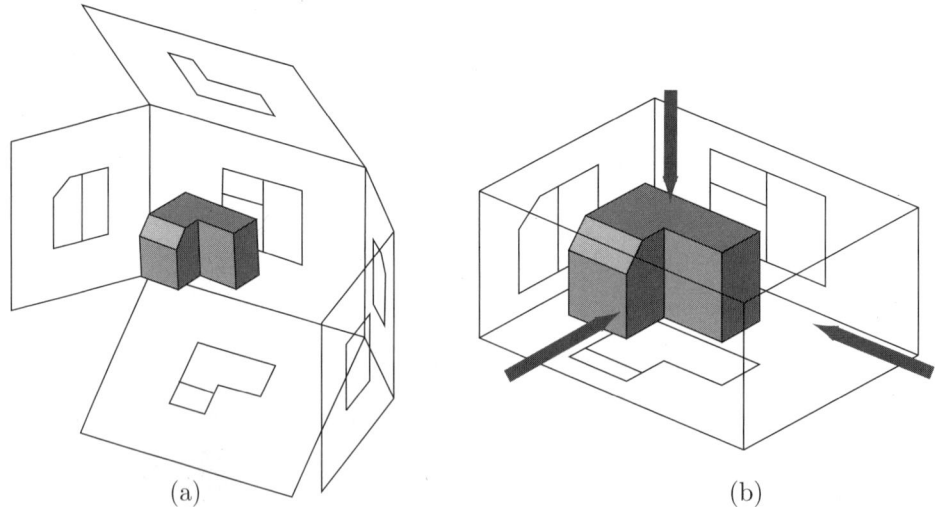

(a)   (b)

Figure 2.2: Six and Three Orthographic Projections.

If one side of the box is the xy plane, then a point $\mathbf{P} = (x, y, z)$ is projected on this side by removing its z coordinate to become $\mathbf{P}^* = (x, y)$. This operation can be carried out formally by multiplying $\mathbf{P}$ by matrix $\mathbf{T}_z$ of Equation (2.1). Similarly, matrices $\mathbf{T}_x$

and $\mathbf{T}_y$ project points orthographically on the yz and the xz planes, respectively.

$$\mathbf{T}_x = \begin{pmatrix} 0 & 0 & 0 \\ 0 & 1 & 0 \\ 0 & 0 & 1 \end{pmatrix}, \ \mathbf{T}_y = \begin{pmatrix} 1 & 0 & 0 \\ 0 & 0 & 0 \\ 0 & 0 & 1 \end{pmatrix}, \ \mathbf{T}_z = \begin{pmatrix} 1 & 0 & 0 \\ 0 & 1 & 0 \\ 0 & 0 & 0 \end{pmatrix}. \tag{2.1}$$

The object of Figure 2.2 has two properties that make it especially easy to project. It is similar to a cube, and its edges are aligned with the coordinate axes. In general, if the main edges of the object are not aligned with the coordinate axes, its orthographic projections along the axes may look unfamiliar and confusing, and it is preferable to rotate the object, if at all possible, and align it before it is projected. If the object is not cubical, the best option is to select on the object three axes that are judged the "main" ones and align them with the coordinate axes. The object is then surrounded by a bounding box (Figure 2.3) and the box is projected. Once this is done, the object is transferred into the projected bounding box in a process similar to that described in Section 3.3. If the object is so complex that it is impossible to find three such axes, then the designer should consider projecting several sectional views of the object or using a nonorthographic projection.

Figure 2.3: Orthographic Projection of a Curved Object.

◇ **Exercise 2.1:** Try to interpret the three orthographic projections of Figure 2.4.

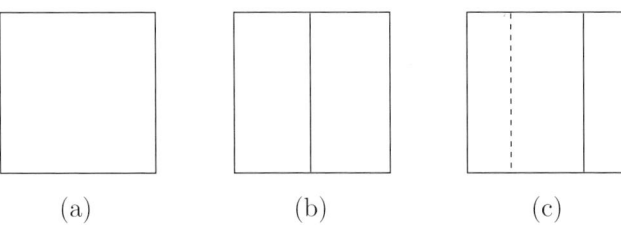

(a)   (b)   (c)

Figure 2.4: Three Orthographic Projections for Exercise 2.1.

The main advantage of orthographic projections is the ease of measuring dimensions. The projection of a segment of length l on the object is a segment of length l (or of a length related to l in a simple way) on the projection plane. This helps in manufacturing an object directly from a drawing and is the main reason orthographic projections are used in technical drawing.

Figure 2.5 shows a side view and the top view of a thin hexagon. It is easy to see that a segment of length l on side a becomes a segment of the same length on the projection, while a segment of length l on side b becomes a segment of length $l \cos \beta$ on the projection (where $\beta = 270° - \alpha$).

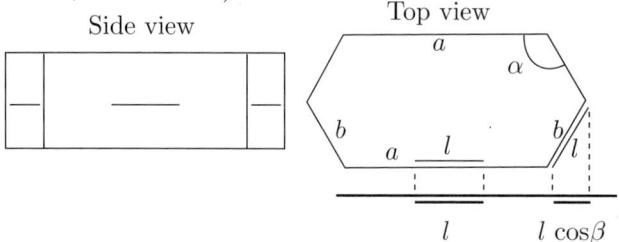

Figure 2.5: Segments on the Sides of a Hexagon.

> I feel like I am diagonally parked in a parallel universe.
>
> —Unknown

2.2 Axonometric Projections

The term axonometric is derived from the Greek $\alpha\xi\omega\nu$ or $\alpha\xi o\nu\alpha\varsigma$ (axon, axis) and $\mu\epsilon\tau\rho o\nu$ (metron, a measure). We approach this type of parallel projections from two points of view.

Approach 1: Linear perspective, the topic of Chapter 3, was developed in the West during the Renaissance and is based on geometric optics. The observer is considered a point that receives straight rays of light and senses only the color, the intensity, and the direction of a ray but not the distance it has traveled. Oriental art, in contrast, has developed in a different direction and has adopted a different system of perspective, one that is suitable for scroll paintings.

A Chinese scroll painting is normally executed on a horizontal rectangle about 40 cm high and several meters long. The painting is viewed slowly from right to left while unrolling the scroll, and it tells a story in time. As the eye moves to the left, we see later occurrences of the same scene, not new views. We can call this approach to art "narrative," in contrast to Western art, which is situational. Figure C.4 (page 236) is an example of this type of art. It is a 33-foot-long scroll titled *A City of Cathay* that was painted by artists of the Qing court (1662–1795).

Because of the temporal approach to scroll art, Chinese (and other Oriental artists) had to develop a system of perspective with no vanishing points, no explicit light sources, and no shadows. The result was a special type of parallel perspective, known today as "Chinese perspective" or axonometric projection. If we imagine the scroll to be the xy plane and we view it along the z axis, then lines that are parallel to the z axis are drawn parallel on the scroll instead of converging to a vanishing point.

Approach 2: An orthographic projection of an object shows the details of only one of its main faces, which is why three or even six projections are needed. Each projection may be detailed and it may show the true shape of that face with the correct dimensions, but it shows little or nothing of the rest of the object. Thus, interpreting and understanding orthographic projections requires experience. Viewing an object from above, from below, and from four sides tends to confuse an inexperienced person. Engineers, architects, and designers may be familiar with orthographic projections, but they have to draw plans that will be viewed and comprehended by their superiors and customers, and this suggests a projection method that will include some perspective, will show more than one face of the object, and will also make it easy to compute dimensions from the drawing. Linear perspective is easy to visualize and understand, but for engineers and designers it has at least three disadvantages: (1) it is complex to compute and draw, (2) the relation between dimensions on the diagram and real dimensions of the object is complex, and (3) distant objects look small. A common compromise is a drawing in one of the three varieties of axonometric projections.

Axonometric projections show more of the object in each projection but at the price of having wrong dimensions and angles. An axonometric projection typically shows three or more faces of the object, but it shrinks some of the dimensions. When a dimension is measured on the drawing, some computations are needed to convert it to a true dimension on the object. This is an easy, albeit nontrivial, procedure. An axonometric projection shows the true shape of a face of the object (with true dimensions) only if the face happens to be parallel to the projection plane. Otherwise, the shape of the face is distorted and its dimensions are shrunk.

Before we get to the details, here is a summary of the properties of axonometric projections:

- Axonometric projections are parallel, so a group of parallel lines on the object will appear parallel in the projection.

- There are no vanishing points. Thus, a wide image can be scrolled slowly while different parts of it are observed. At every point, the viewer will see the same perspective.

- Distant objects retain their size regardless of their distance from the observer. If the parameters of the projection are known, then the dimensions of any object, far or nearby, can be computed from measurements taken on the projection.

- There are standards for axonometric projections. A standard may specify the orientation of the object relative to the observer, which makes it easy for the observer to compute distances directly from the projection.

To construct an axonometric projection, the object may first have to be rotated to bring the desired faces toward the projection plane. It is then projected on that plane in parallel. We assume that the projection plane is the xy plane, so the projection is done by clearing the z coordinates of all the points or, equivalently, by multiplying each point, after rotating it, by matrix $\mathbf{T}_z$ of Equation (2.1). Assuming that we first rotate the object ϕ degrees about the y axis and then θ degrees about the x axis, the combined rotation/projection matrix is [see Equation (1.30)]

$$\mathbf{T} = \begin{pmatrix} \cos\phi & 0 & -\sin\phi \\ 0 & 1 & 0 \\ \sin\phi & 0 & \cos\phi \end{pmatrix} \begin{pmatrix} 1 & 0 & 0 \\ 0 & \cos\theta & \sin\theta \\ 0 & -\sin\theta & \cos\theta \end{pmatrix} \begin{pmatrix} 1 & 0 & 0 \\ 0 & 1 & 0 \\ 0 & 0 & 0 \end{pmatrix}$$

$$= \begin{pmatrix} \cos\phi & \sin\phi\sin\theta & 0 \\ 0 & \cos\theta & 0 \\ \sin\phi & -\cos\phi\sin\theta & 0 \end{pmatrix}. \tag{2.2}$$

To find how various dimensions are affected by these transformations, we start with the vector $(1,0,0)$. This is a unit vector in the direction of the x axis. Multiplying it by $\mathbf{T}$ gives another vector, which we denote by $(x_1, x_2, 0)$. Its magnitude is $s_x = \sqrt{x_1^2 + x_2^2}$ and since the original vector had magnitude 1, the quantity s_x expresses the ratio of magnitudes or the factor by which all dimensions in the x direction have shrunk after the transformation/projection $\mathbf{T}$. Similarly, selecting unit vectors $(0,1,0)$ and $(0,0,1)$ in the y and z directions and multiplying them by $\mathbf{T}$ produces vectors $(y_1, y_2, 0)$ and $(z_1, z_2, 0)$ and shrinking factors $s_y = \sqrt{y_1^2 + y_2^2}$ and $s_z = \sqrt{z_1^2 + z_2^2}$ in the y and z directions, respectively.

Figure 2.6a shows a unit cube rotated such that its three sides, which used to be parallel to the coordinate axes, seem to have different lengths. Such an axonometric projection is called *trimetric*.

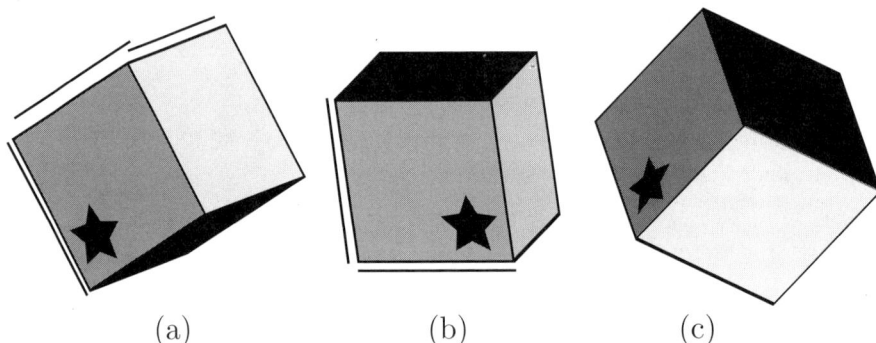

| (a) | (b) | (c) |

Figure 2.6: The Three Types of Axonometric Projections.

Figure 2.6b shows the same unit cube rotated such that two of its three sides seem to have the same length, while the third side looks shorter. Such an axonometric projection is called *dimetric*. Similarly, Figure 2.6c shows the same unit cube rotated such that all its sides seem to have the same length. This type of axonometric projection is called *isometric*.

Matrix $\mathbf{T}$ of Equation (2.2) can be used to calculate the special rotations that produce a dimetric projection. Consider the product of a unit vector in the x direction and $\mathbf{T}$:

$$(1,0,0) \begin{pmatrix} \cos\phi & \sin\phi\sin\theta & 0 \\ 0 & \cos\theta & 0 \\ \sin\phi & -\cos\phi\sin\theta & 0 \end{pmatrix} = (\cos\phi, \sin\phi\sin\theta, 0). \qquad (2.3)$$

This product shows that any vector in the x direction shrinks, after being rotated by matrix $\mathbf{T}$, by a factor s_x given by Equation (2.4). The same equation also produces the shrink factors s_y and s_z of any vector in the y and z directions.

$$s_x = \sqrt{\cos^2\phi + \sin^2\phi\sin^2\theta}, \quad s_y = \sqrt{\cos^2\theta}, \quad s_z = \sqrt{\sin^2\phi + \cos^2\phi\sin^2\theta}. \qquad (2.4)$$

If we want a dimetric projection where equal-size segments in the x and y directions will have equal sizes after the projection, we set $s_x = s_y$ or, equivalently,

$$\cos^2\phi + \sin^2\phi\sin^2\theta = \cos^2\theta,$$

which produces the relation

$$\sin^2\phi = \frac{\sin^2\theta}{1 - \sin^2\theta}. \qquad (2.5)$$

Equation (2.5) together with the expression for s_z^2 yields

$$\begin{aligned}
s_z^2 &= \sin^2\phi + \cos^2\phi\sin^2\theta = \sin^2\phi + (1 - \sin^2\phi)\sin^2\theta \\
&= \sin^2\phi(1 - \sin^2\theta) + \sin^2\theta \\
&= \frac{\sin^2\theta}{1 - \sin^2\theta}(1 - \sin^2\theta) + \sin^2\theta,
\end{aligned}$$

or $2\sin^4\theta - (2 + s_z^2)\sin^2\theta + s_z^2 = 0$, a quadratic equation in $\sin^2\theta$ whose solutions are $\sin^2\theta = s_z^2/2$ and $\sin^2\theta = 1$. The second solution cannot be used in Equation (2.5) and has to be discarded. The first solution produces

$$\theta = \sin^{-1}\left(\pm\frac{s_z}{\sqrt{2}}\right) \quad \text{and} \quad \phi = \sin^{-1}\left(\pm\frac{s_z}{\sqrt{2 - s_z^2}}\right). \qquad (2.6)$$

Since the sine function has values in the range $[-1, 1]$, the argument of $\sin^{-1}$ must be in this range. The expression $s_z/\sqrt{2}$ is in this range when $-\sqrt{2} \leq s_z \leq +\sqrt{2}$, and the expression $s_z/\sqrt{2 - s_z^2}$ is in this range when $-1 \leq s_z \leq +1$. Since s_z is a shrinking factor, it is nonnegative, which implies that it must be in the interval $[0, 1]$. Also, since Equation (2.6) contains a $\pm$, any value of s_z produces four solutions.

Example: Given $s_z = 1/2$, we calculate θ and ϕ:

$$\theta = \sin^{-1}\left(\pm\frac{0.5}{\sqrt{2}}\right) = \sin^{-1}(\pm 0.35355) = \pm 20.7°,$$

$$\phi = \sin^{-1}\left(\pm\frac{0.5}{\sqrt{2 - 0.5^2}}\right) = \sin^{-1}(\pm 0.378) = \pm 22.2°.$$

The two rotations are illustrated in Figure 2.7.

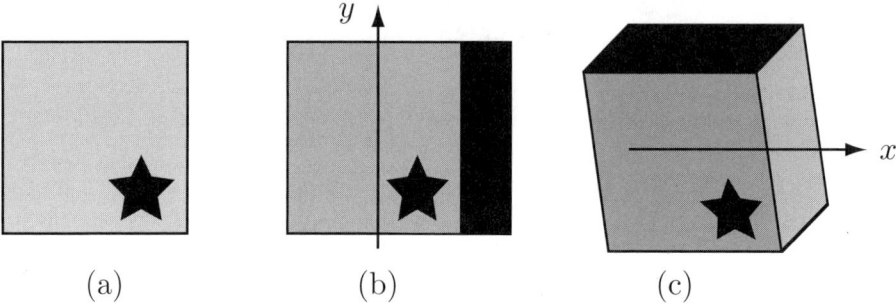

(a)   (b)   (c)

Figure 2.7: Rotations for Dimetric Projection.

◇ **Exercise 2.2:** Repeat the example for $s_z = 0.625$.

◇ **Exercise 2.3:** Calculate θ and ϕ for $s_x = s_z$ (equal shrink factors in the x and z directions).

The condition for an isometric projection (Figure 2.6c) is $s_x = s_y = s_z$. We already know that $s_x = s_y$ results in Equation (2.5). Similarly, it is easy to see that $s_y = s_z$ results in $\cos^2 \theta = \sin^2 \phi + \cos^2 \phi \sin^2 \theta$, which can be written

$$\sin^2 \phi = \frac{1 - 2\sin^2 \theta}{1 - \sin^2 \theta}. \tag{2.7}$$

Equations (2.5) and (2.7) result in $\sin^2 \theta = 1 - 2\sin^2 \theta$ or $\sin^2 \theta = 1/3$, yielding $\theta = \pm 35.26°$. The rotation angle ϕ can now be calculated from Equation (2.5):

$$\sin^2 \phi = \frac{1/3}{1 - 1/3} = 1/2, \quad \text{yielding } \phi = \pm 45°.$$

The shrink factors can be calculated from, for example, $s_y = \cos^2 \theta = \sqrt{2/3} \approx 0.8165$.

We conclude that the isometric projection is the most useful but also the most restrictive of the three axonometric projections. Given a diagram with the isometric projection of an object, we can measure distances on the diagram and divide them by

0.8165 to obtain actual dimensions on the object. However, the diagram must show the object (whose main edges are assumed to be originally aligned with the coordinate axes) after being rotated by $\pm 45°$ about the y axis and by $\pm 35.26°$ about the x axis. If these rotations result in obscuring important object features, a less restrictive projection, such as dimetric or trimetric, must be used.

Standards for Axonometric Projections

Several common standards for axonometric projections exist and are described here. We start with a simple $30°$ standard for isometric projections whose principle is illustrated in Figure 2.8. Part (a) of the figure shows a cube projected in this standard after it has been rotated $\phi = 45°$ about the y axis and $\theta = 35°$ about the x axis. Part (b) shows the same cube with dimensions and angles. It is not difficult to see that α satisfies $\tan\alpha = h/w$, which is why $\alpha = \arctan(h/w)$. The standard specifies the ratio $h/w = 1/\sqrt{3}$, which results in $\alpha \approx 30°$. The $30°$ angle is convenient because $\sin 30° = 1/2$. This part of the figure also shows that $\theta = \arcsin(h/w)$, a quantity that happens to be close to $35°$. This projection is attributed by [Krikke 00] to William Farish, who developed it in 1822.

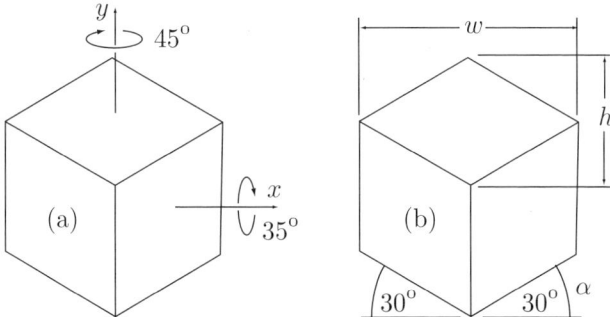

Figure 2.8: The $30°$ Standard for Isometric Projections.

A $30°$ angle is convenient for drafters because $\sin 30° = 1/2$. However, in our age of computers and computer-aided design, virtually all graphics output devices (monitors, plotters, and printers) use a raster scan and are based on pixels. A line is drawn as a set of individual pixels, and even a little experience with such lines shows that a line at $30°$ to the horizontal looks bad. Much better results are obtained when drawing a line at about $27°$ because the tangent of this angle is 0.5, resulting in a line made of identical sets of pixels (Figure 2.9).

Figure 2.9: Pixels for $30°$ and $27°$ Lines.

As a result, the 27° standard for axonometric projections (Figure 2.10) makes more sense. This standard is sometimes also called the 1:2 isometric projection because it is based on the ratio $h/w = 1/2$.

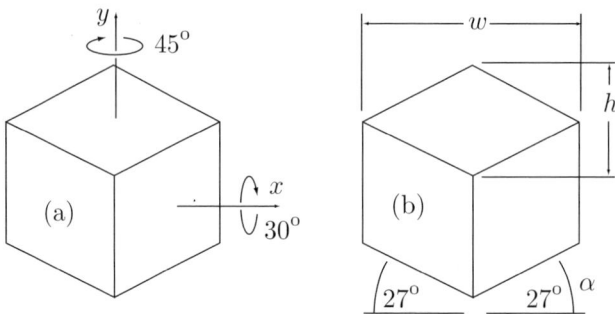

Figure 2.10: The 27° Isometric Projection.

A similar standard is based on the ratio $h/w = 1$, which leads to $\alpha = 45°$. This case is also known as the military isometric projection. This projection is suitable for applications where the horizontal faces of the projected object are important. Figure 2.11 shows that the xz plane becomes a regular rhombus in this projection, which makes it easy to read details and measure distances on this plane.

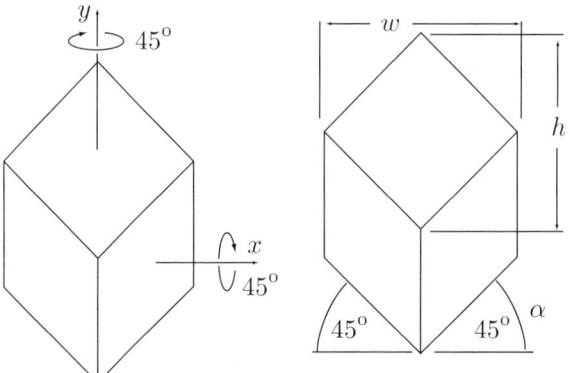

Figure 2.11: The 45° Isometric Projection.

A Dutch standard for dimetric projections is based on the ratio $h/w = 0.33$. It is known as the 42°/7° standard because it results in angles α and β of these sizes (Figure 2.12). The z axis (the one that's drawn at 42°) is scaled by a factor of 1/2.

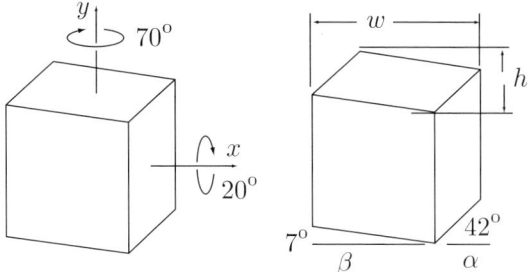

Figure 2.12: The 42°/7° Dimetric Projection.

> What recurrent impressions of the same were possible by hypothesis?
> Retreating, at the terminus of the Great Northern Railway, Amiens Street, with con-
> stant uniform acceleration, along parallel lines meeting at infinity, if produced: along
> parallel lines, reproduced from infinity, with constant uniform retardation, at the
> terminus of the Great Northern Railway, Amiens Street, returning.
>
> —James Joyce, *Ulysses*

2.3 Oblique Projections

An oblique projection is a special case of a parallel projection (i.e., with a center of projection at infinity) where the projecting rays are not perpendicular to the projection plane. We have already seen that axonometric projections show more object details than orthographic projections but make it more cumbersome to compute object dimensions from the flat projection. Similarly, oblique projections generally show more object details than axonometric projections but distort angles and dimensions even more. In an oblique projection, only those faces of the object that are parallel to the projection plane are projected with their true dimensions. Other faces are distorted such that measuring dimensions on them requires calculations.

Figure 2.13 illustrates the principle of oblique projections. A three-dimensional point $\mathbf{P} = (x, y, z)$ is projected obliquely onto a point $\mathbf{P}^*$ on the xy plane. We denote the point $(x, y, 0)$ by $\mathbf{Q}$ and examine the angle θ between the two segments $\mathbf{PP}^*$ and $\mathbf{P}^*\mathbf{Q}$. A cavalier projection is obtained when $\theta = 45°$ and a cabinet projection is the result of $\theta = 63.43°$.

Because of the special 45° angle, the three shrink factors of a cavalier projection are equal, as will be shown later. In a cabinet projection, the shrink factors in the x and y directions (assuming that the object is projected on the xy plane) equal $1/2$.

Figure 2.14a illustrates the geometry of oblique projections and can be used to derive their transformation matrix. We assume that the projection plane is $z = 0$ (the xy plane) and that all the projecting rays hit this plane at an angle θ. Two projecting rays are shown, one projecting the special point $\mathbf{P} = (0, 0, 1)$ to a point $(a, b, 0)$ and the other projecting $\mathbf{Q} = (0, 0, z)$, a general point on the z axis, to a point $(A, B, 0)$. The origin $(0, 0, 0)$ is projected onto itself, so the projection of the unit segment from the

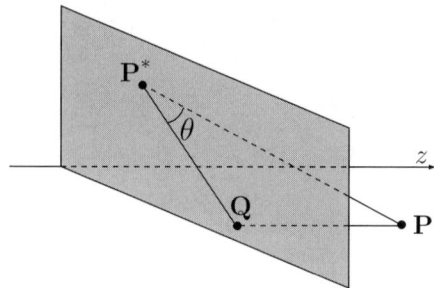

Figure 2.13: Oblique Projections.

origin to $\mathbf{P}$ is the segment of size s from the origin to $(a, b, 0)$. The value s is therefore the shrink factor of the oblique projection. The three quantities a, b, and s are related by $a = s \cos \phi$ and $b = s \sin \phi$, where ϕ is measured on the projection plane. The shrink factor s is also related to the projection angle θ by $\tan \theta = 1/s$ or $s = \cot \theta$.

> The diagram can be drawn quite quickly because the designer used a style of drawing called oblique projection. So long as basic rules are followed, oblique projection is quite easy to master and it may be a suitable style for you to use in a design project. The basic rules are outlined below.
> http://www.technologystudent.com/designpro/oblique1.htm

We now consider the projecting ray from $\mathbf{Q}$ to $(A, B, 0)$. Since $\mathbf{Q}$ is at a distance z from the origin, the distance on the projection plane between the origin and point $(A, B, 0)$ is sz. From this we obtain the relations $A = sz \cos \phi$ and $B = sz \sin \phi$. The next step is to consider the projection of a general point (x, y, z). All the projecting rays are parallel, so a little thinking shows that moving a point from $(0, 0, z)$ to $(x, 0, z)$ moves its projection from $(A, B, 0)$ to $(x + A, B, 0)$. Similarly, moving a point from $(0, 0, z)$ to $(0, y, z)$ moves its projection from $(A, B, 0)$ to $(A, y + B, 0)$. A general point located at (x, y, z) is therefore projected to a point at $(x + A, y + B, 0)$. Thus, the rule of oblique projections is

$$(x, y, z) \longrightarrow (x + sz \cos \phi, y + sz \sin \phi, 0), \tag{2.8}$$

which can be written in terms of a transformation matrix

$$\mathbf{P}^* = \mathbf{PT} = (x, y, z) \begin{pmatrix} 1 & 0 & 0 \\ 0 & 1 & 0 \\ s \cos \phi & s \sin \phi & 0 \end{pmatrix}. \tag{2.9}$$

With the help of this matrix we examine the following special cases.

1. A cavalier projection. It is defined as the case where the projection angle is $45°$, which implies $s = \cot(45°) = 1$. Thus, all edges and segments have shrink factors of 1.

2. A projection angle of $90°$. A value $\theta = 90°$ implies a shrink factor $s = \cot(90°) = 0$. Matrix $\mathbf{T}$ of Equation (2.9) reduces to matrix $\mathbf{T}_z$ of Equation (2.1), showing how the oblique projection reduces in this case to an orthographic projection.

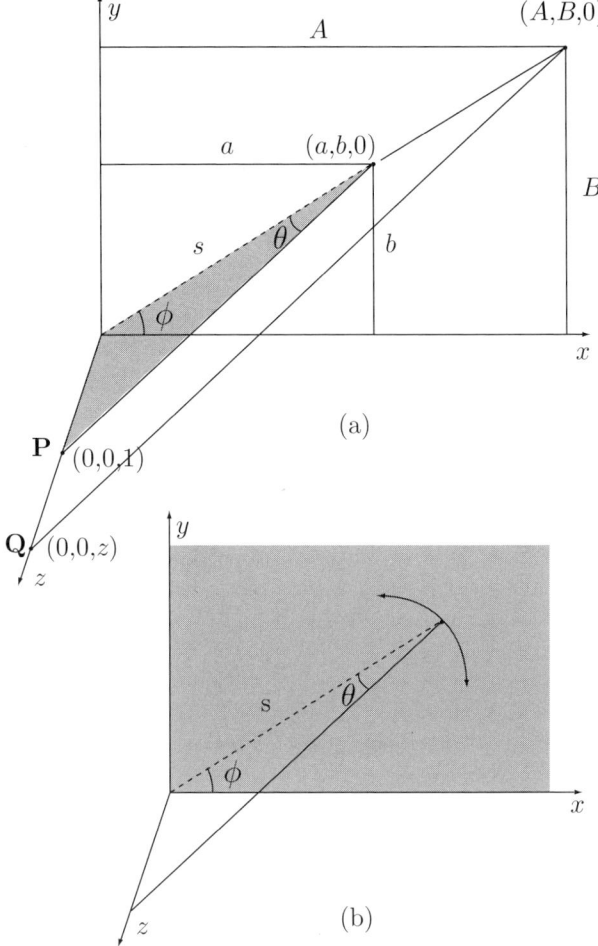

Figure 2.14: Oblique Projections.

3. A cabinet projection. It is defined as the case where the projection angle is 63.43°, which implies $s = \cot(63.43°) = 1/2$. All edges and segments perpendicular to the projection plane have shrink factors of $1/2$.

Figure 2.14b shows how ϕ and θ are independent. For a given projection angle θ, it is possible to assign ϕ any value by rotating the triangle in the figure. In practice, this means that an object can be projected several times, with different values of ϕ but with the same projection angle θ. Such projections may give all the necessary visual information about the object while having the same shrink factors.

Shrink: To become constricted from heat, moisture, or cold.
(A typical dictionary definition)

Axonometric and oblique projections are generally considered different, but Figure 2.15 shows that the difference between them is a matter of taste and terminology. If we rotate the object and light rays of the oblique projection 45° counterclockwise, the result on the projection plane is identical to the axonometric projection.

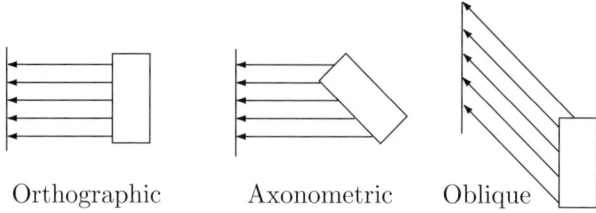

Orthographic   Axonometric   Oblique

Figure 2.15: Comparing Parallel Projections.

She could afterward calmly discuss with him such
blameless technicalities as hidden line algorithms and
buffer refresh times, cabinet versus cavalier projections
and Hermite versus Bézier parametric cubic curve forms.

—John Updike, *Roger's Version* (1986)

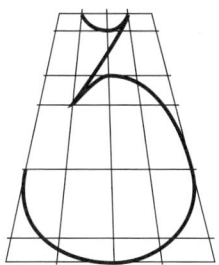

3
Perspective Projection

The term *perspective* refers to several techniques that create the illusion of depth (three dimensions) on a two-dimensional surface. *Linear perspective* is one of these methods. It can be defined as a method for correctly placing objects in a painting or a drawing so they appear closer to the observer or farther away from him. The keyword in this definition is *correctly*. It implies that a flat picture in linear perspective creates in the viewer's brain the same sensation as the original three-dimensional scene. The main tool employed by linear perspective is vanishing points.

This chapter starts by explaining vanishing points. This is followed, in Section 3.2, by a short history of perspective in art. The remainder of the chapter develops simple mathematical tools to compute the two-dimensional perspective projection of any given three-dimensional point.

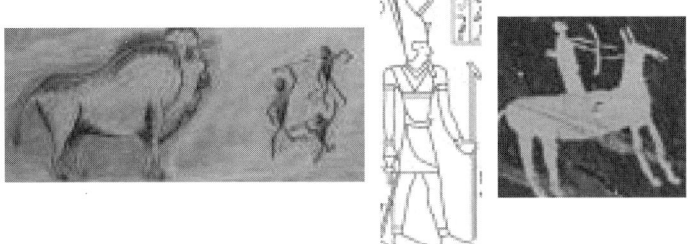

Figure 3.1: Ancient Art.

The Bible is eternal and is always the same, but most other objects and processes around us change and develop continually. Hot air balloons, cheese making, and bicycles are familiar examples of items that constantly develop and improve. Art is another example. Ancient art tends to be flat, as illustrated by Figure 3.1. The Lascaux cave drawings, Navajo rock drawings, and ancient Egyptian art shown in the figure are two-dimensional. They are flat and do not attempt to create a sensation of depth.

Flatness is also a common feature of modern art. The abstract art and cartoons of Figure 3.2 look flat and use the painter's algorithm to create the barest hints of depth. (The painter's algorithm is simply the way painters work. The first objects painted may be partly or fully covered and obscured by objects painted later.)

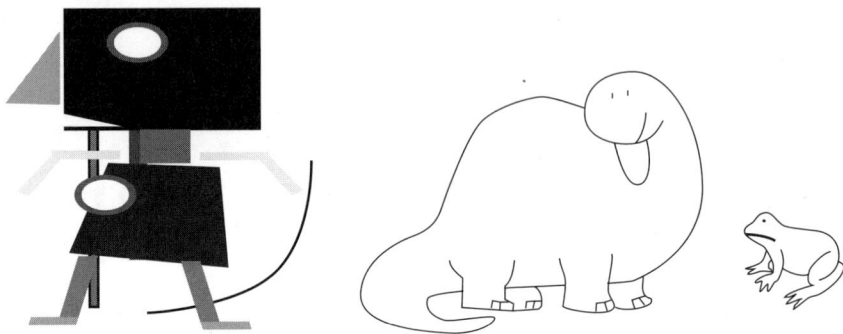

Figure 3.2: Modern Art (Color Version on Page 231).

Art, especially painting and drawing, went through a revolution during the Italian renaissance in the late Middle Ages. An important part of this revolution was the technique of perspective. Almost overnight it became possible to create the illusion of a three-dimensional scene in a flat, two-dimensional picture. Section 3.2 surveys the historical developments that led to an understanding of perspective, but Figure 3.3 illustrates the basic idea. Part (a) of the figure shows a small, flat plane defined by two sets of parallel lines. In part (b), some lines are made to converge to a *vanishing point*, thereby creating the sensation of depth. Part (c) maintains this feeling even though the vanishing point itself has been removed. Finally, part (d) illustrates how four copies of this plane can be connected to form an object that appears to us as a cube, a box, or a room, even though we know that it is only a collection of lines on a flat surface.

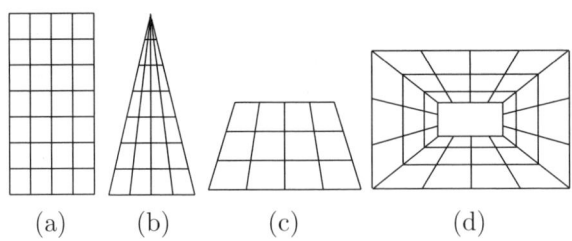

(a)   (b)   (c)   (d)

Figure 3.3: Converging Lines.

Figure 3.4 is another illustration of the same concept. It is easy to see that the railway tracks of part (a) are wrong, while part (b) looks realistic.

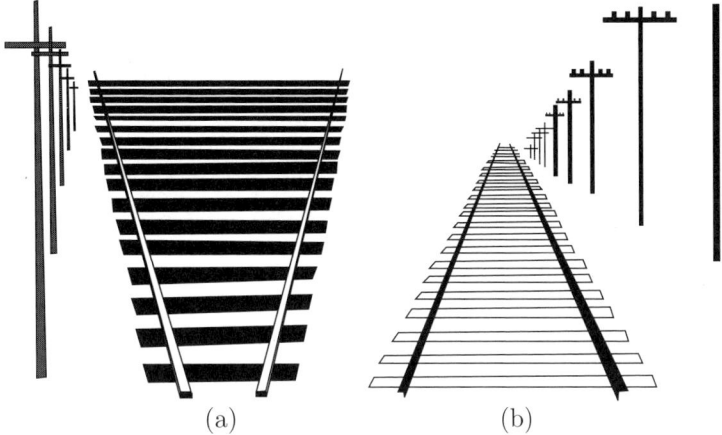

(a)   (b)

Figure 3.4: (a) Wrong and (b) Correct Perspective.

◇ **Exercise 3.1:** Search the works of art (modern or otherwise) for examples of wrong or reversed perspective.

> Simply stated, sound perspective means that something seen happening in the foreground of the shot must make a louder noise than something seen to be further away. Most failures to respect the rule are instinctively heard as "bad sound," as imperfect or amateur use of recording technology.
>
> —David Bellos, *Jacques Tati* (1999).

3.1 One Two Three...Infinity

The first step toward understanding perspective is an understanding of converging lines and vanishing points. Imagine a simple house shaped like a cube. If we stand in front of it, we see only its front wall, a square, much like the one depicted in Figure 3.5a. If, however, we imagine the house to be transparent, it would look like part (b) of the figure. Its back wall is farther away from us, so it looks smaller than its front wall, which is why the four parallel lines connecting the front and back walls do not look parallel; they seem to converge to an imaginary point called a *vanishing point*. The vanishing point exists only in our imagination, and we can imagine it only if we extend the four lines in question. Thus, the vanishing point is a result of the way the brain interprets what the eyes see.

We now walk around our transparent, cubic house and turn to the left, such that our line of sight is aimed at one of the corners, as shown in Figure 3.5c. The house is the same: it hasn't moved or changed shape. We, the viewers, are also the same, only our position and orientation have changed. Yet, when we look at the house, we see two groups of lines converging at two vanishing points (Figure 3.5d).

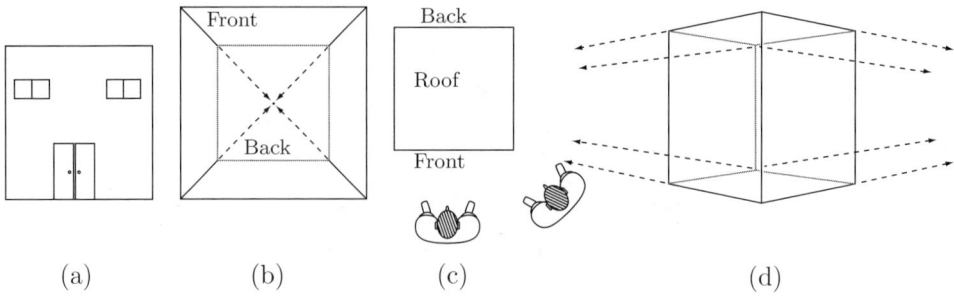

(a)   (b)   (c)   (d)

Figure 3.5: Vanishing Points.

Figure 3.6 shows examples of perspective with three vanishing points. Imagine a person standing in front of a corner of a skyscraper, craning his neck in an attempt to see all the way to the top of the building. Because of the height of the building, its top seems smaller than its bottom, so the straight, parallel lines connecting top to bottom also seem to converge to a vanishing point. Even a small object, such as a cube, can feature three vanishing points if it is hoisted up and we are positioned under it. Even a small, one-story house can feature three vanishing points if it has a traditional pitched roof.

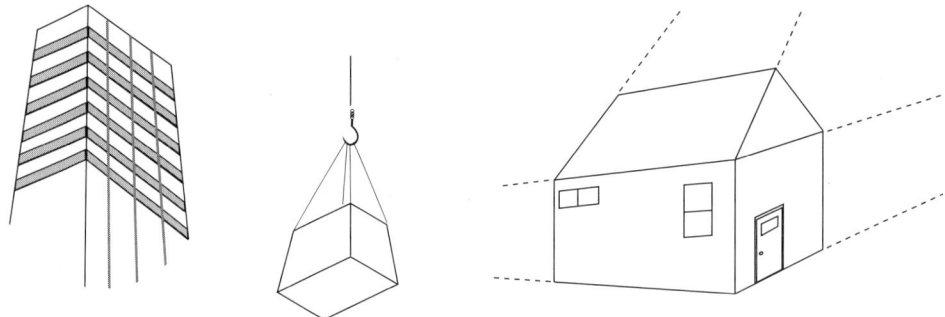

Figure 3.6: Three Vanishing Points.

We live in a three-dimensional world, which is why we can visualize objects with one, two, or three dimensions, but not more. A line or a curve has one dimension. A flat plane or a curved surface is two-dimensional. A solid object has three dimensions, so the question is, can an object feature more than three vanishing points? The answer, which may come as a surprise to some, is yes, as illustrated by Figures C.1 (page 232) and 3.7. Everyday objects, such as a chest of drawers (if you have messed yours up in order to prove my point, please take the time to put it back in order) and a circular staircase, can feature any number of vanishing points.

⋄ **Exercise 3.2:** Come up with other common objects or scenes that feature many vanishing points.

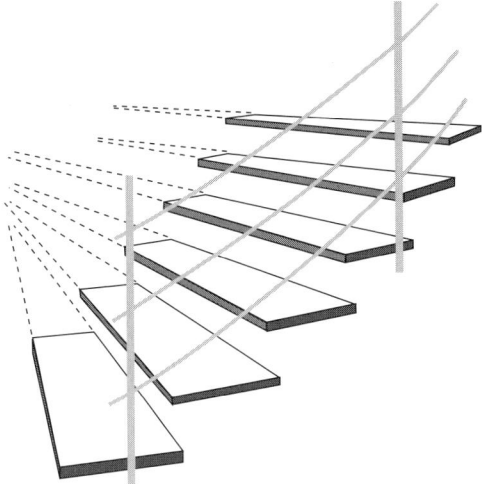

Figure 3.7: Many Vanishing Points.

We therefore conclude that an object seen in perspective can have any number of vanishing points, even zero. In addition, the number and positions of those points vary when the object is moved or changes its orientation and when the viewer moves, turns, bends, tilts his head, or cranes his neck. The rule governing the number and position of the vanishing points is simple and can be considered the main principle of perspective. Before this rule is stated, let's take another look at Figure 3.5d. It features two vanishing points, each created by a group of parallel lines. The point is that originally (i.e., in Figure 3.5b) these lines seem parallel, but when the viewer moves to a different location, looking at the same object from a different direction, these same lines no longer look parallel and seem to converge.

Figure 3.8: Effects of a Small Rotation.

Figure 3.8 serves to further illustrate the behavior of the vanishing points. Part (a) of the figure shows the cube of Figure 3.5b with the four parallel edges marked with an asterisk. In part (b), the cube is rotated through a small angle, which slightly changes the orientation of these four edges relative to the viewer. They no longer appear parallel, and seem to converge to a distant vanishing point on the far left of the figure. In addition, the four lines that originally converged at the center of the cube now converge to a vanishing point slightly to the right of center.

This small rotation has resulted in the same cube featuring two vanishing points. It can be interpreted by saying that the rotation has moved the original vanishing point slightly to the right and the new vanishing point isn't really new. It was originally located at infinity and has moved by the rotation to a finite (albeit distant) location.

These observations should help the reader to understand and agree with the following statement: In order for an object to feature vanishing points, it must have groups of straight parallel lines. The lines may be generated by the intersection of two planes on the object, as in the case of a cube, or they may be painted or scribed on the surface of the object. They may even be located inside the object, if it is transparent. Any such group of lines results in a vanishing point, except if the lines are perpendicular to the line of sight of the viewer. Rotating the cube of Figure 3.8 has changed the orientation of a group of four parallel lines that were originally perpendicular to the line of sight but are no longer so. The new orientation has therefore added a vanishing point.

The conclusion is that an object may have any number of vanishing points depending on its shape and orientation, on groups of parallel lines that happen to be on it, and on the direction from which it is viewed.

The statement above is the rule governing vanishing points. It should be stressed that the vanishing points are not real. They exist only in our imagination and we imagine them because of the particular way our brain interprets the signals sent from our eyes.

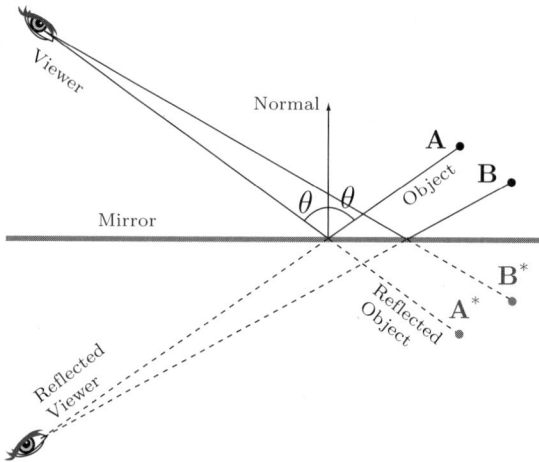

Figure 3.9: The Rule of Reflection.

An interesting example of vanishing points is a reflection in a mirror. A ray of light that strikes a mirror is reflected in a direction determined by the normal to the mirror. The rule of reflection (Figure 3.9) is that the angle of incidence equals the angle of reflection. Points "A" and "B" in the figure are seen by the viewer as if they are deep in the mirror, and any group of parallel lines on a reflected object seems to converge in the mirror to a new vanishing point.

Figure 3.10 shows a cube (in two-point perspective) reflected in a mirror. The two real vanishing points are vp_1 and vp_2. The cube seen in the mirror also has two virtual vanishing points, vp_1^* and vp_2^*, and it is easy to see the symmetric relation between the real and virtual points.

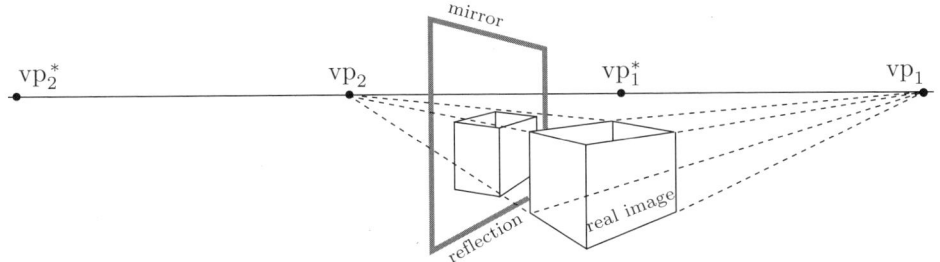

Figure 3.10: Real and Virtual Vanishing Points.

Note. This section discusses straight lines and their convergence, which is why the examples here employ cubes and other objects with large, flat surfaces and straight lines. However, curved objects with no straight, parallel lines can also be seen (and drawn) in perspective, and techniques for achieving this are described in Section 3.3.

Vanishing points and converging lines are important in perspective, but perspective has another important aspect. When an object is moved away from the viewer, it appears smaller, but it also features less perspective. The amount of perspective seen depends on the relation between the size of the viewed object and its distance from the viewer. To see why this is so, we go back to the cube of Figure 3.5b, duplicated in Figure 3.11a. Assuming that this cube is 10 cm on a side and that it is viewed from a distance of 10 cm, its back face is 20 cm from the viewer, twice the distance of the front face. The back face therefore seems to the viewer much smaller than the front face, and the object is seen with a lot of perspective. If this cube is moved 90 cm away from the viewer, its front face ends up at 100 cm and its back face at 110 cm from the viewer. The difference between front and back is much smaller compared with the distance from the viewer, causing the back face to appear just a shade smaller than the front, with the result that the object appears to have much less perspective (Figure 3.11b).

◊ **Exercise 3.3:** In addition to featuring less perspective, a distant object also looks small. Can we bring such an object closer without increasing its perspective?

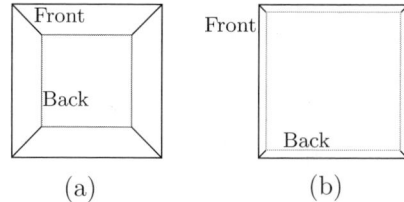

Figure 3.11: (a) More and (b) Less Perspective.

3.2 History of Perspective

In art, the term "perspective" refers to a technique for depicting a three-dimensional scene on a two-dimensional flat surface. The result is similar, but not identical, to the way we perceive three-dimensional objects and scenes in space. Our eyes are separated by a few centimeters and therefore see slightly different views of the same scene. The brain combines these views in a complex way to generate the sensation of depth. When we move, turn, or raise or lower our head, the image we see changes continuously. A painting or drawing in perspective, on the other hand, is based on a fixed viewpoint and is equivalent to looking at the scene through a peephole with one eye.

The principles of perspective were known to the ancients. Many Greek vase paintings indicate a grasp of the principles of perspective. Roman wall paintings show lines converging to vanishing points, and the Roman architect Vitruvius describes perspective in his writings [Vitruvius 06]. In the Middle Ages, especially in the 13th and 14th centuries, several artists in Italy, France, and Holland (and perhaps also in the East) independently discovered (or rediscovered) some of the principles of perspective, especially the concept of lines converging to a vanishing point. However, none came up with a complete and consistent theory of perspective. Such a theory had to wait until the second decade of the 15th century, when it was developed first experimentally by Filippo Brunelleschi and later in more detail by Leon Battista Alberti. (Some experts also credit the painter Paolo Uccello with major contributions to the understanding of perspective.)

> [Uccello] would remain the long night in his study to work out the vanishing points of his perspective, and when summoned to his bed by his wife replied in the celebrated words: "How fair a thing is this perspective." Being endowed by nature with a sophisticated and subtle disposition, he took pleasure in nothing save in investigating difficult and impossible questions of perspective.... When engaged in these matters, Paolo would remain alone in his house almost like a hermit, with hardly any intercourse, for weeks and months, not allowing himself to be seen.... By using up his time on these fancies he remained more poor than famous during his lifetime.
>
> —Giorgio Vasari, *The Lives of the Artists* (1567)

The remainder of this section discusses the contributions made by three Renaissance figures, Brunelleschi, Masaccio, and Alberti, to the understanding of perspective.

Brunelleschi

Filippo Brunelleschi, known to his contemporaries as "Pippo," was born in Florence in 1377. His father, Ser Brunellesco di Lippo Lapi, was a prosperous notary, but young Filippo showed an interest in machines and in solving mechanical problems. (The term "ser" was a title of respect, while "di Lippo Lapi" indicates that Brunellesco's father was named Lippo and was from the Lapi family.) Filippo was therefore apprenticed, at age 15, to a local goldsmith (perhaps one Benincasa Lotti). For the next six years he learned to cast metals, work with enamel, engrave and emboss silver, and use precious metals to decorate manuscripts with gold leaf and to make jewels and religious artifacts.

After completing his apprenticeship in 1398 at age 21, Brunelleschi was sworn as a master goldsmith and became a well-known goldsmith in Florence and other cities. From 1401 to 1416 or 1417, he seems to have spent most of his time in Rome (although this is uncertain), working as a goldsmith, making clocks, and surveying the many ruins of the eternal city. Returning to Florence after 13 years of absence, Brunelleschi, then 40, became involved in the competition for the great dome of the Santa Maria del Fiore Cathedral. This was to be both the largest dome ever attempted, with a diameter of more than 143 feet, and the tallest one, starting at a height of about 170 feet off the ground and reaching about 280 feet. (The lantern on top of it adds more than 70 feet to that.)

Even though known as a goldsmith, not an architect, Brunelleschi won the 1418 competition because of his original approach to the problem. The novel aspect of his plan for the dome was to build it without any scaffolding. (The term "centering" was then used.) This idea, and the 1:12 model of the dome that he built in brick to demonstrate his method, helped convince the committee of judges to give him the commission. He then spent the years from 1420 to 1436 supervising the construction while also designing and building ingenious machines to haul heavy loads to the top.

Brunelleschi, a true Renaissance man both because of his interests and achievements and because of his time period, died in 1446. Like Donatello, Masaccio, da Vinci, and Michelangelo, he never married. For more information on Brunelleschi, his work, and his times, see [King 00] and [Walker 02].

A biography of Brunelleschi [Manetti 88] was written in the 1480s, four decades after the death of its subject, by his pupil Antonio Manetti, which brings us to Brunelleschi's contribution to perspective. In this biography, Manetti describes Brunelleschi's panel drawing, a trompe l'oeil that was then used by Brunelleschi in an experiment that fuses nature and art, similar to an optical trick. This historically important painting has since been lost, but it (and the experiment) are described in detail by Manetti.

> trompe l'oeil.
> 1. A style of painting that gives an illusion of photographic reality.
> 2. A painting or effect created in this style.

The peepshow experiment. Brunelleschi placed himself at a point three braccia (about six feet) inside the doorway of the not yet completed cathedral of Santa Maria del Fiore. His idea was to specify a precise viewing point at which a viewer could compare a real scene with a perspective painting of the same scene. Looking outside across the

Piazza del Duomo, he clearly saw, about 115 feet away, the Baptistery of San Giovanni, one of Florence's most well-known landmarks. This structure was a good choice for the study of perspective because it is shaped like an octagon, so someone standing in front of it sees its three front walls in two-point perspective. (It also features left–right symmetry, so reflecting it horizontally does not change its shape.) Brunelleschi then painted what he saw through the doorframe—the Baptistery and some of the surrounding streets—in perspective on a small panel about 12 inches wide. Finally, he drilled a small hole in the panel at the center of the Baptistery's door (Figure 3.12a) because this point of the Baptistery would be directly opposite the eye of a viewer standing at the specified viewing point.

> The world having so long been without artists of lofty soul or inspired talent, heaven ordained that it should receive from the hand of Filippo the greatest, the tallest, and the finest edifice of ancient and modern times, demonstrating that Tuscan genius, although moribund, was not yet dead.
>
> —Giorgio Vasari, *The Lives of the Artists* (1567)

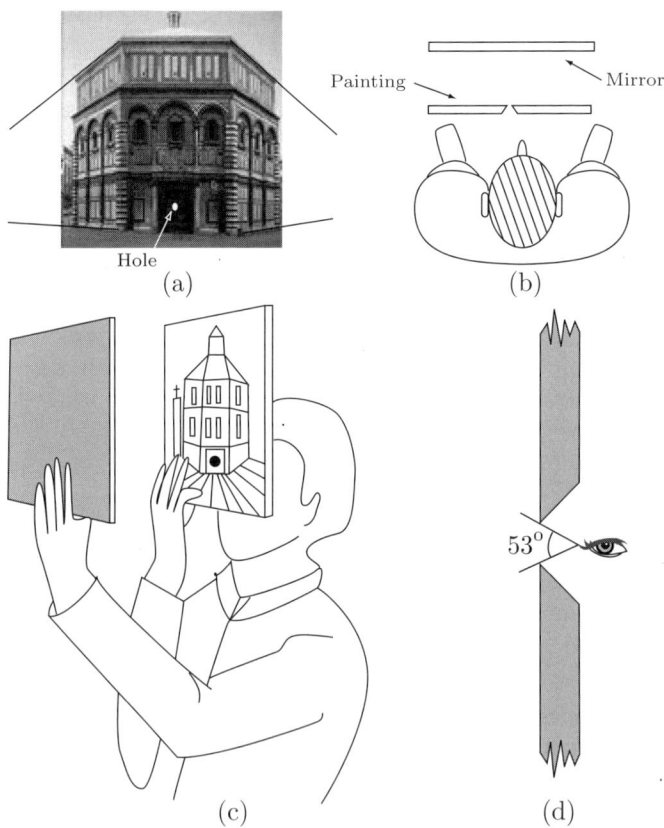

Figure 3.12: Brunelleschi's Experiment in Perspective.

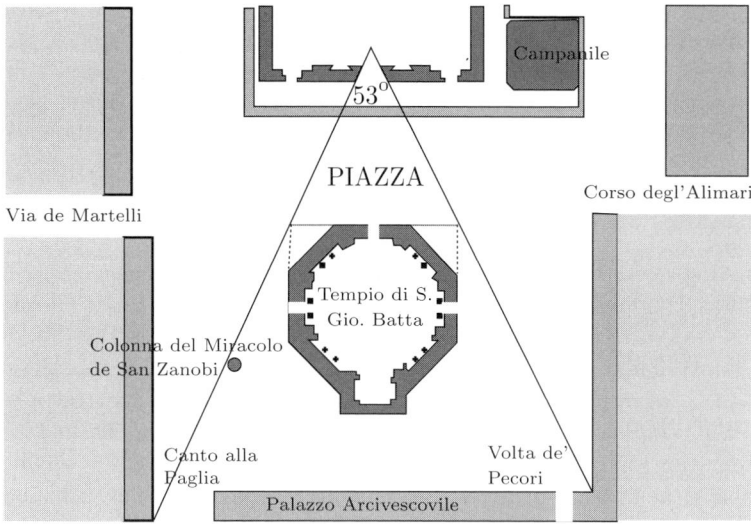

Figure 3.13: Plan of the Piazza del Duomo, Florence (After [Sgrilli 33]).

Brunelleschi then rotated the panel 180° and looked through the hole at the Baptistery. He then inserted a mirror and held it at arm's length as shown in Figure 3.12bc and looked at his painting reflected in the mirror. This became Brunelleschi's celebrated peepshow experiment, which proved the lifelike qualities of perspective. In his biography, Manetti claims to have held this painting in his hands and to have repeated the experiment. He was unable to tell the difference between the image reflected in the mirror and the real scene (without the mirror). (However, modern travelers to Florence recommend the use of a pair of heavy-duty tripods to hold the image and the mirror at their precise locations.)

> [Brunelleschi] had made a hole in the panel on which there was this painting;... which hole was as small as a lentil on the painting side of the panel, and on the back it opened pyramidally, like a woman's straw hat, to the size of a ducat or a little more. And he wished the eye to be placed at the back, where it was large, by whoever had it to see, with the one hand bringing it close to the eye, and with the other holding a mirror opposite, so that there the painting came to be reflected back... which on being seen,... it seemed as if the real thing was seen: I have had the painting in my hand and have seen it many times in these days, so I can give testimony.
>
> —Antonio Manetti, *The Life of Brunelleschi* (1480s)

Manetti mentions another interesting fact. The painting was about 12 inches wide and Brunelleschi recommended watching it from a distance of 6 inches, so the reflection seen in the mirror appears to be at a distance of 12 inches from the viewer. We know that $\tan 26.6° = 0.5$, which implies that the apex angle of an isosceles triangle whose height equals its base is $2 \times 26.6 \approx 53°$ (see also Exercise 3.29). This trigonometric fact suggests that, as seen from the viewing point specified by Brunelleschi, the Baptistery spans a viewing angle of about 53°, and this is verified by Figure 3.13, which follows the site plan given by [Sgrilli 33]. Finally, Manetti mentions that the diameter of the hole on the painted side of the panel was about the thickness of a bean (6–7 mm). Figure 3.12d illustrates how the same angle of 53° is obtained if the eye of the viewer is glued to the back of the panel (where according to Manetti the hole was bigger, about the size of a ducat, 20 mm) and the thickness of the panel is the same 6–7 mm.

Masaccio

Perhaps the first great Renaissance painter to use the ideas of Brunelleschi in a serious work of art was Tommaso di ser Giovanni di Mone (or Tommaso di ser Giovanni cassai), known to us as Masaccio, a nickname that can be translated as Big Thomas, Rough Thomas, Clumsy Thomas, Sloppy Thomas, Bad Thomas, or even the Messy Thomas. He died in 1428, at age 27, and in his last two years he painted a fresco, today titled *Trinity* (or *Holy Trinity*), in the church of Santa Maria della Novella in Florence. The accurate execution of one-point perspective in this picture creates the illusion of a sculpture placed in a cavity in the wall, although the picture is flat. This large picture (approximately 6.7×3.2 m, or 21 ft $10\frac{1}{2}$ in by 10 ft 5 in) has a sad history of incompetent restoration and a 19th century attempt to cut it off the wall and move it to another wall in the same church. Figure C.2 (page 233) is a small replica showing how the single vanishing point was placed by the artist at the viewer's position.

The architectural setting of this fresco [the *Trinity*] is so accurate in its perspective and so Brunelleschian in style that some scholars have suggested Brunelleschi drew the sinopia, or cartoon, on the wall for Masaccio to paint. This is certainly possible, but it is also quite possible that Masaccio—a master draftsman as well as an inspired painter—could have done the whole work himself. Perhaps it doesn't matter. The important fact for the future of Western art is that Masaccio met Brunelleschi and gained such a deep knowledge of perspective that he set a standard for every painter to follow.

—Paul Robert Walker, *The Feud that Sparked the Renaissance* (2002)

Alberti

In 1435–36, Leon Battista Alberti wrote and published (in Latin and Italian) *Il Trattato della Pittura e I Cinque Ordini Archittonici* ("On Painting"), where he describes a simple geometric method for constructing a correct one-point perspective of a horizontal grid on a vertical picture plane. This method was later simplified by Piero della Francesca in his 1478 mathematical treatise *De prospectiva pingendi* and is illustrated in Figure 3.14.

The left part of the figure shows a side view where the picture plane is intercepted by a family of visual rays that emanate from the viewer's eye. Each ray connects the eye to one of the transversals (or divisions) of the grid on the ground. The point where the ray intercepts the picture plane is then transferred to the front view (on the right part of the figure) to indicate where to place the particular transversal in the picture. It is easy to see how the transversals, which are equally spaced on the ground, become

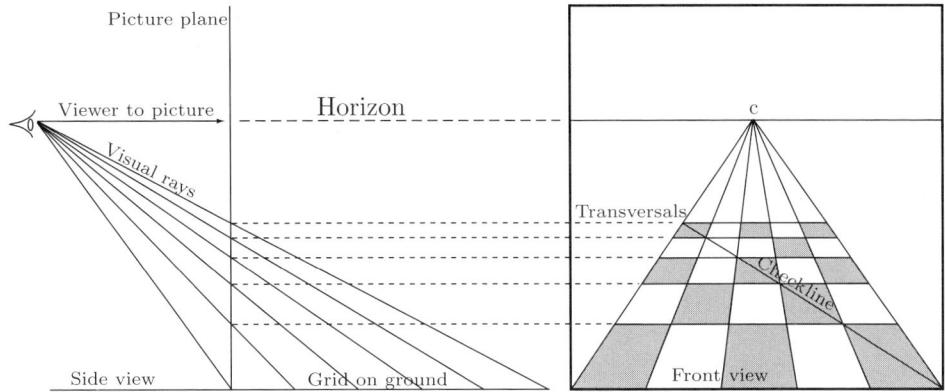

Figure 3.14: Alberti's Method of Traversals in One-Point Perspective.

closer and closer in the picture. The last step is to draw a diagonal line in the front view to check for the accuracy of this geometric construction.

> The canvas is an open window through which I see what I want to paint.
> —Leon Battista Alberti

In his book, Alberti also shows how such a floor, accurately drawn in perspective, can serve to determine the correct dimensions (both horizontal and vertical) of objects positioned on the floor and elsewhere in the picture. Figure 3.15 illustrates how a grid on a floor is used to determine the height of a large, box-like object placed on the floor. Alberti used the braccio (plural braccia), a length unit that equals approximately 58 cm (or 23 in, roughly the length of a man's arm), and a length of four braccia, measured on the floor, is employed to determine the heights of the box at its front and back.

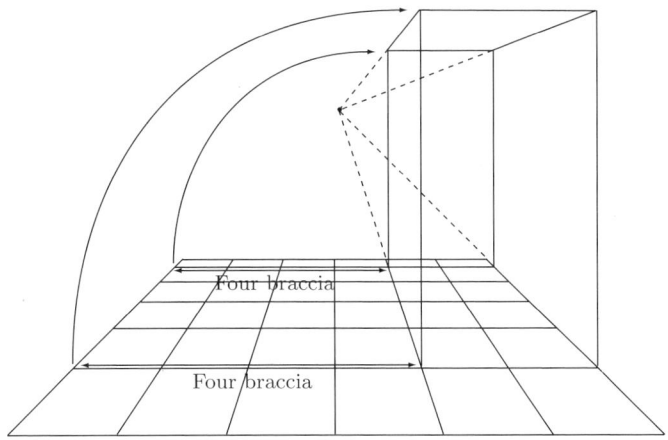

Figure 3.15: Determining Vertical Dimensions from the Floor.

It is such precisely described methods and techniques that distinguish Alberti from his predecessors and justify the title "pioneer" or "originator" of perspective.

◇ **Exercise 3.4:** Given the simple two-point perspective of Figure 3.16, show how the equally spaced vertical lines (labeled "e") were constructed.

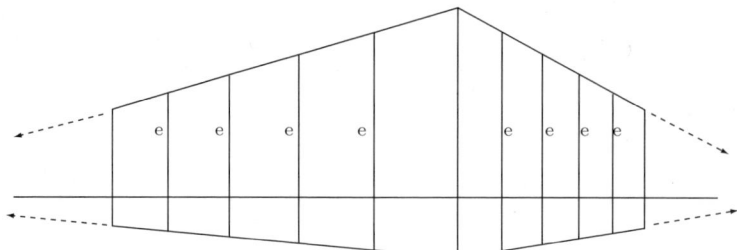

Figure 3.16: Two-Point Perspective with Equally Spaced Lines.

Leonardo da Vinci, who certainly knew about perspective, developed his own projection, now known as aerial or atmospheric perspective. This method of adding depth to a two-dimensional painting is based on the perception that contrasts of color and shade appear greater in nearby objects than in those far away, and that warm colors (such as red, orange, and yellow) appear to advance, while cool colors (blue, violet, and green) appear to recede. Aerial perspective is also used in East Asian art, where zones of mist are sometimes used to separate near and distant parts of the scene.

3.3 Perspective in Curved Objects

Up until now, we have discussed perspective, converging lines, and vanishing points in cubes or other objects with large flat surfaces on which it is easy to draw straight lines. Our accumulated life experience, however, teaches us that even curved objects—objects without flat parts and with no groups of straight, parallel lines—are seen in perspective. This section shows how to extend the principles of perspective discussed earlier to arbitrary surfaces.

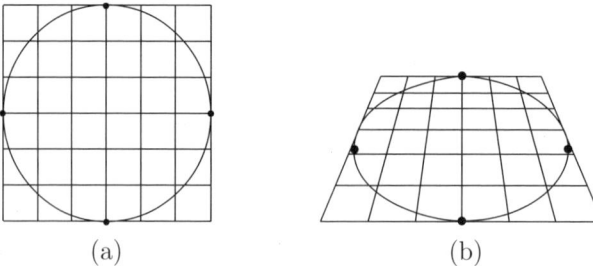

(a)   (b)

Figure 3.17: Alberti's Method of Perspective Drawing.

The main idea was already proposed by Alberti and is illustrated in Figure 3.17 for a circle. Start with a flat, nonperspective drawing of a curved object and place a regular rectangular grid on it [part (a)]. Redraw the grid in perspective, with a vanishing point [part (b)], and go over the two grids box by box. For each box, copy that part of the object seen in the first grid and modify it according to the shape of the box in the second grid. The final result (the circle in perspective) looks like an ellipse, but notice how the left and right extreme points of the projected circle (i.e., the ellipse's major axis) no longer lie on the central horizontal line but have moved below it.

⋄ **Exercise 3.5:** Explain why.

A variant of this method starts by locating key points on the curved object (points that make it easy to draw the entire object), assigning them coordinates, and locating them on the perspective grid. Figure 3.18 shows an example of a large digit **5** where $5 \times 7 = 35$ key points have been located. The digit is placed in a rectangle, and grid lines are added and labeled 1 through 5 and "A" through "G," resulting in a nonuniform grid. This grid is then transformed in perspective (one-point or two-point) and the key points located in the new grid, which makes it easy to draw the large **5** in perspective.

⋄ **Exercise 3.6:** Show the geometric construction that transfers the 35 key points to a grid in one-point perspective.

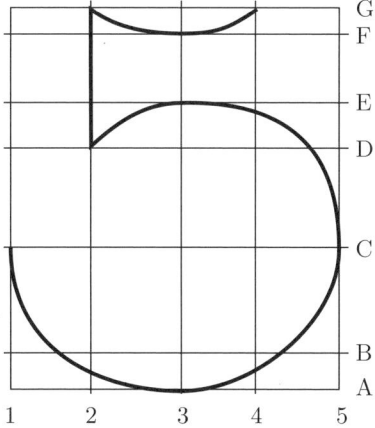

Figure 3.18: A Large Digit "5."

The great German painter Albrecht Dürer showed how to extend Alberti's approach to three-dimensional objects (Figure 3.19). Lay the object (a lute in the figure) on a table behind a frame and attach a string with a pulley and a weight to the wall in front of the frame. A wooden leaf is attached to the frame with hinges, and a sheet of blank paper is mounted on the leaf. Now move the free end of the string to an arbitrary point on the object and determine the point where the string intercepts the frame. (This is done by two moveable wires or threads, as shown in the upper part of the figure.)

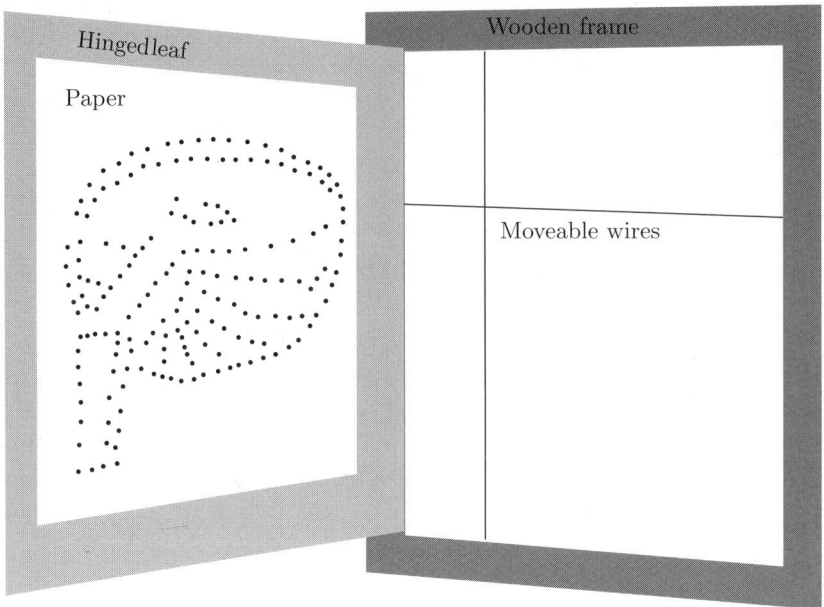

Figure 3.19: Dürer's Method of Perspective Drawing.

Remove the string temporarily, close the hinged leaf, and mark the intersection point of the wires on the paper. This is repeated for many points on the object, which later permits the artist to interpolate the points and complete the drawing.

In contrast with Renaissance and classical artists, who mostly tried to create works true to nature, many impressionist and modern artists consider the use of color and technique more important than accurate perspective. Figure C.3 (page 234) is a classic example of this approach. It shows the famous yellow chair painted by Vincent van Gogh several times during his short stay in Arles. Even a quick glance at it creates the impression that something is wrong. However, van Gogh fans (this author not numbered among them) claim that his mastery of color, combined with his technique and style, resulted in paintings full of appeal and charm, in spite of the crude perspective (or even because of it). Another example that some may call *divergent perspective* is *The Chair* by David Hockney (1985).

3.4 The Mathematics of Perspective

The mathematics of linear perspective is easy to derive and to apply to various situations. The mathematical problem involves three entities, a (three-dimensional) object to be projected, a projection plane, and a viewer watching the projection on this plane. The object and the viewer are located on different sides of the projection plane, and the problem is to determine what the viewer will see on the plane. It is like having a transparent plane and looking through it at an object. Specifically, given an arbitrary point $\mathbf{P} = (x, y, z)$ on the object, we want to compute the two-dimensional coordinates (x^*, y^*) of its projection $\mathbf{P}^*$ on the projection plane. Once this is done for all the points of the object, the perspective projection of the object appears on the projection plane. Thus, the problem is to find a transformation $\mathbf{T}$ that will transform $\mathbf{P}$ to $\mathbf{P}^*$. We use the notation $\mathbf{P}^* = \mathbf{PT}$ from Chapter 1.

Often, there is no need to compute the projections of all the points of the object. If $\mathbf{P}_1$ and $\mathbf{P}_2$ are the two endpoints of a straight line on the object, then only their projections $\mathbf{P}_1^*$ and $\mathbf{P}_2^*$ need be computed and a straight line is then drawn between them on the plane. In the case of a curve, it is enough to compute the projections of several points on the curve and either interpolate them on the projection plane or simply connect them with short, straight segments.

It is obvious that what the viewer will see on the projection plane depends on the position and orientation of the viewer. The viewer and the object have to be located on different sides of the plane, and the viewer should look at the plane. If the viewer moves, turns, or tilts his head, he will see something else on the projection plane and may not even see this plane at all. Similarly, if the object is moved or if the projection plane is moved or is rotated, the projection will change. Thus, the mathematical expressions for perspective must depend on the location and orientation of the viewer and the projection plane, as well as on the location of each point $\mathbf{P}$ on the object.

We start with a special case—where the viewer is positioned at a special location, looking in a special direction at a specially placed projection plane—and show how to project any three-dimensional point to a two-dimensional point on the projection plane.

There is no need to consider the orientation of the object because each point **P** on the object is projected separately. Starting in Section 3.5, this treatment is generalized and we show how to project an object on any projection plane and with the viewer located anywhere and looking in an arbitrary direction.

The discussion of perspective and of converging lines earlier in this chapter implies that we are looking for a transformation **T** that satisfies the following conditions:

1. As the object is moved away from the projection plane, its projection shrinks. This corresponds to the well-known fact that distant objects appear small.

2. The projection of a distant object features less perspective, as illustrated by Figure 3.11. The reader may claim that the projection of a distant object is too small to be seen, so the loss of perspective may not matter, but the point is that we can look at a distant object through a telescope. This brings the object closer, so it looks big, but there is still loss of perspective.

3. Any group of straight parallel lines on the object seems to converge to a vanishing point, except if the lines are perpendicular to the line of sight of the viewer. This rule of vanishing points is stated and discussed in Section 3.1.

The remainder of this section derives the special case of perspective projection in four steps as follows:

1. We describe the special case and state the rule of projection.
2. The mathematical expressions are derived using just similar triangles.
3. We show that this rule satisfies the three requirements above.
4. We include this rule in the general three-dimensional transformation matrix. This produces a 4×4 matrix that can be used to transform the points of an object and also project them on a plane.

Step 1. The special case discussed in this section places the viewer at point $(0, 0, -k)$, where k, a positive real number, is a parameter selected by the user. The viewer looks in the positive z axis, so the line of sight is the vector $(0, 0, 1)$. Finally, the projection plane is the xy plane. In order for the projection to make sense, we state again that the viewer and the object must be on different sides of the projection plane, which implies that all the points of the object must have nonnegative z coordinates. [The points will normally have positive z coordinates, but they may also be of the form $(x, y, 0)$; i.e., located on the projection plane.]

This special case is referred to as the *standard position* (Figure 3.20a) and is mentioned often in this book. The rule of perspective projection is a special case of the general rule of projection (page 2) where the center of projection is at the viewer. Thus, in order to project point **P**, we compute the line segment that connects **P** to the viewer at point $(0, 0, -k)$ and place the projected point $\mathbf{P}^*$ where this segment intercepts the xy plane. (The segment always intercepts the xy plane because the object and the viewer are located on opposite sides of the plane.) Because the projection plane is the xy plane, the coordinates of the projected point are $(x^*, y^*, 0)$, indicating that it is two-dimensional.

It is important to realize that the viewer and the projection plane constitute a single unit and have to be moved and rotated together. This is illustrated in Figure 3.20b and especially in Figure 3.21a, which shows the viewer-plane unit moving around the

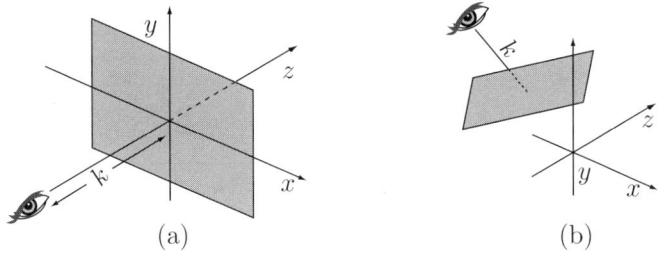

Figure 3.20: (a) Standard and (b) Nonstandard Positions.

object and the viewer looking at the object from different directions, examining various projections of it on the plane. It is pointless to move the viewer around the object while the projection plane stays at the same location (Figure 3.21b) because such a viewer will generally not even be looking at the plane. Thus, the projection plane must move with the viewer and must remain perpendicular to the line of sight of the viewer and at a distance of k units from him (although k may be varied by the user).

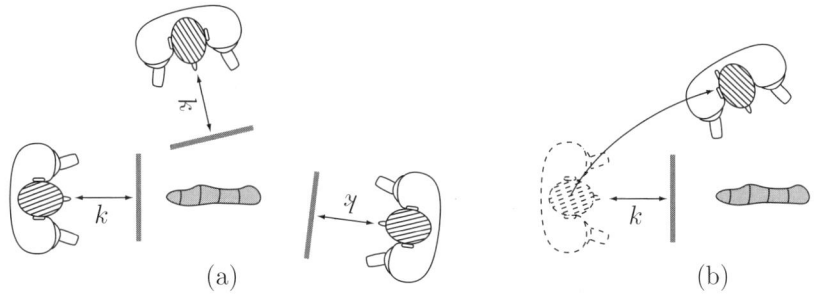

Figure 3.21: Moving the Viewer and the Projection Plane.

Step 2. The two similar triangles of Figure 3.22 yield the simple relations

$$\frac{x^*}{k} = \frac{x}{z+k} \quad \text{and} \quad \frac{y^*}{k} = \frac{y}{z+k},$$

from which we obtain

$$x^* = \frac{x}{(z/k)+1} \quad \text{and} \quad y^* = \frac{y}{(z/k)+1}. \tag{3.1}$$

(Some authors assign the x coordinate a negative sign. This is a result of the difference between left-handed and right-handed coordinate systems as discussed in Section 1.3. See also Exercise 3.27.) The +1 in the denominator of Equation (3.1) is important. It guarantees that the denominator will never be zero. The denominator can be zero only if $z/k = -1$, but k is positive and z is nonnegative.

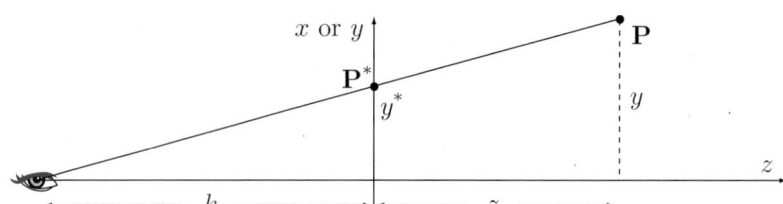

Figure 3.22: Perspective by Similar Triangles.

Step 3. Equation (3.1) can be employed to show that the projection rule of Step 1 results in a projection that satisfies the three conditions above and can therefore be called perspective. Condition 1 says that a distant object should appear small. The object can become distant in three ways:

1. increasing the z coordinates of its points;
2. increasing the x or y coordinates;
3. increasing the value of k.

For large values of z, Equation (3.1) yields small values for x^* and y^*. Specifically

$$\lim_{z \to \infty} x^* = 0 \quad \text{and} \quad \lim_{z \to \infty} y^* = 0.$$

For large values of x or y, imagine two points, $\mathbf{P}_1 = (x_1, y_1, z_1)$ and $\mathbf{P}_2 = (x_2, y_1, z_1)$, on the object that differ only in their x coordinates. They are projected to the two points $\mathbf{P}_1^* = (x_1^*, y_1^*)$ and $\mathbf{P}_2^* = (x_2^*, y_1^*)$, which have identical y coordinates, and the ratio of their x coordinates is

$$\frac{x_1^*}{x_2^*} = \frac{x_1}{(z_1/k) + 1} \bigg/ \frac{x_2}{(z_1/k) + 1} = \frac{x_1}{x_2}. \tag{3.2}$$

Thus, when both x_1 and x_2 grow, the ratio x_1^*/x_2^* approaches 1, which implies that the two projected points $\mathbf{P}_1^*$ and $\mathbf{P}_2^*$ get closer. Since $\mathbf{P}_1$ and $\mathbf{P}_2$ are any points with the same y and z coordinates, this implies that all the points with the same y and z coordinates produce projections that are very close. The object seems to have shrunk in the x dimension (Figure 3.23a).

The case where k increases (i.e., the viewer moves away from the projection plane) is different. Figure 3.23b shows how the projection of the object becomes bigger and bigger in this case until, at the limit, when the viewer is at infinity, the projection reaches the actual size of the object. The perspective projection is reduced in this limit to a parallel projection. However, even though the projection itself gets bigger, the viewer sees a small projected object because the projection plane and everything on it look small to a distant viewer.

Condition 2 demands that a distant object feature less perspective. We already know that an object can become distant in three ways each of which is individually treated here.

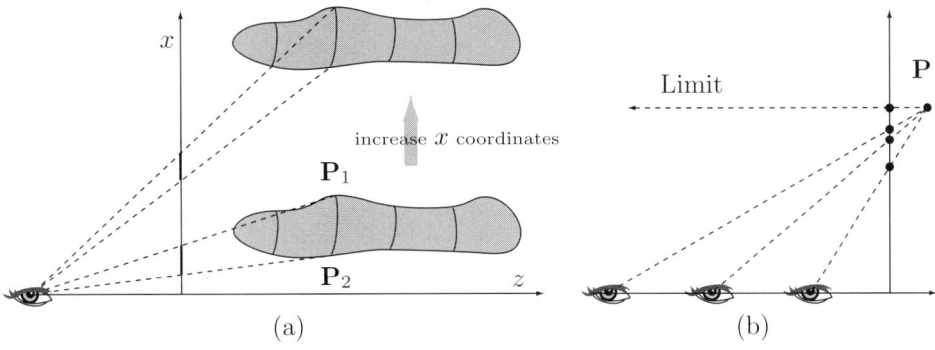

Figure 3.23: (a) Large x Dimensions. (b) Large Values of k.

1. The z coordinates are increased. We select two object points $\mathbf{P}_1 = (x_1, y_1, z_1)$ and $\mathbf{P}_2 = (x_1, y_1, z_2)$ with the same x and y coordinates and different z coordinates. We denote their projected points by $\mathbf{P}_1^* = (x_1^*, y_1^*)$ and $\mathbf{P}_2^* = (x_2^*, y_2^*)$ and compute the ratio x_1^*/x_2^*:

$$\frac{x_1^*}{x_2^*} = \frac{x_1}{(z_1/k) + 1} \bigg/ \frac{x_1}{(z_2/k) + 1} = \frac{z_2 + k}{z_1 + k}. \tag{3.3}$$

When the z coordinates are increased, this ratio approaches 1, thereby showing that the distance between the projected points is decreased, resulting in less perspective.

2. The x or y coordinates are increased. Equation (3.2) shows that the projected points get closer in this case, too.

3. The value of k is increased. In this case, Equation (3.3) shows that the projected points get closer, again implying less perspective.

Condition 3 is also easy to verify, at least in the case of lines parallel to the z axis. Figure 3.24 shows how a group of lines parallel to the z axis are projected to line segments that converge at the origin.

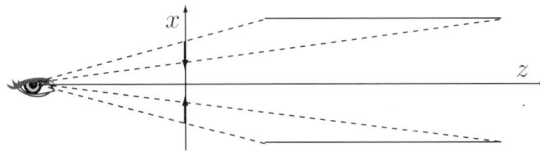

Figure 3.24: Lines Parallel to the z Axis.

Step 4. The projection expressed by Equation (3.1) can be included in the general 4×4 transformation matrix in three dimensions [Equation (1.23)]. The result is

$$\mathbf{T}_p = \begin{pmatrix} 1 & 0 & 0 & 0 \\ 0 & 1 & 0 & 0 \\ 0 & 0 & 0 & r \\ 0 & 0 & 0 & 1 \end{pmatrix}. \tag{3.4}$$

A simple test verifies that the product $(x, y, z, 1)\mathbf{T}_p$ yields $(x, y, 0, rz + 1)$ or, after dividing by the fourth coordinate, $(x/(rz+1), y/(rz+1), 0, 1)$. This agrees with Equation (3.1) if we assume that $r = 1/k$. (Recall that k is strictly positive and is never zero. The viewer never presses his eyes to the projection plane.)

It is now clear that there are two more special cases that are geometrically equivalent to our standard position. These are the cases where the viewer is positioned on the negative side of the x axis (or the y axis) at a certain distance from the origin and the projection plane is the yz (or xz) plane. The object is located on the positive side of the x (or y) axis. These cases correspond to the transformation matrices

$$\mathbf{T}_x = \begin{pmatrix} 0 & 0 & 0 & p \\ 0 & 1 & 0 & 0 \\ 0 & 0 & 1 & 0 \\ 0 & 0 & 0 & 1 \end{pmatrix} \quad \text{and} \quad \mathbf{T}_y = \begin{pmatrix} 1 & 0 & 0 & 0 \\ 0 & 0 & 0 & q \\ 0 & 0 & 1 & 0 \\ 0 & 0 & 0 & 1 \end{pmatrix},$$

where both $1/p$ and $1/q$ are the distances of the viewer from the origin.

The general case, where the viewer can be positioned anywhere and looking in any direction, is covered in Section 3.5. Before we get to this material, here are some examples of points projected in the standard position.

Linear example. We arbitrarily select the two points $\mathbf{P}_1 = (2, 3, 1)$ and $\mathbf{P}_2 = (3, -1, 2)$ and the distance $k = 1$. Notice that the z coordinates of these points are nonnegative. The points are projected to

$$\mathbf{P}_1^* = \left[\frac{2}{(1/1) + 1}, \frac{3}{(1/1) + 1} \right] = (1, 3/2) \text{ and } \mathbf{P}_2^* = \left[\frac{3}{(2/1) + 1}, \frac{-1}{(2/1) + 1} \right] = (1, -1/3).$$

We now select the midpoint $\mathbf{P}_m = (\mathbf{P}_1 + \mathbf{P}_2)/2 = (5/2, 1, 3/2)$ and project it to

$$\mathbf{P}_m^* = \left[\frac{5/2}{\frac{3/2}{1} + 1}, \frac{1}{\frac{3/2}{1} + 1} \right] = (1, 2/5).$$

Point $\mathbf{P}_m$ is located on the straight segment connecting $\mathbf{P}_1$ to $\mathbf{P}_2$ (it is the midpoint of the segment) and $\mathbf{P}_m^*$ is on the segment connecting $\mathbf{P}_1^*$ to $\mathbf{P}_2^*$ (although it isn't the midpoint, because it is easy to see that $\mathbf{P}_m^* = 0.4\mathbf{P}_1^* + 0.6\mathbf{P}_2^*$). The perspective projection of a straight segment is a straight segment, which is why it is done in practice by projecting the two endpoints and connecting them on the projection plane with a straight segment.

Converging lines. We now select an arbitrary point $\mathbf{P}_3 = (0, 2, 3)$ and compute a new point $\mathbf{P}_4 = (1, -2, 4)$ from the relation $\mathbf{P}_4 - \mathbf{P}_3 = \mathbf{P}_2 - \mathbf{P}_1$. The difference of two points is a vector, so this relation guarantees that the vector from $\mathbf{P}_3$ to $\mathbf{P}_4$ equals the vector from $\mathbf{P}_1$ to $\mathbf{P}_2$, or, equivalently, that the two line segments $\mathbf{P}_1\mathbf{P}_2$ and $\mathbf{P}_3\mathbf{P}_4$ are parallel. The two new points are projected to yield

$$\mathbf{P}_3^* = \left[0, \frac{2}{(3/1) + 1} \right] = (0, 1/2) \quad \text{and} \quad \mathbf{P}_4^* = \left[\frac{1}{(4/1) + 1}, \frac{-2}{(4/1) + 1} \right] = (1/5, -2/5).$$

The parametric equation of the straight segment connecting $\mathbf{P}_3^*$ to $\mathbf{P}_4^*$ is (see Equation (Ans.7))

$$L_2(w) = w(\mathbf{P}_4^* - \mathbf{P}_3^*) + \mathbf{P}_3^* = w(1/5, -9/10) + (0, 1/2) \quad \text{for} \quad 0 \le w \le 1,$$

and the parametric equation of the straight segment connecting $\mathbf{P}_1^*$ to $\mathbf{P}_2^*$ is

$$L_1(u) = u(\mathbf{P}_2^* - \mathbf{P}_1^*) + \mathbf{P}_1^* = u(0, -4/3) + (1, 3/2) \quad \text{for} \quad 0 \le u \le 1,$$

the point is that although the original segments $\mathbf{P}_1\mathbf{P}_2$ and $\mathbf{P}_3\mathbf{P}_4$ are parallel, the two projected segments are not parallel. They meet at point $L_1(33/8) = L_2(5) = (1, -4)$.

Another way to prove that the two projected line segments converge is to show that they are not parallel by computing and comparing their directions (or slopes). It's easy to see that $\mathbf{P}_2^* - \mathbf{P}_1^* = (0, -4/3)$ but $\mathbf{P}_4^* - \mathbf{P}_3^* = (1/5, -9/10)$. Line segment L_1 moves straight down, whereas L_2 has a slope of $(-9/10)/(1/5) = -4.5$.

◇ **Exercise 3.7:** Select two line segments that are perpendicular to the line of sight of the viewer, and show that their projections on the xy plane are parallel.

Projecting curves. We select the three points $\mathbf{P}_1 = (-1, 0, 1)$, $\mathbf{P}_2 = (0, 1, 2)$, and $\mathbf{P}_3 = (1, 1, 3)$ and compute the Bézier curve $\mathbf{P}(t)$ defined by them

$$\mathbf{P}(t) = (1 - t)^2(-1, 0, 1) + 2t(1 - t)(0, 1, 2) + t^2(1, 1, 3).$$

The midpoint of this curve is

$$\mathbf{P}(0.5) = (-1/4, 0, 1/4) + (0, 1/2, 1) + (1/4, 1/4, 3/4) = (0, 3/4, 2).$$

We now project the three original points and obtain

$$\mathbf{P}_1^* = \left[\frac{-1}{(1/1) + 1}, 0 \right] = (-1/2, 0), \quad \mathbf{P}_2^* = \left[0, \frac{1}{(2/1) + 1} \right] = (0, 1/3),$$

$$\mathbf{P}_3^* = \left[\frac{1}{(3/1) + 1}, \frac{1}{(3/1) + 1} \right] = (1/4, 1/4).$$

The Bézier curve defined by these points is

$$\mathbf{P}^*(t) = (1 - t)^2(-1/2, 0) + 2t(1 - t)(0, 1/3) + t^2(1/4, 1/4).$$

The point of this example is that the projection of $\mathbf{P}(0.5)$, which is $(0, 1/4)$, is not located on $\mathbf{P}^*(t)$. This illustrates the nonlinear nature of the Bézier curve (as well as most other curves).

⋄ **Exercise 3.8:** Show why point $(0, 1/4)$ is not located on $\mathbf{P}^*(t)$.

Transforming and projecting. This example illustrates the advantage of the projection matrix $\mathbf{T}_p$ of Equation (3.4). Given an object, we might want to transform it before we project its points. In such a case, all we have to do is prepare the individual 4×4 transformation matrices, multiply them together in the order of the transformations, and multiply the result by $\mathbf{T}_p$. Assume that we want to apply the following transformations to our object: (1) Rotate it about the x axis by $90°$ from the direction of positive y to the direction of positive z (Figure 3.25a). (2) Translate it by 3 units in the positive z direction. (3) Scale it by a factor of $1/2$ (i.e., shrink it to half its size) in the y dimension. The three transformation matrices are

$$\mathbf{T}_R = \begin{bmatrix} 1 & 0 & 0 & 0 \\ 0 & 0 & 1 & 0 \\ 0 & -1 & 0 & 0 \\ 0 & 0 & 0 & 1 \end{bmatrix}, \quad \mathbf{T}_T = \begin{bmatrix} 1 & 0 & 0 & 0 \\ 0 & 1 & 0 & 0 \\ 0 & 0 & 1 & 0 \\ 0 & 0 & 3 & 1 \end{bmatrix}, \quad \mathbf{T}_S = \begin{bmatrix} 1 & 0 & 0 & 0 \\ 0 & 1/2 & 0 & 0 \\ 0 & 0 & 1 & 0 \\ 0 & 0 & 0 & 1 \end{bmatrix}$$

and their product with $\mathbf{T}_p$ (we assume $k = 1$, so $r = 1$) produces

$$\mathbf{T} = \mathbf{T}_R \mathbf{T}_T \mathbf{T}_S \begin{bmatrix} 1 & 0 & 0 & 0 \\ 0 & 1 & 0 & 0 \\ 0 & 0 & 0 & 1 \\ 0 & 0 & 0 & 1 \end{bmatrix} = \begin{bmatrix} 1 & 0 & 0 & 0 \\ 0 & 0 & 0 & 1 \\ 0 & -1/2 & 0 & 0 \\ 0 & 0 & 0 & 4 \end{bmatrix}. \tag{3.5}$$

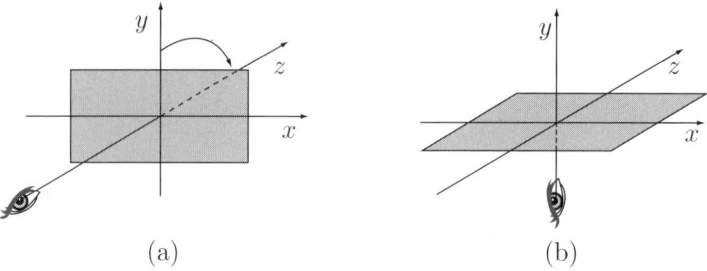

(a)   (b)

Figure 3.25: Rotation about the x Axis.

We can now pick any point on the object, write it as a 4-tuple in homogeneous coordinates, and multiply it by $\mathbf{T}$ to obtain its projection after applying the three transformations to it. Notice that a point cannot be scaled, but the effect of scaling is to move points such that the scaled object will shrink to half its size in the y dimension. As an example, multiplying point $(0, 1, -4, 1)$ by $\mathbf{T}$ results in $(0, 2, 0, 5)$, which, after dividing by the fourth coordinate, produces the two-dimensional point $(0, 2/5)$.

⋄ **Exercise 3.9:** Multiply point $(0, 1, -4, 1)$ by the product $\mathbf{T}_R \mathbf{T}_T \mathbf{T}_S$ and explain the result.

⋄ **Exercise 3.10:** The previous paragraph has mentioned *scaling*, so let's consider another subtle effect of this simple transformation. The transformation matrix for scaling is

$$\begin{pmatrix} T_1 & 0 & 0 & 0 \\ 0 & T_2 & 0 & 0 \\ 0 & 0 & T_3 & 0 \\ 0 & 0 & 0 & 1 \end{pmatrix}.$$

When combined with perspective projection, it yields

$$\begin{pmatrix} T_1 & 0 & 0 & 0 \\ 0 & T_2 & 0 & 0 \\ 0 & 0 & T_3 & 0 \\ 0 & 0 & 0 & 1 \end{pmatrix} \begin{pmatrix} 1 & 0 & 0 & 0 \\ 0 & 1 & 0 & 0 \\ 0 & 0 & 0 & r \\ 0 & 0 & 0 & 1 \end{pmatrix} = \begin{pmatrix} T_1 & 0 & 0 & 0 \\ 0 & T_2 & 0 & 0 \\ 0 & 0 & 0 & T_3r \\ 0 & 0 & 0 & 1 \end{pmatrix}.$$

Hence, a point $(x, y, z, 1)$ is transformed to $(T_1x, T_2y, 0, T_3rz + 1)$, which implies

$$x^* = \frac{T_1x}{T_3rz + 1}, \qquad y^* = \frac{T_2y}{T_3rz + 1}.$$

In the special case of uniform scaling, $T_1 = T_2 = T_3 = T$, we get $x^* = x/(rz + 1/T)$, $y^* = y/(rz + 1/T)$. The problem is that when T gets large (large magnification), $1/T$ becomes small, resulting in

$$x^* \approx \frac{x}{rz} = \frac{xk}{z}, \qquad y^* \approx \frac{y}{rz} = \frac{yk}{z}.$$

We don't seem to get the expected magnification. What's the explanation?

The rightmost column of matrix **T** of Equation (3.5) is important and will serve (on page 110) to illuminate the properties of the general perspective projection. The three top elements of this column are 0, 1, and 0. The reader may remember that the general transformation matrix [Equation (1.23)] denotes these elements by p, q, and r. Thus, element q of matrix **T** is nonzero. It has already been mentioned that element r of matrix $\mathbf{T}_p$ is nonzero because the viewer is positioned on the z axis. The reason that element q of matrix **T** is nonzero is the rotation about the x axis. We can interpret this rotation either as a rotation of the point or as a rotation of the coordinate system. In the latter case, this rotation has changed the projection plane from the xy plane to the xz plane and has also moved the viewer (because the viewer and the projection plane constitute one unit) from his standard position on the z axis to a new location on the y axis (Figure 3.25b). The fact that q is nonzero tells us that the y axis now intercepts the projection plane. Page 110 sheds more light on the function of matrix elements p, q, and r.

⋄ **Exercise 3.11:** Compute the coordinates of the object point **P** that happens to be projected to the origin after the three transformations.

Negative z coordinates. It has already been mentioned several times that the viewer and the object have to be located on different sides of the projection plane. In the standard position, this means that all the object points have to have nonnegative z coordinates. This example shows what happens when object points have invalid coordinates. (See also Exercise 3.19.) Figure 3.26a shows the two points $\mathbf{P}_1 = (0, 1, -1)$ and $\mathbf{P}_2 = (0, 1, 1)$ and a viewer located at $(0, 0, -3)$. When Equation (3.1) is used to project the two points, the results are

$$\mathbf{P}_1^* = \left[0, \frac{1}{(-1/3) + 1}, 0\right] = (0, 3/2, 0) \quad \text{and} \quad \mathbf{P}_2^* = \left[0, \frac{1}{(1/3) + 1}, 0\right] = (0, 3/4, 0).$$

The result seems to make sense, but Figure 3.26b shows that when $\mathbf{P}_1$ is moved to the left (i.e., toward larger negative z values), its projection climbs up the y axis quickly and without limit, thereby creating a distorted projection of the entire object. When $\mathbf{P}_1$ is located right over the viewer [when it is moved to $(0, 1, -3)$], its projection is undefined, and when it is moved farther to the left, its projection becomes negative. In such a case, those parts of the object that are in front of the viewer are projected right side up but distorted, and those parts that are behind the viewer are projected upside down.

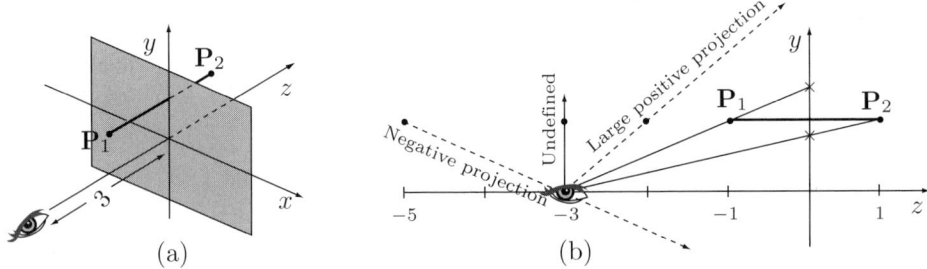

Figure 3.26: Perspective Projection with Negative z Coordinates.

3.5 General Perspective

The standard position is just a special case of perspective projection. It simplifies the computations of the projected points and should be used whenever possible. There are cases, however, where the viewer has to be positioned at different points and has to look in different directions. A common example is computer animation. In a typical animation sequence, there is an object or a scene and we imagine a camera moving around or above the scene, taking snapshots much like a real movie camera. While the camera is moving, the object or objects in the scene may also move along a path, rotate, shrink, or become distorted by shearing.

An animation sequence is therefore done in steps, where each step starts by moving, rotating, or otherwise transforming the object (if necessary), moving the camera (which becomes the viewer) to its appropriate position for the step, orienting it, so it looks in the right direction, and finally taking a snapshot. The last operation, taking a snapshot, is done by computing the perspective projections of all the object's points and plotting the points on the projection plane. The resulting image on the projecting plane then becomes the next animation frame, and the final animation is screened at a fast rate (typically 18 to 24 frames per second) to create the illusion of smooth animation.

Because animation is such an important application of perspective projection, we often use the term "screen" instead of "projection plane." The main difference between a screen and a plane is that the former has a finite size, whereas the latter is infinitely large. In order to derive the mathematics of general perspective, we need to know at least (1) the location **B** of the viewer, (2) the direction **D** of the viewer's line of sight, and (3) the coordinates of all the points **P** on the object. Figure 3.27 illustrates another complication that often arises. The figure shows viewers located at the same point and looking in the same direction, but with screens that have different orientations (although each is perpendicular to the line of sight). Thus, in order to fully specify the viewer-screen unit, we sometimes also need to specify the direction **T** of the top of the screen.

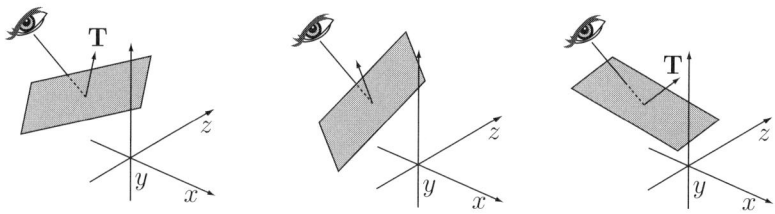

Figure 3.27: General Perspective with Different Screen Orientations.

We start this section with a simple example that illustrates how rotation and translation, combined with basic concepts from geometry, can be applied to the computation of perspective projection. Similar computations can be carried out in other cases, but they are normally very messy. Future sections of this chapter illustrate better approaches to the problem of general perspective.

In this example, we assume that the viewer has been moved from the standard position by a translation and his line of sight has been rotated. (It is also possible to first rotate the viewer and then translate him.) We compute the new location and direction of the viewer and use this information to compute the equation of the projection plane. (Alternatively, we can determine the new equation of the projection plane by applying to it the same transformations applied to the viewer.) Once this equation is known, we compute the straight segment $\mathbf{P}(t)$ that connects the object point **P** to the viewer. The final step is to calculate the point $\mathbf{P}(t_0)$ where this segment intercepts the projection plane. This point is the projection $\mathbf{P}^*$ of **P**.

In the example, we rotate the viewer θ degrees counterclockwise about the y axis from the positive z to the positive x direction (Figure 3.28a). The viewer ends up at point

$$(0,0,-k) \begin{pmatrix} \cos\theta & 0 & -\sin\theta \\ 0 & 1 & 0 \\ \sin\theta & 0 & \cos\theta \end{pmatrix} = (-k\sin\theta, 0, -k\cos\theta) = (-k\alpha, 0, -k\beta), \qquad (3.6)$$

where $\alpha = \sin\theta$ and $\beta = \cos\theta$ (notice that $\alpha^2 + \beta^2 = 1$). We select a general point $\mathbf{P} = (l, m, n)$ on the object and compute its projection $\mathbf{P}^*$ on the new projection plane. Notice that the new projection plane is still perpendicular to the line of sight of the viewer and is still at a distance of k units. It is no longer identical to the xy plane, but it still contains the origin.

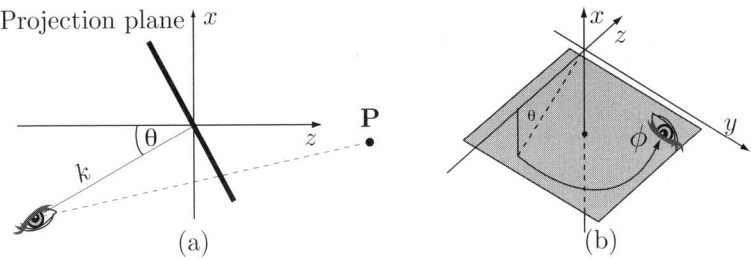

Figure 3.28: Viewer Rotated About the y Axis.

◇ **Exercise 3.12:** The previous paragraph talks about rotating the viewer counterclockwise, but Equation (3.6) looks like Equation (1.4), which generates clockwise rotation. What's the explanation?

■ The first task is to find the equation of the projection plane. Vector $(-k\alpha, 0, -k\beta)$ is perpendicular to the plane (it is the *normal* to the plane), so it is perpendicular to any general vector (x, y, z) on the plane. This is why their dot product is zero. From $(-k\alpha, 0, -k\beta) \bullet (x, y, z) = 0$, we obtain the plane equation $\alpha x = -\beta z$.

◇ **Exercise 3.13:** Why doesn't this equation involve y?

An alternate way to derive the plane equation is to start with the equation of the original plane and transform it by means of Equation (3.6). The original plane was the xy plane, whose equation is $z = 0$. A general point on this plane has coordinates $(a, b, 0)$. When multiplied by the rotation matrix of Equation (3.6), the point is transformed to $(\beta a, b, -\alpha a)$. Thus, a general point (x, y, z) on the new plane has an x coordinate that's the product of an arbitrary number a and $\cos\theta$, a z coordinate that's the product of the same number a and $-\sin\theta$, and an arbitrary y coordinate. The relation between the coordinates can therefore be expressed as $z = -\alpha a = -\alpha(x/\beta)$ or $\alpha x = -\beta z$.

■ Next, we find the equation of the line segment from the viewer to point $\mathbf{P}$. We use the parametric representation $\mathbf{P}(t) = (\mathbf{P}_2 - \mathbf{P}_1)t + \mathbf{P}_1$ [Equation (Ans.7)]. When applied to the viewer (denoted by $\mathbf{P}_1$) and to point $\mathbf{P} = (l, m, n)$ (denoted by $\mathbf{P}_2$), it yields

$$\mathbf{P}(t) = (l + k\alpha, m, n + k\beta)t + (-k\alpha, 0, -k\beta)$$
$$= ((l + k\alpha)t - k\alpha, mt, (n + k\beta)t - k\beta)$$
$$= \big(P_x(t), P_y(t), P_z(t)\big).$$

■ Our next task is to find the intersection point of the line and the projection plane. This is obtained at the value t_0 that satisfies $\alpha P_x(t_0) = -\beta P_z(t_0)$ or

$$\alpha\big((l + k\alpha)t_0 - k\alpha\big) = -\beta\big((n + k\beta)t_0 - k\beta\big).$$

The solution is

$$t_0 = \frac{k(\alpha^2 + \beta^2)}{\alpha l + \beta n + k(\alpha^2 + \beta^2)} = \frac{k}{\alpha l + \beta n + k}.$$

The intersection point is $\mathbf{P}(t_0)$.

■ The next task is to find the three coordinates of the projected point $\mathbf{P}^* = \mathbf{P}(t_0)$. The x coordinate is

$$x^* = P_x(t_0) = (l + k\alpha)t_0 - k\alpha = (l + k\alpha)\frac{k}{\alpha l + \beta n + k} - k\alpha = \frac{lk\beta^2 - nk\alpha\beta}{\alpha l + \beta n + k}.$$

The y coordinate is

$$y^* = P_y(t_0) = mt_0 = \frac{mk}{\alpha l + \beta n + k},$$

and the z coordinate is

$$z^* = P_z(t_0) = (n + k\beta)t_0 - k\beta = (n + k\beta)\frac{k}{\alpha l + \beta n + k} - k\beta = \frac{-lk\alpha\beta + nk\alpha^2}{\alpha l + \beta n + k}.$$

From $(x^*, y^*, z^*) = (X/H, Y/H, Z/H)$, we obtain

$$X = lk\beta^2 - nk\alpha\beta,$$
$$Y = mk,$$
$$Z = -lk\alpha\beta + nk\alpha^2,$$
$$H = \alpha l + \beta n + k.$$

■ Using the four expressions above and keeping in mind that (l, m, n) are the coordinates of point $\mathbf{P}$, it is easy to figure out the transformation matrix that projects $\mathbf{P}$ to $\mathbf{P}^*$:

$$(l, m, n, 1)\mathbf{T} = (X, Y, Z, H) \quad \text{implies} \quad \mathbf{T} = \begin{pmatrix} k\beta^2 & 0 & -k\alpha\beta & \alpha \\ 0 & k & 0 & 0 \\ -k\alpha\beta & 0 & k\alpha^2 & \beta \\ 0 & 0 & 0 & k \end{pmatrix}. \quad (3.7)$$

The three quantities α, 0, and β that appear at the top of the rightmost column of matrix **T** correspond to elements p, q, and r of the general 4×4 transformation matrix. They tell us which of the three coordinate axes is intercepted by the projection plane. In our case, the first and third quantities are nonzero (except for $\theta = 0$ and $\theta = 90°$), which implies that the new projection plane intercepts the x and z axes. Page 110 has more to say about elements p, q, and r.

⋄ **Exercise 3.14:** Calculate the values of matrix (3.7) for the three special cases $\theta = 0°$, 45°, and 90°.

⋄ **Exercise 3.15:** Given the point $\mathbf{P} = (\beta l, m, -\alpha l)$, calculate its projection. Explain the result!

⋄ **Exercise 3.16:** Imagine rotating the viewer, who is now at $(-k\alpha, 0, -k\beta)$, a second time, by an angle ϕ about the x axis (Figure 3.28b). The new position of the viewer is

$$
(-k\alpha, 0, -k\beta)
\begin{pmatrix}
1 & 0 & 0 \\
0 & \cos\phi & -\sin\phi \\
0 & \sin\phi & \cos\phi
\end{pmatrix}
$$
$$
= (-k\sin\theta, -k\cos\theta\sin\phi, -k\cos\theta\cos\phi)
$$
$$
= (k\alpha, -k\beta\gamma, -k\beta\delta),
$$

where $\gamma = \sin\phi$ and $\delta = \cos\phi$. Derive the projection matrix for this case using steps similar to the ones above.

⋄ **Exercise 3.17:** After two rotations, the viewer may be located at any point in space. This is still not the most general case because there is another constraint. What is it?

It is important to realize that matrix (3.7) isn't as useful as it may seem at first. It generates the coordinates of projected points, but those coordinates are on the plane $\alpha x = -\beta z$. In practice, we want to display the projected points on the screen, which is two-dimensional, so we have to go through another step. We have to define two local axes on $\alpha x = -\beta z$ and then figure out the coordinates of the projected points relative to those axes. This is why the approaches discussed in the remainder of this chapter are preferable. They project points onto the xy plane, where they effectively have just two coordinates. Before looking at these approaches, however, here is a short summary of the method used in this section.

Summary. The method of this section proceeds in the following steps:

1. Derive the equation of the projection plane.
2. Determine the equation of the line segment connecting an arbitrary point $\mathbf{P}$ on the object to the viewer (see Equation (Ans.7)).
3. Locate the intersection point of the line and the plane.
4. Convert the coordinates of the intersection point to screen coordinates.

It is possible to use these steps to figure out the projection matrix for the general case where the viewer may be located at any point $\mathbf{B}$, looking in an arbitrary given direction **D**. This approach to the computation, however, is messy because in addition to **B**, **D**, and k, another vector is needed to define the "up" direction of the projection plane. In this section, we started with the "up" direction in the positive y direction. After the two rotations, that direction has changed, but it is fully determined by the rotations and does not need to be explicitly specified. Another drawback of this approach is that points are projected on a three-dimensional plane, so they have three dimensions. In practice, the projected image should be displayed on the computer monitor, which is two-dimensional, so we would like the computations to produce two-dimensional points. The following two sections show how to project points on the xy plane, which effectively makes them two-dimensional.

> Perspective, as its inventor remarked, is a beautiful thing. What horrors of damp huts, where human beings languish, may not become picturesque through aerial distance! What hymning of cancerous vices may we not languish over as sublimest art in the safe remoteness of a strange language and artificial phrase! Yet we keep a repugnance to rheumatism and other painful effects when presented in our personal experience.
>
> —George Eliot, *Daniel Deronda* (1876)

3.6 Transforming the Object

The theory of special relativity teaches that movement (at a constant speed and in a straight line) is relative, which suggests the following idea. Instead of transforming the viewer to a new location, computing the new equation of the projection plane, and going through all the computations of the previous section, why not leave the viewer at the standard position and transform the object instead? After all, we are interested only in what the viewer sees on the screen. The absolute locations of the viewer and the object are irrelevant. If the viewer is left at the standard position, then any point on the (transformed) object can be projected by means of matrix $\mathbf{T}_p$ of Equation (3.4), which greatly simplifies the computations.

This approach is ideal for cases where the viewer is located at the standard position and has to be transformed by means of translations and/or rotations (or even reflections, but no scaling or shearing) to a new location, where he can observe the object from a different direction. This approach is useful, for example, in computer animation. Suppose that we have to transform the viewer from the standard position to a new location by means of a transformation $\mathbf{A}$ that consists of several translations, rotations, and/or reflections (thus, $\mathbf{A} = \mathbf{T}_1 \cdot \mathbf{T}_2 \cdots \mathbf{T}_n$). Instead of this, we leave the viewer in the standard position and apply the inverse transformation $\mathbf{A}^{-1}$ to the object. Direct multiplication proves that the inverse $\mathbf{A}^{-1}$ of our product matrix $\mathbf{A}$ is given by $\mathbf{A}^{-1} = \mathbf{T}_n^{-1} \cdots \mathbf{T}_2^{-1} \cdot \mathbf{T}_1^{-1}$ (where $\mathbf{T}_i^{-1}$ is the reverse of transformation $\mathbf{T}_i$). A nice feature of this approach is that the individual $\mathbf{T}_i$ transformations are only translations, rotations, and reflections, and these transformations have simple inverses.

The point is that transforming the viewer with $\mathbf{A}$ or transforming the object with $\mathbf{A}^{-1}$ will bring them to the *same relative position*. Once the object has been transformed, we can use matrix $\mathbf{T}_p$ [Equation (3.4)] to compute the perspective projection because the viewer is still located at the standard position. In practice, there is, of course, no need to actually transform the object. All that we have to do is compute matrix $\mathbf{T} = \mathbf{A}^{-1} \cdot \mathbf{T}_p$ and multiply each point of the object by $\mathbf{T}$.

Example: A viewer located at the standard position and an object close to the origin (Figure 3.29). Suppose that we want to translate the viewer to the origin, rotate him $45°$ counterclockwise and then translate him k units in both the negative x and negative z directions (Figure 3.29a,b,c). The transformation matrices are

$$\mathbf{T}_1 = \begin{pmatrix} 1 & 0 & 0 & 0 \\ 0 & 1 & 0 & 0 \\ 0 & 0 & 1 & 0 \\ 0 & 0 & k & 1 \end{pmatrix}, \quad \mathbf{T}_2 = \begin{pmatrix} \cos 45° & 0 & -\sin 45° & 0 \\ 0 & 1 & 0 & 0 \\ \sin 45° & 0 & \cos 45° & 0 \\ 0 & 0 & 0 & 1 \end{pmatrix},$$

$$\mathbf{T}_3 = \begin{pmatrix} 1 & 0 & 0 & 0 \\ 0 & 1 & 0 & 0 \\ 0 & 0 & 1 & 0 \\ -k & 0 & -k & 1 \end{pmatrix}.$$

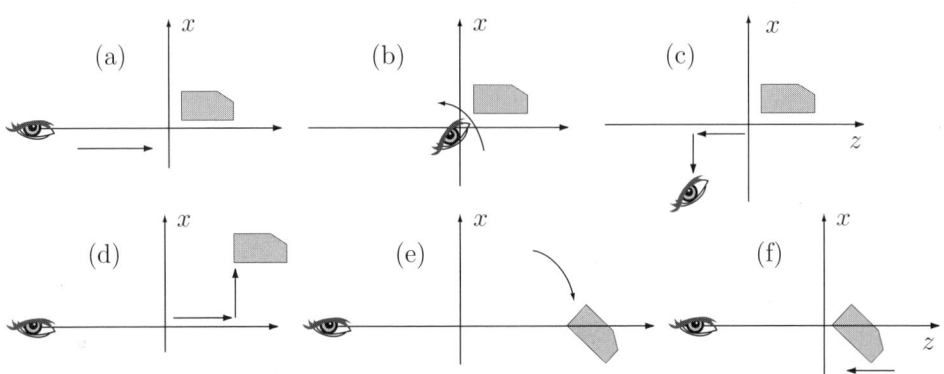

Figure 3.29: Transforming Viewer or Object.

The reverse transformations, performed in reverse order, are (Figure 3.29d,e,f)

$$\mathbf{A}^{-1} = \begin{pmatrix} 1 & 0 & 0 & 0 \\ 0 & 1 & 0 & 0 \\ 0 & 0 & 1 & 0 \\ k & 0 & k & 1 \end{pmatrix} \begin{pmatrix} \cos 45° & 0 & \sin 45° & 0 \\ 0 & 1 & 0 & 0 \\ -\sin 45° & 0 & \cos 45° & 0 \\ 0 & 0 & 0 & 1 \end{pmatrix} \begin{pmatrix} 1 & 0 & 0 & 0 \\ 0 & 1 & 0 & 0 \\ 0 & 0 & 1 & 0 \\ 0 & 0 & -k & 1 \end{pmatrix}$$

$$= \begin{pmatrix} \sin 45° & 0 & \sin 45° & 0 \\ 0 & 1 & 0 & 0 \\ -\sin 45° & 0 & \sin 45° & 0 \\ 0 & 0 & -k + 2k\sin 45° & 1 \end{pmatrix}.$$

Any point $\mathbf{P} = (x, y, z, 1)$ on the object can be projected to a two-dimensional point $\mathbf{P}^*$ on the screen by

$$\mathbf{P}^* = \mathbf{PA}^{-1}\mathbf{T}_p = (x, y, z, 1) \begin{pmatrix} a & 0 & 0 & a/k \\ 0 & 1 & 0 & 0 \\ -a & 0 & 0 & a/k \\ 0 & 0 & 0 & 2a \end{pmatrix}$$

$$= (a(x - z), y, 0, a(2k + x + z)/k),$$

resulting in

$$x^* = \frac{k(x - z)}{2k + x + z}, \quad y^* = \frac{yk}{a(2k + x + z)},$$

where $a = \sin 45°$. A comparison of parts (c) and (f) in Figure 3.29 shows how the viewer and the object end up in the same relative positions.

If transforming the viewer involves only translations and rotations (and no reflections), it is possible to transform the viewer from the standard position to any location in space by means of (1) a translation to the origin, (2) a general rotation about the origin, and (3) another translation from the origin to the final location. The two translations are easy to express, and Section 3.7 shows how to derive the transformation matrix that will rotate the viewer so his line of sight becomes any given direction $\mathbf{D}$.

The following example serves to illustrate this claim. Suppose that we want to translate the viewer from the standard position $(0, 0, -k)$ to an arbitrary location $\mathbf{B} = (a, b, c)$ and then rotate him about some axis that goes through the origin (or, equivalently, first rotate him and then translate him to $\mathbf{B}$). A rotation about the origin requires a temporary translation from $\mathbf{B}$ to the origin, a rotation, and a translation back to $\mathbf{B}$. Thus, we need the four transformation matrices

$$\mathbf{T}_1 = \begin{bmatrix} 1 & 0 & 0 & 0 \\ 0 & 1 & 0 & 0 \\ 0 & 0 & 1 & 0 \\ a & b & c+k & 1 \end{bmatrix}, \quad \mathbf{T}_2 = \begin{bmatrix} 1 & 0 & 0 & 0 \\ 0 & 1 & 0 & 0 \\ 0 & 0 & 1 & 0 \\ -a & -b & -c & 1 \end{bmatrix},$$

$$\mathbf{T}_3 = \begin{bmatrix} . & . & . & 0 \\ . & . & . & 0 \\ . & . & . & 0 \\ 0 & 0 & 0 & 1 \end{bmatrix}, \quad \mathbf{T}_4 = \begin{bmatrix} 1 & 0 & 0 & 0 \\ 0 & 1 & 0 & 0 \\ 0 & 0 & 1 & 0 \\ a & b & c & 1 \end{bmatrix},$$

where the elements of the rotation matrix $\mathbf{T}_3$ are irrelevant and are not shown. Direct multiplication verifies that the product $\mathbf{T}_1\mathbf{T}_2$ is a transformation matrix that translates from the standard position $(0, 0, -k)$ to the origin. Thus, instead of the four matrices above, we need only three transformation matrices, a translation to the origin, a rotation about the origin, and a translation to point $\mathbf{B}$.

⋄ **Exercise 3.18:** Suppose that we first want to rotate the viewer about the origin and then translate him to point $\mathbf{B} = (a, b, c)$. The rotation requires three transformations, a translation $\mathbf{T}_1$ to the origin, a rotation $\mathbf{T}_2$ about the origin, and a translation $\mathbf{T}_3$ back to $(0, 0, -k)$. This must be followed by a translation $\mathbf{T}_4$ from the standard position to $\mathbf{B}$. Show that the last two translations, $\mathbf{T}_3$ and $\mathbf{T}_4$, can be replaced by one translation.

When reflections are included in addition to translations and rotations, more than four transformation matrices may be needed. Figure 3.30a shows a simple example. Given a viewer at $(0, 0, -2)$, we want to reflect it about the plane $(x, 0, x - 1)$ and rotate it 45° about the y axis (Figure 3.30b). The viewer can be considered a point which has no dimensions and no "left" and "right" directions. Thus, the reflection moves the viewer to another location but does not "reverse" him. However, the viewer and the screen have to be treated and moved as a single unit, which is why a full treatment of perspective projection should include a "top" vector that points in the direction of the top of the screen. When the viewer-screen unit is reflected, the left and right sides of the screen are reversed and the "top" vector changes direction (Figure 3.30c).

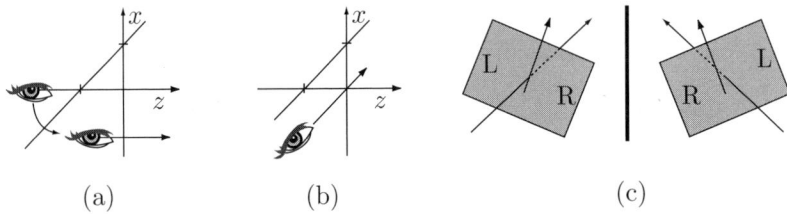

(a)   (b)   (c)

Figure 3.30: Reflecting the Viewer.

In general, a reflection about an arbitrary plane in three dimensions requires five transformations: (1) a translation that brings one point of the plane to the origin, (2) a rotation about the origin that brings the plane to one of the three coordinate planes, (3) a reflection about that plane, (4) a reverse rotation, and (5) the reverse translation. In many special cases, such as a plane parallel to one of the coordinate planes, this process can be simplified, but in general a reflection followed by a rotation requires eight $(5 + 3)$ transformations. In order to apply the inverse transformations to points on the object, we have to determine the inverses of all the transformation matrices involved, but fortunately the inverses of translation, rotation, and reflection about one of the coordinate planes are trivial to figure out.

It should again be emphasized that the viewer and the projection plane constitute a single unit and should be transformed together. Even though the approach discussed in this section transforms the object and not the viewer, it is still important to make sure that the object remains on the other side of the projection plane from the viewer after all the transformations. Thus, after an object point is transformed and before it is projected, it is important to verify that its z coordinate is still nonnegative. It is also important to make sure that enough points are selected on the object, because otherwise it may happen that two points with nonnegative z coordinates are connected on the object with a curve, some of whose points may have negative z coordinates when projected. Figure 3.31a is an example of an object where $\mathbf{P}_3$ initially is not included as an object point. The transformations move the object to the left such that part of the curve between points $\mathbf{P}_1$ and $\mathbf{P}_2$ ends up to the left of the xy plane and $\mathbf{P}_3$ has a negative z coordinate. Once $\mathbf{P}_3$ is included as an object point, the software discovers that its projection has a negative z coordinate of, say, a units. The software then moves all the object points a units to the right (Figure 3.31b) to obtain the correct projection on the xy plane.

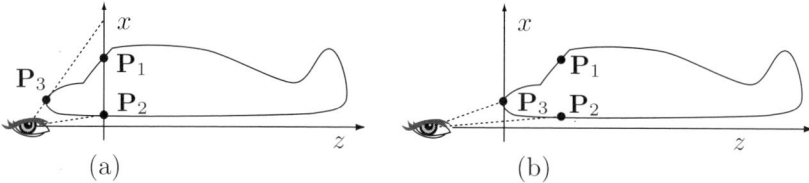

Figure 3.31: An Object with Negative z Coordinates.

On the other side of the screen, it all looks so easy.
—Jeff Bridges (as Kevin Flynn) in *Tron* (1982)

3.7 Viewer at an Arbitrary Location

The previous section dealt with the case where the viewer is initially located at the standard position. This section looks at the more general problem where the viewer is located at an arbitrary point $\mathbf{B} = (a, b, c)$, looking in a given direction $\mathbf{D} = (d, e, f)$ (Figure 3.32a). The approach taken here is to transform the viewer to the standard position in three simple steps: (1) translate the viewer from $\mathbf{B}$ to the origin (the screen is also translated by the same amount; Figure 3.32b); (2) rotate the viewer-screen unit in three dimensions until $\mathbf{D}$ coincides with $(0, 0, 1)$ (i.e., it points in the positive z direction, Figure 3.32c), and (3) translate the viewer and screen from the origin to point $(0, 0, -k)$ (Figure 3.32d). These three transformations bring the viewer to the standard position and the screen to the xy plane. The same transformations are then applied to every point $\mathbf{P}$ of the image, thereby bringing the viewer and the image to the same relative positions they had before the transformations. One way to understand this approach is to imagine that the viewer and all the image points are transformed as one unit, such that the viewer ends up at the standard position. Another way to look at this approach is to imagine that we transform the coordinate axes (Section 1.5), while the viewer and the image are not moved.

Now that the viewer is located at the standard position, matrix $\mathbf{T}_p$ [Equation (3.4)] can be used to project image points. This approach has the advantage that all the image points are projected on the xy plane, so that the projected points are effectively two-dimensional. In practice, there is no need to actually transform the viewer and the

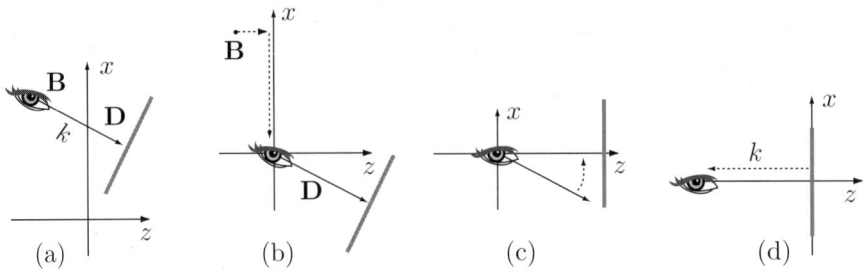

Figure 3.32: Transforming the Viewer-Screen Unit.

screen. We simply use the coordinates (a, b, c) of point $\mathbf{B}$ and the components (d, e, f) of vector $\mathbf{D}$ to derive the three transformation matrices $\mathbf{T}_1$ (translation), $\mathbf{T}_2$ (rotation), and $\mathbf{T}_3$ (second translation) and multiply $\mathbf{T} = \mathbf{T}_1 \mathbf{T}_2 \mathbf{T}_3 \mathbf{T}_p$. Any point $\mathbf{P}$ on the object is then transformed and projected in a single step by the multiplication $\mathbf{P}^* = \mathbf{PT}$.

This approach is developed here for the general case but is first illustrated by two examples where the coordinates of $\mathbf{B}$ and the components of $\mathbf{D}$ are known numbers.

Example 1. The viewer is located at $\mathbf{B} = (1, 1, 1)$ and is looking in direction $\mathbf{D} = (1, 0, 1)$ (i.e., midway between the directions of positive x and positive z). Matrix $\mathbf{T}_1$ below translates from $(1, 1, 1)$ to the origin. Matrix $\mathbf{T}_2$ rotates by $45°$ from the positive x to the positive z direction. Matrix $\mathbf{T}_3$ translates from the origin to point $(0, 0, -k)$. The result is (we denote $s = \cos 45° = \sin 45° = 1/\sqrt{2}$)

$$
\mathbf{T} = \mathbf{T}_1 \mathbf{T}_2 \mathbf{T}_3 \mathbf{T}_p
$$

$$
= \begin{pmatrix} 1 & 0 & 0 & 0 \\ 0 & 1 & 0 & 0 \\ 0 & 0 & 1 & 0 \\ -1 & -1 & -1 & 1 \end{pmatrix} \begin{pmatrix} s & 0 & s & 0 \\ 0 & 1 & 0 & 0 \\ -s & 0 & s & 0 \\ 0 & 0 & 0 & 1 \end{pmatrix} \begin{pmatrix} 1 & 0 & 0 & 0 \\ 0 & 1 & 0 & 0 \\ 0 & 0 & 1 & 0 \\ 0 & 0 & -k & 1 \end{pmatrix} \begin{pmatrix} 1 & 0 & 0 & 0 \\ 0 & 1 & 0 & 0 \\ 0 & 0 & 0 & r \\ 0 & 0 & 0 & 1 \end{pmatrix}
$$

$$
= \begin{pmatrix} s & 0 & s & 0 \\ 0 & 1 & 0 & 0 \\ -s & 0 & s & 0 \\ 0 & -1 & -2s - k & 1 \end{pmatrix} \begin{pmatrix} 1 & 0 & 0 & 0 \\ 0 & 1 & 0 & 0 \\ 0 & 0 & 0 & r \\ 0 & 0 & 0 & 1 \end{pmatrix}
$$

$$
= \begin{pmatrix} s & 0 & 0 & sr \\ 0 & 1 & 0 & 0 \\ -s & 0 & 0 & sr \\ 0 & -1 & 0 & 1 - kr - 2rs \end{pmatrix}
$$

$$
= \begin{pmatrix} s & 0 & 0 & sr \\ 0 & 1 & 0 & 0 \\ -s & 0 & 0 & sr \\ 0 & -1 & 0 & -2rs \end{pmatrix}. \tag{3.8}
$$

(Recall that $k = 1/r$.) The projection of any point $\mathbf{P} = (x, y, z)$ is calculated by

$\mathbf{P}^* = \mathbf{PT}$. We illustrate this for two points.

1: Point $\mathbf{P} = (1,1,1)$ is projected to $\mathbf{P}^* = (0,0,0)$ because

$$(1,1,1,1) \begin{pmatrix} s & 0 & 0 & sr \\ 0 & 1 & 0 & 0 \\ -s & 0 & 0 & sr \\ 0 & -1 & 0 & -2rs \end{pmatrix} = (0,0,0,0).$$

2: Point $\mathbf{P} = (2k,0,2k)$ is projected to $\mathbf{P}^* = (0,-1/\sqrt{2}(2-r),0)$ because

$$(2k,0,2k,1) \begin{pmatrix} s & 0 & 0 & sr \\ 0 & 1 & 0 & 0 \\ -s & 0 & 0 & sr \\ 0 & -1 & 0 & -2rs \end{pmatrix} = (0,-1,0,2s(2-r)).$$

◇ **Exercise 3.19:** The product

$$(0,0,0,1) \begin{pmatrix} s & 0 & 0 & sr \\ 0 & 1 & 0 & 0 \\ -s & 0 & 0 & sr \\ 0 & -1 & 0 & -2rs \end{pmatrix}$$

equals $(0,-1,0,-2sr)$, which suggests that the origin $(0,0,0)$ is projected on the screen at point $\mathbf{P}^* = (0,k/\sqrt{2},0)$. This, however, does not make sense since point $(0,0,0)$ was originally "behind" the viewer and should remain behind it after all the transformations. What's the explanation?

Mighty is geometry; joined with art, resistless.
— Euripides

Note. Notice the rightmost column of matrix $\mathbf{T}$ [Equation (3.8)]. The first and third elements of that column are nonzero, which indicates that the projection plane intercepts the x and z axes. This is discussed in detail on page 110.

Example 2. The viewer is located at $\mathbf{B} = (-k\sin\theta, 0, -k\cos\theta) = (-k\alpha, 0, -k\beta)$ and is looking in direction $\mathbf{D} = (\alpha, 0, \beta)$ (i.e., toward the origin). Matrices $\mathbf{T}_1$, $\mathbf{T}_2$, $\mathbf{T}_3$, and $\mathbf{T}_p$ below are similar to the ones from the previous example. The result is

$$\mathbf{T} = \mathbf{T}_1\mathbf{T}_2\mathbf{T}_3\mathbf{T}_p$$

$$= \begin{pmatrix} 1 & 0 & 0 & 0 \\ 0 & 1 & 0 & 0 \\ 0 & 0 & 1 & 0 \\ k\alpha & 0 & k\beta & 1 \end{pmatrix} \begin{pmatrix} \beta & 0 & \alpha & 0 \\ 0 & 1 & 0 & 0 \\ -\alpha & 0 & \beta & 0 \\ 0 & 0 & 0 & 1 \end{pmatrix} \begin{pmatrix} 1 & 0 & 0 & 0 \\ 0 & 1 & 0 & 0 \\ 0 & 0 & 1 & 0 \\ 0 & 0 & -k & 1 \end{pmatrix} \begin{pmatrix} 1 & 0 & 0 & 0 \\ 0 & 1 & 0 & 0 \\ 0 & 0 & 0 & r \\ 0 & 0 & 0 & 1 \end{pmatrix}$$

$$= \begin{pmatrix} \beta & 0 & \alpha & 0 \\ 0 & 1 & 0 & 0 \\ -\alpha & 0 & \beta & 0 \\ 0 & 0 & 0 & 1 \end{pmatrix} \begin{pmatrix} 1 & 0 & 0 & 0 \\ 0 & 1 & 0 & 0 \\ 0 & 0 & 0 & r \\ 0 & 0 & 0 & 1 \end{pmatrix}$$

$$= \begin{pmatrix} \beta & 0 & 0 & \alpha r \\ 0 & 1 & 0 & 0 \\ -\alpha & 0 & 0 & \beta r \\ 0 & 0 & 0 & 1 \end{pmatrix}. \tag{3.9}$$

It is easy to see that for $\theta = 0$ (where $\alpha = 0$ and $\beta = 1$), matrix (3.9) reduces to matrix (3.4).

⋄ **Exercise 3.20:** Assuming a viewer positioned as in the example above, calculate the projection of point $\mathbf{P} = (\beta l, m, -\alpha l)$.

⋄ **Exercise 3.21:** Projection matrices (3.9) and (3.7) correspond to the same geometry, so one would think that they should be identical. Why are they different?

We now develop this approach for the general case where a viewer is located at an arbitrary point $\mathbf{B} = (a, b, c)$ looking in an arbitrary direction $\mathbf{D} = (d, e, f)$, where vector $\mathbf{D}$ is assumed to be normalized (i.e., $d^2 + e^2 + f^2 = 1$). Translating the viewer to the origin is done, as usual, by matrix $\mathbf{T}_1$:

$$\mathbf{T}_1 = \begin{pmatrix} 1 & 0 & 0 & 0 \\ 0 & 1 & 0 & 0 \\ 0 & 0 & 1 & 0 \\ -a & -b & -c & 1 \end{pmatrix}. \tag{3.10}$$

The main task is to rotate vector $\mathbf{D}$ so it coincides with the positive z direction. The rotation should be about an axis that's perpendicular to both $\mathbf{D}$ and the z axis. A general vector in this direction is obtained by the cross product

$$\mathbf{D} \times (0, 0, 1) = (d, e, f) \times (0, 0, 1) = (e, -d, 0).$$

Normalizing this vector yields

$$\mathbf{u} = \frac{(e, -d, 0)}{\sqrt{e^2 + d^2}} = \left(\frac{e}{\sqrt{1 - f^2}}, \frac{-d}{\sqrt{1 - f^2}}, 0 \right).$$

Vector $\mathbf{u}$ is a unit vector in the direction of rotation. The rotation angle θ is the angle between vectors $\mathbf{D}$ and $z = (0, 0, 1)$. Since both are unit vectors, we can employ the dot product to obtain $\cos\theta = \mathbf{D} \bullet (0, 0, 1) = f$ and $\sin\theta = \sqrt{1 - \cos^2\theta} = \sqrt{1 - f^2}$. Notice that $\sin\theta$ is nonnegative because the angle between vector $\mathbf{D}$ and the z axis is measured between the direction of $\mathbf{D}$ and the positive z direction and is consequently always in the interval $[0, \pi]$.

The rotation matrix is obtained from Equation (1.32)

$$\mathbf{T}_2 = \begin{pmatrix} \frac{e^2+f-f^3-e^2f}{1-f^2} & \frac{-ed}{1+f} & d & 0 \\ \frac{-ed}{1+f} & \frac{d^2+f-f^3-d^2f}{1-f^2} & e & 0 \\ -d & -e & f & 0 \\ 0 & 0 & 0 & 1 \end{pmatrix}. \tag{3.11}$$

The two other tasks are to translate the viewer from the origin to point $(0, 0, -k)$ by means of $\mathbf{T}_3$ and to use matrix $\mathbf{T}_p$ to project from the standard position:

$$\mathbf{T}_3 = \begin{pmatrix} 1 & 0 & 0 & 0 \\ 0 & 1 & 0 & 0 \\ 0 & 0 & 1 & 0 \\ 0 & 0 & -k & 1 \end{pmatrix}, \quad \mathbf{T}_p = \begin{pmatrix} 1 & 0 & 0 & 0 \\ 0 & 1 & 0 & 0 \\ 0 & 0 & 0 & r \\ 0 & 0 & 0 & 1 \end{pmatrix}. \tag{3.12}$$

The result is the matrix product

$$\mathbf{T}_g = \mathbf{T}_1\mathbf{T}_2\mathbf{T}_3\mathbf{T}_p \tag{3.13}$$

$$= \begin{bmatrix} \frac{e^2+f+f^2}{1+f} & \frac{-de}{1+f} & 0 & dr \\ \frac{-de}{1+f} & \frac{d^2+f+f^2}{1+f} & 0 & er \\ -d & -e & 0 & fr \\ \frac{cd+bde-ae^2-af+cdf-af^2}{1+f} & \frac{-bd^2+ce+ade-bf+cef-bf^2}{1+f} & 0 & -(ad+be+cf)r \end{bmatrix}.$$

For the special case of a viewer located at $\mathbf{B} = (-k\sin\theta, 0, -k\cos\theta) = (-k\alpha, 0, -k\beta)$ and looking in direction $\mathbf{D} = (\alpha, 0, \beta)$, this reduces to matrix (3.9).

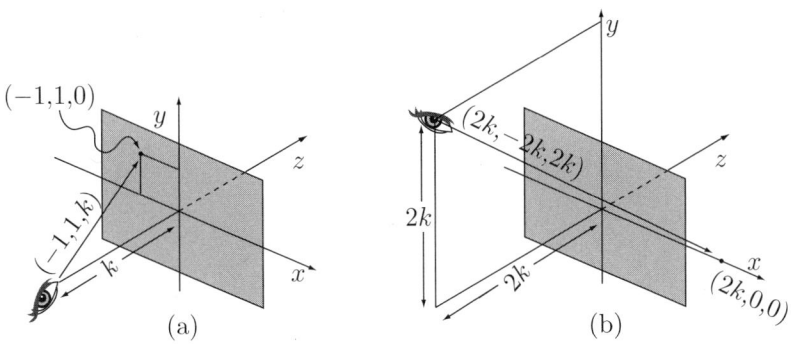

Figure 3.33: Two Tests of Matrix $\mathbf{T}_g$.

Matrix $\mathbf{T}_g$ is now tested twice. The first test (Figure 3.33a) assumes that the viewer is at the standard location $(0, 0, -k)$ but looking in direction $(-1, 1, k)$. (These components still have to be normalized.) We compute the projection of point $(-1, 1, 0)$

and the figure shows that this projection should be at the origin because the viewer is looking directly at the point. The *Mathematica* code

```
<< LinearAlgebra'Orthogonalization'
k = 3.; r = 1/k;
{a, b, c} = {0, 0, -k}; {d, e, f} = Normalize[{-1, 1, k}]
T = {{(e^2 + f + f^2)/(1 + f), -d e/(1 + f), 0, d r},
  {-d e/(1 + f), (d^2 + f + f^2)/(1 + f), 0, e r},
  {-d, -e, 0, f r},
  {(c d + b d e - a e^2 - a f + c d f - a f^2)/(1 + f),
  (-b d^2 + c e + a d e - b f + c e f - b f^2)/(1 + f),
  0, -(a d + b e + c f)  r}};
{-1, 1, 0, 1}.T
```

computes the normalized components of $\mathbf{D}$ as $(-0.3015, 0.3015, 0.9045)$ and the projected point as the 4-tuple $(0, 0, 0, 1.1)$ (i.e., the origin).

The second test (Figure 3.33b) assumes that the viewer is located at $\mathbf{B} = (0, 2k, -2k)$ looking in (the still unnormalized) direction $(2k, -2k, 2k)$. We compute the projection of point $(2k, 0, 0)$, and the figure again suggests that this projection should be at the origin because the viewer is looking directly at the point. Code similar to the above yields the normalized direction vector $\mathbf{D}$ as $(0.577, -0.577, 0.577)$ and the projected point as $(0, 0, 0, 3.5)$, again the origin.

⋄ **Exercise 3.22:** Perform a similar test for $\mathbf{B} = (0, 2k, -k)$ and unnormalized $\mathbf{D} = (0, -1, -1)$. Use mathematical software to compute the projection of point $(0, 0, -4k)$. Notice that the viewer is looking at the z axis a little "past" this point.

The rightmost column of $\mathbf{T}_g$ is especially interesting. Its three top elements are dr, er, and fr, where $r = 1/k$ is the inverse of the (strictly positive) distance k of the viewer from the screen and (d, e, f) are the components of vector $\mathbf{D}$. If any of these components is zero, the corresponding element of $\mathbf{T}_g$ will also be zero, which implies that there is a simple relationship between these three matrix elements and the direction $\mathbf{D}$ of the viewer's line of sight. Since the screen is perpendicular to the line of sight, we end up with the following interesting result.

The three matrix elements dr, er, and fr indicate which of the three coordinate axes is intercepted by the screen before the screen is transformed to the standard position.

For example, if $e = 0$ and d and f are nonzero, then $\mathbf{D}$ is a vector in the xz plane (and is not in the x or z direction), so the projection plane intercepts the x and z axes but is parallel to the y axis and does not intercept it. This result has already been mentioned several times in the past and is often referred to as *n-point perspective*, where n can be 1, 2, or 3. Figure 3.34 illustrates the justification for this term. The figure shows a cube centered on the origin and three viewers looking at it. Viewer 1 is located on the z axis and sees one vanishing point. Viewer 2 is located on the xz plane and therefore sees two vanishing points, and viewer 3 is located above the xz plane and so sees three vanishing points. However, the term "n-point perspective" refers to the number of coordinate axes, 1, 2, or 3, intercepted by the projection plane, not to the number of vanishing points actually observed by the viewer. The viewer can observe any number

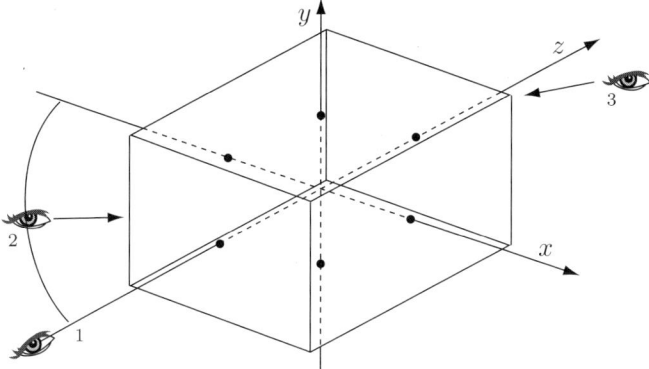

Figure 3.34: n-Point Perspective.

of vanishing points, depending on the existence of groups of straight, parallel lines on the object (page 76).

$\diamond$ **Exercise 3.23:** Calculate matrix (3.13) twice, first for the case where $\mathbf{D} = (0, 0, 1)$ (viewer looking in the positive z direction) and then for $\mathbf{D} = (0, 0, 1)$ and $\mathbf{B} = (0, 0, -k)$ (the standard position).

$\diamond$ **Exercise 3.24:** Assuming a viewer at point $\mathbf{B} = (0, 1, 0)$ looking in direction $\mathbf{D} = (0, 1, 1)$, calculate the projection of point $\mathbf{P} = (0, 1, 10)$.

Matrix $\mathbf{T}_g$ of Equation (3.13) contains the expression $1 + f$ in the denominators of certain elements, which may cause undefined values when $f = -1$. Since we assume that vector $\mathbf{D}$ is normalized, $d^2 + e^2 + f^2$ must be equal to 1, so the case $f = -1$ implies $d = e = 0$, which, in turn, implies $\mathbf{D} = (0, 0, -1)$ (i.e., a viewer looking in the negative z direction). It turns out that $\mathbf{T}_g$ can be used even in this case. When $d = e = 0$, we can write

$$T_g[1, 1] = \frac{e^2 + f + f^2}{1 + f} = \frac{f(1 + f)}{1 + f} = f = -1,$$

$$T_g[2, 2] = \frac{d^2 + f + f^2}{1 + f} = \frac{f(1 + f)}{1 + f} = f = -1.$$

$$T_g[4, 1] = \frac{cd + bde - ae^2 - af + cdf - af^2}{1 + f} = -af\frac{1 + f}{1 + f} = a,$$

$$T_g[4, 2] = \frac{-bd^2 + ce + ade - bf + cef - bf^2}{1 + f} = -bf\frac{1 + f}{1 + f} = b.$$

Matrix elements $T_g[1, 2] = T_g[2, 1] = -de/(1 + f)$ have the indefinite form $0/0$, but we artificially set them to zero. Matrix $\mathbf{T}_g$ becomes

$$\mathbf{T}_g = \begin{pmatrix} -1 & 0 & 0 & 0 \\ 0 & -1 & 0 & 0 \\ 0 & 0 & 0 & -r \\ a & b & 0 & cr \end{pmatrix}. \tag{3.14}$$

This is a matrix that transforms a point $\mathbf{P} = (x, y, z, 1)$ to point

$$\mathbf{P}^* = \left(\frac{-x+a}{(c-z)r}, \frac{-y+b}{(c-z)r}, 0 \right).$$

Following are two quick tests of this matrix. They were performed with the following *Mathematica* code:

```
(* code to check matrix T_g for the case 1 + f = 0 *)
r = 1/k; {a, b, c} = {0, 0, -k};
T = {{-1, 0, 0, 0}, {0, -1, 0, 0}, {0, 0, 0, -r}, {a, b, 0, c r}};
{x, y, z, 1}.T
```

1. When the viewer $\mathbf{B}$ is located at the standard location $(0, 0, -k)$, matrix $\mathbf{T}_g$ of Equation (3.14) transforms an arbitrary point $\mathbf{P} = (x, y, z)$ to the point

$$\mathbf{P}^* = \left(\frac{x}{(k+z)r}, \frac{y}{(k+z)r}, 0 \right) = \left(\frac{x}{1+z/k}, \frac{y}{1+z/k}, 0 \right),$$

which is the familiar Equation (3.1).

2. When the viewer $\mathbf{B}$ is located at $(1, 1, 1)$, point $(x, y, z) = (1, 1, -1)$ is transformed to

$$\left(\frac{1-1}{(1+1)r}, \frac{1-1}{(1+1)r}, 0 \right) = (0, 0, 0).$$

The reader should visualize this situation with the help of a diagram to see why the result is correct.

The Top Vector. This section's approach to general perspective moves the viewer from an arbitrary location $\mathbf{B}$ to the standard position while rotating his line of sight from an arbitrary direction $\mathbf{D}$ to the positive z direction. This is done in the following three steps: (1) a translation from $\mathbf{B}$ to the origin, (2) a rotation, and (3) a translation to point $(0, 0, -k)$. However, Figures 3.27 and Ans.8b illustrate why another rotation is sometimes needed after step 3 in order to correct the orientation of the screen. Figure 3.35 shows a viewer moved from a general location to the standard position and how the extra rotation serves to align the top of the screen with the y axis in a new step 4.

The software normally has no idea how the screen is oriented initially and how it should be oriented when the viewer is brought to the standard position. If this orientation is important, the user should specify the direction $\mathbf{Q}$ of the top of the screen, and step 4 should be added to rotate the viewer-screen unit about the z axis until $\mathbf{Q}$ is aligned with the x or y axis or any other desired direction.

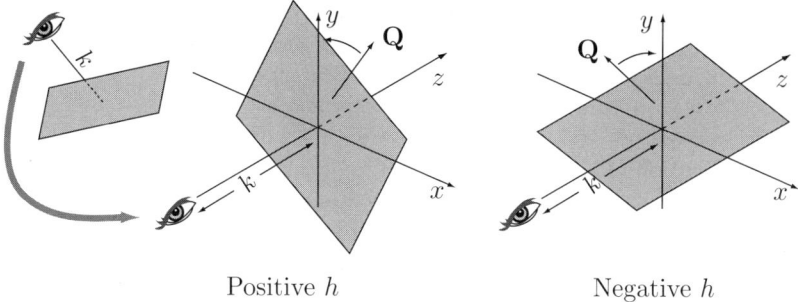

Figure 3.35: The Top Vector.

This extra step can be ignored in cases where the projection plane is rotated through a small angle or is infinitely large. In practice, however, the projection plane is the screen on which the three-dimensional scene is projected. This screen has a finite size and should normally be oriented such that its top points in the positive y direction.

The rotation matrix of step 4 is easy to derive. We assume that the first three steps have brought the screen to the xy plane and have transformed the original top vector $\mathbf{Q}$ to $\mathbf{Q} = (h, i, 0)$. [We assume that $(h, i, 0)$ is already normalized, so $h^2 + i^2 = 1$ or $h = \pm\sqrt{1 - i^2}$.] We further assume that the rotation of step 4 should align $\mathbf{Q}$ with the positive y axis $(0, 1, 0)$. The rotation is about the z axis, and the angle ϕ of rotation is determined by $\cos\phi = \mathbf{Q} \bullet (0, 1, 0) = i$ and $\sin\phi = \sqrt{1 - i^2} = h$. The rotation matrix is therefore given by

$$\mathbf{T}_4 = \begin{bmatrix} i & h & 0 & 0 \\ -h & i & 0 & 0 \\ 0 & 0 & 1 & 0 \\ 0 & 0 & 0 & 1 \end{bmatrix}.$$

Matrix $\mathbf{T}_4$ rotates vector $(h, i, 0, 1)$ to the positive y axis. The following *Mathematica* code verifies this for the four special normalized vectors $(a, a, 0, 0)$, $(a, -a, 0, 0)$, $(-a, a, 0, 0)$, and $(-a, -a, 0, 0)$, where $a = 1/\sqrt{2}$. In each case, once the values of h and i on line 1 are set to a or -a, the result is $(0, 1, 0, 0)$.

```
1 a = 1/Sqrt[2]; h = a; i = a;
2 T = {{i, h, 0, 0},{-h, i, 0, 0},{0, 0, 1, 0},{0, 0, 0, 1}};
3 {h, i, 0, 1}.T
```

⋄ **Exercise 3.25:** Assume a viewer located at $\mathbf{B} = (0, 2k, -2k)$ looking in (unnormalized) direction $\mathbf{D} = (0, -1, -1)$, as in Exercise 3.22 and a value $k = \sqrt{2}$. Figure Ans.8a illustrates the geometry of this case.

   1. Derive the equation of the projection plane.

   2. Multiply the transformation matrices of Equations (3.10), (3.11), and (3.12) to obtain one transformation $\mathbf{T}_{123}$ that brings the viewer to the standard position.

   3. Pick up a point on the projection plane and compute its coordinate on the xy plane after transformation $\mathbf{T}_{123}$.

3.8 A Coordinate-Free Approach: I

The discussion of general perspective in the previous sections is based on points and their coordinates relative to a three-dimensional coordinate system. This section presents a coordinate-free approach to the same problem that is based on vectors and vector operations. The location of the origin and the directions of the coordinate axes are not needed, although they may serve to illuminate the particular geometry of the examples presented here. The term "point" is still used, but we refer to a point in terms of the vector connecting it with the origin instead of as a triplet (x, y, z) of coordinates.

Figure 3.36a shows a viewer at point **B** looking in an arbitrary direction **a**. The screen is, as always, perpendicular to the line of sight **a**, and we assume that $|\mathbf{a}| = k > 0$. The center of the screen is at point **C**. Note that vector **a** gives both the direction of view of the viewer and the distance between the viewer and the screen.

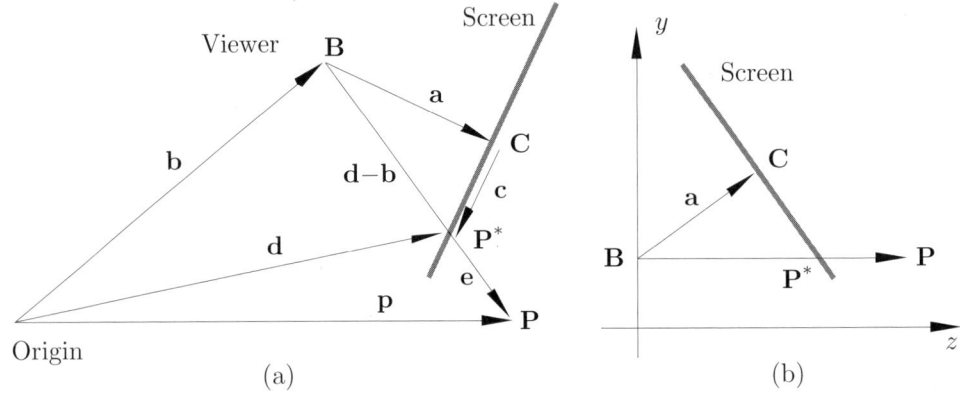

Figure 3.36: (a) General Perspective with Vectors. (b) Example.

The derivation of the projection is surprisingly easy. We select an arbitrary point **P** on the other side of the screen from the viewer and connect it with the viewer. The intersection of line **BP** and the screen is the projected point $\mathbf{P}^*$. Vector **b** indicates the position of the viewer. Vector **c** indicates the direction $\mathbf{CP}^*$ on the screen. Vector **d** is the position vector of point $\mathbf{P}^*$. Vector **e** connects **B** to **P**. Vector **p** points from the origin to point **P**. Vector $\mathbf{d} - \mathbf{b}$ connects point **B** to point $\mathbf{P}^*$.

Vector **p** is the sum $\mathbf{p} = \mathbf{b} + \mathbf{e}$, which implies $\mathbf{e} = \mathbf{p} - \mathbf{b}$. From $\mathbf{d} = \mathbf{b} + \mathbf{a} + \mathbf{c}$, we get $\mathbf{c} = \mathbf{d} - \mathbf{b} - \mathbf{a}$. Vector $\mathbf{d} - \mathbf{b}$ is in the direction of **e**, so we can write $\mathbf{d} - \mathbf{b} = \alpha \mathbf{e} = \alpha(\mathbf{p} - \mathbf{b})$, where α is a real number. This implies $\mathbf{c} = \alpha(\mathbf{p} - \mathbf{b}) - \mathbf{a}$ or

$$\mathbf{d} = \mathbf{b} + \mathbf{a} + \mathbf{c} = \mathbf{b} + \alpha(\mathbf{p} - \mathbf{b}). \qquad (3.15)$$

Since the line of sight is perpendicular to the screen, we can write $\mathbf{a} \bullet \mathbf{c} = 0$, which implies $\mathbf{a} \bullet [\alpha(\mathbf{p} - \mathbf{b}) - \mathbf{a}] = 0$, or $\alpha \mathbf{a} \bullet (\mathbf{p} - \mathbf{b}) = \mathbf{a} \bullet \mathbf{a}$, or

$$\alpha = \frac{|\mathbf{a}|^2}{\mathbf{a} \bullet (\mathbf{p} - \mathbf{b})}. \qquad (3.16)$$

Before we continue with the analysis, the following cases should be discussed:

1. α is positive. This is the normal case. It means that the viewer and point **P** are on different sides of the screen and the projection is meaningful.

2. α is zero. This implies a vector **a** of magnitude zero (i.e., a viewer positioned at the screen). Either the viewer or the screen should be moved before anything can be meaningfully displayed.

3. α is negative. This implies that **P** and the viewer are on the same side of the screen, so **P** should not be projected.

4. α is undefined. This occurs when $\mathbf{a} \bullet (\mathbf{p} - \mathbf{b}) = 0$, implying that **a** is perpendicular to $\mathbf{p} - \mathbf{b}$ and therefore to **e**. Vector **e** is therefore parallel to the screen, making it impossible to project **P**.

After α is computed and checked, we can proceed in one of two ways: (1) We can use Equation (3.15) to calculate vector **d**, which points directly to $\mathbf{P}^*$ on the screen, or (2) we can calculate the screen coordinates of vector **c**. In the latter case, we consider the center of the screen (point **C**) a local origin and we define two unit vectors **u** and **w** to serve as local axes on the screen. The screen coordinates of **c** are, in this case, the projections $\mathbf{u} \bullet \mathbf{c}$ and $\mathbf{w} \bullet \mathbf{c}$ of **c** on these axes.

In order to compute **u** and **w**, we recall that they should be on the screen (and therefore perpendicular to **a**) and also perpendicular to each other. We can therefore write $\mathbf{a} \bullet \mathbf{u} = \mathbf{a} \bullet \mathbf{w} = \mathbf{u} \bullet \mathbf{w} = 0$. It also makes sense to require that **u** be in the xy plane (which will cause **w** to point in the z direction as much as possible). Solving these equations results in

$$\mathbf{u} = (a_y, -a_x, 0) \qquad \text{and} \qquad \mathbf{w} = (a_x a_z, a_y a_z, -a_x^2 - a_y^2). \tag{3.17}$$

Vectors **u** and **w** should then be normalized.

Note that **u** and **w** are undefined if **a** points in the z direction [if $\mathbf{a} = (0, 0, a_z)$, then $\mathbf{u} = \mathbf{w} = (0, 0, 0)$, an undefined direction]. However, in this case the screen is parallel to the xy plane, so we can simply define the local coordinate axes as $\mathbf{u} = (1, 0, 0) = \mathbf{i}$ and $\mathbf{w} = (0, 1, 0) = \mathbf{j}$.

This novel approach to general perspective is illustrated by two examples.

Example 1. This is a simple example (Figure 3.36b) where all the points lie on the yz plane.

We assume a viewer at $\mathbf{B} = (0, 1, 0)$, looking in direction $(0, 1, 1)$ (i.e., 45° in the yz plane). Vector **a** must point in this direction, and we assume $\mathbf{a} = (0, 2, 2)$ (i.e., the center of the screen is at a distance of $|\mathbf{a}| = \sqrt{2^2 + 2^2} = \sqrt{8}$ units from the viewer). We further assume that the point **P** to be projected is at $(0, 1, 10)$. The center of the screen (point **C**) is easily seen to be at $\mathbf{b} + \mathbf{a} = (0, 1, 0) + (0, 2, 2) = (0, 3, 2)$. The first step is to determine α

$$\alpha = \frac{|\mathbf{a}|^2}{\mathbf{a} \bullet (\mathbf{p} - \mathbf{b})} = \frac{8}{(0, 2, 2) \bullet (0 - 0, 1 - 1, 10 - 0)} = \frac{2}{5}.$$

The next step is to compute $\mathbf{d} = \mathbf{b} + \alpha(\mathbf{p} - \mathbf{b}) = (0, 1, 0) + (2/5)(0, 0, 10) = (0, 1, 4)$. The projected point $\mathbf{P}^*$ is therefore at $(0, 1, 4)$. (See the diagram to convince yourself that the precise value of the z coordinate of **P** is irrelevant in this case.)

Next, we calculate the local coordinates of this point on the screen. Vector $\mathbf{c}$ is first obtained by $\mathbf{c} = \alpha(\mathbf{p} - \mathbf{b}) - \mathbf{a} = (2/5)(0,0,10) - (0,2,2) = (0,-2,2)$. The local axes on the screen are computed next from Equation (3.17). They are $\mathbf{u} = (2,0,0)$ and $\mathbf{w} = (0,4,-4)$. We normalize them by dividing each by its magnitude, obtaining $\mathbf{u} = (1,0,0)$ and $\mathbf{w} = (0,1/\sqrt{2},-1/\sqrt{2})$. (Note that $\mathbf{u}$ is in the x direction and $\mathbf{w}$ is in the yz plane.)

Thus, the screen coordinates of $\mathbf{c}$ are $\mathbf{u} \bullet \mathbf{c} = (1,0,0) \bullet (0,-2,2) = 0$ and $\mathbf{w} \bullet \mathbf{c} = (0,1/\sqrt{2},-1/\sqrt{2}) \bullet (0,-2,2) = -\sqrt{8}$. The projected point is therefore $\sqrt{8}$ units away from the center of the screen $\mathbf{C}$. Note that this equals the absolute value of vector $\mathbf{c}$.

As an added bonus, we compute the plane equation of the screen. Let (x,y,z) be a general point on the screen. The vector from the center (point $\mathbf{C}$) to (x,y,z) is $(x-0,y-3,z-2)$. This vector must be perpendicular to the normal to the screen (vector $\mathbf{a}$), which implies

$$0 = \mathbf{a} \bullet (x,y-3,z-2) = (0,2,2) \bullet (x,y-3,z-2), \quad \text{or} \quad y+z=5.$$

This equation relates the y and z coordinates of all the points on the screen. Any point with coordinates $(x,y,5-y)$ is therefore on the screen regardless of the value of x. Note that the projected point $\mathbf{P}^*$ also satisfies this relation.

⋄ **Exercise 3.26:** Generalize the previous example to the case of a general point $\mathbf{P} = (x,y,z)$.

Example 2. Again we give a simple example, illustrated in Figure 3.37. The screen is centered on the origin at a 45° angle, and the viewer is at point $(-k/\sqrt{2},0,-k/\sqrt{2})$, a distance of k units from the screen. To simplify the notation, we introduce the quantity $\psi = k/\sqrt{2}$. From Figure 3.36a it is clear that $\mathbf{a} = (\psi,0,\psi)$ and $\mathbf{b} = -\mathbf{a} = (-\psi,0,-\psi)$. The center of the screen is, as always, at $\mathbf{a} + \mathbf{b}$, which is point $(0,0,0)$.

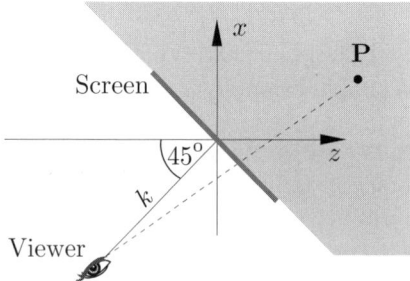

Figure 3.37: Viewer Rotated About the y Axis.

The first step is to determine α:

$$\alpha = \frac{|\mathbf{a}|^2}{\mathbf{a} \bullet (\mathbf{P}-\mathbf{b})} = \frac{2\psi^2}{(\psi,0,\psi) \bullet (x+\psi,y,z+\psi)} = \frac{2\psi}{x+z+2\psi}.$$

(Try to convince yourself that α is positive in the gray area above and to the right of the screen because $x + z + 2\psi$ is positive in this area.)

The next step is to compute vector $\mathbf{d}$

$$
\begin{aligned}
\mathbf{d} &= \mathbf{b} + \alpha(\mathbf{P} - \mathbf{b}) \\
&= (-\psi, 0, -\psi) + \frac{2\psi}{x + z + 2\psi}\,(x + \psi, y, z + \psi) \\
&= \frac{\psi}{x + z + 2\psi}\,(x - z, y, z - x)\,.
\end{aligned}
$$

Notice that $\mathbf{P} = (0,0,0)$ is transformed to $\mathbf{P}^* = (0,0,0)$. Also, every point $\mathbf{P} = (x, 0, -x)$ is transformed to $\mathbf{P}^* = (0,0,0)$.

Since the screen is centered at the origin, we have $\mathbf{c} = \alpha(\mathbf{P}-\mathbf{b})-\mathbf{a} = \alpha(\mathbf{P}-\mathbf{b})+\mathbf{b} = \mathbf{d}$. The next step is to calculate the local screen vectors $\mathbf{u}$ and $\mathbf{w}$ from Equation (3.17). This is straightforward and results in $\mathbf{u} = (0, -\psi, 0)$ and $\mathbf{w} = (\psi, 0, -\psi)$. After normalization, these become $\mathbf{u} = (0, -1, 0)$ and $\mathbf{w} = (1/\sqrt{2}, 0, -1/\sqrt{2})$. Notice that $\mathbf{u}$ is the y axis and $\mathbf{w}$ is in the xz plane.

The screen equation is obtained from $\mathbf{a} \bullet (x, y, z) = 0$, which implies $\psi(x + z) = 0$ or $x = -z$. The last step is to derive the transformation matrix. From

$$
x^* = \frac{X}{H} = \frac{\psi(x - z)}{x + z + 2\psi}, \quad y^* = \frac{Y}{H} = \frac{\psi y}{x + z + 2\psi}, \quad z^* = \frac{Z}{H} = \frac{\psi(z - x)}{x + z + 2\psi},
$$

we get

$$
(X, Y, Z, H) = (x, y, z, 1)\begin{pmatrix} \psi & 0 & -\psi & 1 \\ 0 & \psi & 0 & 0 \\ -\psi & 0 & \psi & 1 \\ 0 & 0 & 0 & 2\psi \end{pmatrix}.
$$

(Notice the two 1's in the last column. They indicate that the projection plane intercepts the x and z axes but not the y axis. This is a two-point perspective.)

3.9 A Coordinate-Free Approach: II

This approach to the problem of perspective projection also uses vectors instead of coordinates, but we assume that the following are given (Figure 3.38):

1. the position of the viewer (vector $\mathbf{b}$);
2. the direction and distance from the viewer to the projection plane (vector $\mathbf{a}$);
3. an "up" vector $\mathbf{Z}$, which determines the direction of the local screen vector $\mathbf{w}$;
4. two viewing half-angles h and v, an approach that is handy when we want to limit the projected image to certain viewing angles, as in Figure 3.13.

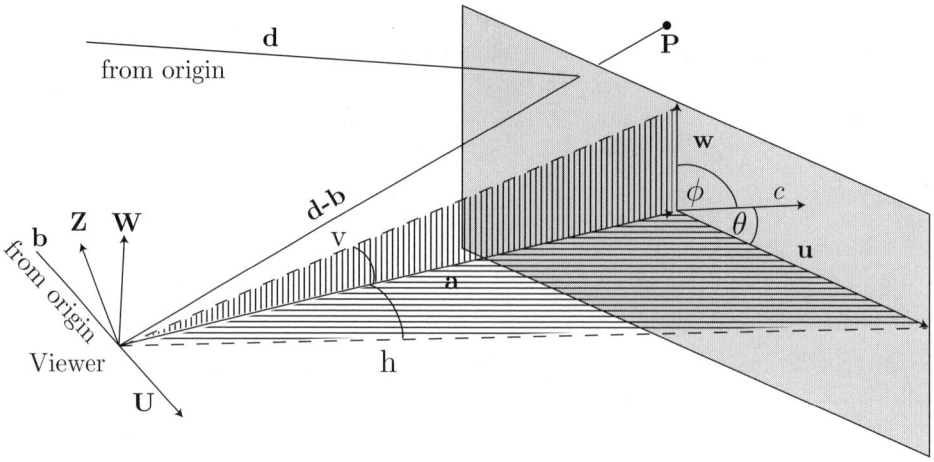

Figure 3.38: A Viewing Geometry.

We proceed in the following simple steps:

1. Calculate vector $\mathbf{U}$ as perpendicular to both $\mathbf{a}$ and $\mathbf{Z}$. $\mathbf{U} = \mathbf{a} \times \mathbf{Z}$.

2. Compute vector $\mathbf{W}$ as perpendicular to both $\mathbf{U}$ and $\mathbf{a}$. $\mathbf{W} = \mathbf{U} \times \mathbf{a}$. Vector $\mathbf{W}$ is in the Za plane and is perpendicular to $\mathbf{a}$. It will serve to determine vector $\mathbf{w}$ on the screen in step 4.

3. Denote $\mathbf{C} = \mathbf{b} + \mathbf{a}$. This points to the center of the screen.

4. Construct the half-screen vectors $\mathbf{u}$ and $\mathbf{w}$. They are in the directions of $\mathbf{U}$ and $\mathbf{W}$, respectively, but their sizes are determined by the viewing angles

$$\mathbf{u} = \frac{\mathbf{U}}{|\mathbf{U}|} |\mathbf{a}| \tan h, \qquad \mathbf{w} = \frac{\mathbf{W}}{|\mathbf{W}|} |\mathbf{a}| \tan v.$$

5. Compute $\alpha = \frac{|\mathbf{a}|^2}{\mathbf{a} \bullet (\mathbf{P} - \mathbf{b})}$ and vectors $\mathbf{d} = \mathbf{b} + \alpha(\mathbf{P} - \mathbf{b})$ and $\mathbf{c} = \alpha(\mathbf{P} - \mathbf{b}) - \mathbf{a}$ in the usual way.

6. Now that $\mathbf{c}$ is known, we use it to determine the two scale factors c_x and c_y:

$$c_x = \frac{|\mathbf{c}| \cos \theta}{|\mathbf{u}|} = \frac{1}{|\mathbf{u}|^2} (\mathbf{c} \bullet \mathbf{u}), \qquad c_y = \frac{|\mathbf{c}| \cos \phi}{|\mathbf{w}|} = \frac{1}{|\mathbf{w}|^2} (\mathbf{c} \bullet \mathbf{w}).$$

These are numbers in the range $[-1, 1]$. Any point $\mathbf{P} = (x, y, z)$ for which either c_x or c_y is greater than 1 or less than -1 is therefore outside the screen and should not be displayed.

The range of values of c_x and c_y assumes that the origin of the screen is at its center. The actual screen coordinates (s_x, s_y) of a pixel depend on the dimensions of the screen (measured in pixels). They are given by

$$s_x = (\text{half the screen width}) \times c_x, \qquad s_y = (\text{half the screen height}) \times c_y.$$

If the origin is at the bottom left corner, then

$$s_x = \text{(half the screen width)} + \text{(half the screen width)} \times c_x,$$
$$s_y = \text{(half the screen height)} + \text{(half the screen height)} \times c_y.$$

If it is at the top left corner,

$$s_x = \text{(half the screen width)} + \text{(half the screen width)} \times c_x,$$
$$s_y = \text{(half the screen height)} - \text{(half the screen height)} \times c_y.$$

Example. We apply the method above to the standard case depicted in Figure 3.22, where the screen is part of the xy plane and is centered on the origin and the viewer is located k units from the origin on the negative z axis. Assuming that the two half-angles h and v are given, we need to compute scale factors c_x and c_y that will make it possible to determine for any given point $\mathbf{P}$ whether its projection on the xy plane is inside or outside the screen.

It is clear that $\mathbf{b} = (0, 0, -k)$ and $\mathbf{a} = (0, 0, k) = -\mathbf{b}$. We also select the positive y direction as our "up" direction, so $\mathbf{Z} = (0, 1, 0)$. To express the final results in a general way, we denote $m = \tan h$ and $n = \tan v$. The calculation is straightforward.

1. $\mathbf{U} = \mathbf{a} \times \mathbf{Z} = (0, 0, k) \times (0, 1, 0) = (-k, 0, 0)$.
2. $\mathbf{W} = \mathbf{U} \times \mathbf{a} = (-k, 0, 0) \times (0, 0, k) = (0, k^2, 0)$.
3. $\mathbf{C} = \mathbf{b} + \mathbf{a} = (0, 0, 0)$. The center of the screen is at the origin.
4. The local screen axes are

$$\mathbf{u} = \frac{\mathbf{U}}{|\mathbf{U}|} |\mathbf{a}| \tan h = (-km, 0, 0), \quad \mathbf{w} = \frac{\mathbf{W}}{|\mathbf{W}|} |\mathbf{a}| \tan v = (0, kn, 0).$$

5. The three quantities α, $\mathbf{d}$, and $\mathbf{c}$ are determined next:

$$\alpha = \frac{|\mathbf{a}|^2}{\mathbf{a} \bullet (\mathbf{P} - \mathbf{b})} = \frac{k^2}{(0, 0, k) \bullet (x, y, z + k)} = \frac{k}{z + k},$$

$$\mathbf{d} = \mathbf{b} + \alpha(\mathbf{P} - \mathbf{b}) = (0, 0, k) + \frac{k}{z + k}(x, y, z + k) = \frac{k}{z + k}(x, y, 0),$$

$$\mathbf{c} = \alpha(\mathbf{P} - \mathbf{b}) - \mathbf{a} = \alpha(\mathbf{P} - \mathbf{b}) + \mathbf{b} = \mathbf{d}.$$

6. The scale factors c_x and c_y can now be obtained:

$$c_x = \frac{\mathbf{c} \bullet \mathbf{u}}{|\mathbf{u}|^2} = \frac{\frac{k}{z+k}(-xkm)}{k^2 m^2} = \frac{-x}{m(z + k)},$$
$$c_y = \frac{\mathbf{c} \bullet \mathbf{w}}{|\mathbf{w}|^2} = \frac{\frac{k}{z+k}(ykn)}{k^2 n^2} = \frac{y}{n(z + k)}.$$

(3.18)

As a simple application of these results, let's select $h = v = 45°$, which implies $m = n = 1$. Let's also assume screen dimensions of 100×100 pixels, a local origin at the center of the screen, and $k = 1$. For point $\mathbf{P} = (1, 2, 1)$, we get the scale factors

$$c_x = \frac{-x}{m(z+k)} = \frac{-1}{1+1} = -0.5, \quad c_y = \frac{y}{n(z+k)} = \frac{2}{1+1} = 1.$$

Thus, the screen coordinates are $s_x = 50 \times (-0.5) = -25$ and $s_y = 1 \times 50 = 50$ (the top of the screen). However, any point with coordinates $(1, y, 1)$ where $y > 2$ would produce a scale factor $c_y > 1$, implying that its projection is outside the screen.

◇ **Exercise 3.27:** Why is Equation (3.18) asymmetric with respect to x and y (i.e., why $-x$ and not $-y$)?

3.9.1 Perspective Depth

The perspective projection converts a three-dimensional point to a two-dimensional point. It completely erases any information about the depth (the z coordinate) of the original point. However, certain algorithms for hidden surface removal need precisely such information. We therefore need to generalize our perspective projection to create a third coordinate z^* with information about the original z coordinate of the projected point. The obvious choice is $z^* = z$, but this has a serious downside: It does not preserve straight lines.

Imagine two three-dimensional points, $\mathbf{P}_1 = (x_1, y_1, z_1)$ and $\mathbf{P}_2 = (x_2, y_2, z_2)$, projected to the points

$$\mathbf{P}_1^* = \left(\frac{x_1 k}{k + z_1}, \frac{y_1 k}{k + z_1}, z_1 \right) \quad \text{and} \quad \mathbf{P}_2^* = \left(\frac{x_2 k}{k + z_2}, \frac{y_2 k}{k + z_2}, z_2 \right).$$

Note that the two projected points are not necessarily on the projection plane. We say that they are located in the *image space*.

The straight segment $\mathbf{P}(t) = \mathbf{P}_1 + (\mathbf{P}_2 - \mathbf{P}_1)t$ (Equation (Ans.7)) connects the two original points, while the segment $\mathbf{P}^*(u) = \mathbf{P}_1^* + (\mathbf{P}_2^* - \mathbf{P}_1^*)u$ connects the two projected ones. It can be shown that an arbitrary point $\mathbf{P}(t_0)$ on $\mathbf{P}(t)$ is projected to a point that's not on $\mathbf{P}^*(u)$.

This is why the perspective depth projection is not chosen simply as $z^* = z$ but as $z^* = z/(k+z)$. This definition preserves depth information, because it has the property $z_1 > z_2 \Rightarrow z_1^* > z_2^*$. It also preserves straight lines.

◇ **Exercise 3.28:** Prove the claim above.

3.10 The Viewing Volume

In order to display realistic images, we have to limit the items that are being displayed to those that would actually be seen by a viewer located at $(0, 0, -k)$ and looking at the image projected on the screen. There are three cases to consider:

1. The viewer and the object being projected should be located on different sides of the projection plane. Any parts of the object located on the same side as the viewer should not be projected. Such parts should be identified and ignored. If the software does not do that, such parts would be projected in a wrong way, upside down and back to front. (See also the discussion of negative z coordinates on page 95.) As an example, consider points $\mathbf{P}_1$ and $\mathbf{P}_2$ in Figure 3.39a. The former is on the other side of the screen from the viewer and is therefore projected correctly. The latter is on the viewer's side of the screen and is projected on the negative side of the x axis. Including points such as $\mathbf{P}_2$ in the projection creates a confusing effect.

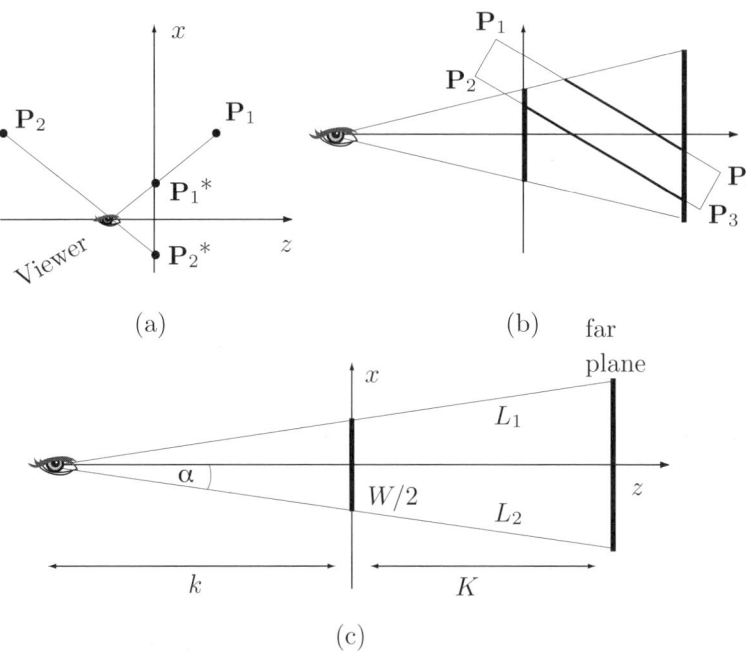

Figure 3.39: The Viewing Volume in Three Dimensions.

2. Those parts of the scene that are located very far away may be too small to be seen by an actual viewer, and we may choose not to project and display them on the screen. User-friendly software should therefore make it possible for the user to select a value K and clip off those parts of the scene whose z coordinates are greater than K. The effect of this is to define a plane located at $z = K$ beyond which nothing is projected.

3. The screen and the far plane now define a truncated pyramid, called the *viewing volume* or *frustum* (Latin for a piece broken off). Those parts of the image that are outside it are either irrelevant or invisible to the viewer and should not be displayed.

Imagine a picture made up of points connected with straight lines. Before displaying the picture, the software should determine which points are outside the viewing volume. Those points should not be displayed but should not be ignored either. Figure 3.39b shows four points connected to form a rectangle. Notice how some of the lines connecting the points should not be displayed and others should be *clipped*. In general, only those parts of the image that are inside the viewing volume should be displayed.

It is easy to determine if a point $\mathbf{P} = (x, y, z)$ is inside the viewing volume. We assume that the screen is a square that is W units on a side. Figure 3.39c shows two of the four lines that bound the pyramid. It is easy to see that $\tan\alpha = (W/2)/k = W/(2k)$. This is also the slope of line L_1. The x-intercept of the line is $W/2$, so the line's equation is $x = (W/2k)z + W/2 = (W/2)(z/k + 1)$. The equation of L_2 is, similarly, $x = -(W/2)(z/k + 1)$. Since the diagram is symmetric with respect to x and y, we conclude that point $\mathbf{P}$ is located inside the pyramid if its coordinates satisfy $|x|, |y| \leq (W/2)(z/k + 1)$.

◇ **Exercise 3.29:** Assume that the distance k of the viewer from the screen equals the size W of the screen. What will be the width of the field of view of the viewer?

Let's assume that two points, $\mathbf{P}$ and $\mathbf{Q}$, are part of the total image and are to be connected with a straight line. The first step is to determine, for each point, whether it is located inside or outside the viewing volume. (If a point is located on the edge of the viewing volume, it is considered to be inside.) In the second step, three cases should be distinguished:

1. Both points are inside the viewing volume. The line connecting them is completely inside the volume and should be fully displayed. This is because the viewing volume is convex. (It is a convex polyhedron.)

2. One point is inside and the other is outside the viewing volume. The line connecting them intercepts the volume at exactly one point. (This, again, is a result of the convexity of the viewing volume.) The interception point should be determined and the line should be clipped.

3. Both points are outside. The line connecting them is either completely outside (and should therefore be ignored) or it intercepts the viewing volume at two points. Both interception points should be calculated and the line segment connecting them should be displayed. (There is also the degenerate case where both interception points are identical; the line is tangent to the viewing volume. In such a case, the line can be ignored or just one pixel displayed.)

3.10.1 Application: Flight Simulation

People have been fascinated by flight since the dawn of history. It is therefore not surprising that simple, inexpensive flight simulators for personal computers appeared as soon as these computers became fast and powerful enough to complete the necessary computations in real time. A flight simulator, even a simple one, is a complex program because it has to simulate the behavior of an airplane and display both the interior (instruments) and exterior (the view from the cockpit) in real time. This section is concerned with displaying the view from the cockpit, and we show that this task is an application of the important concept of *viewing volume*.

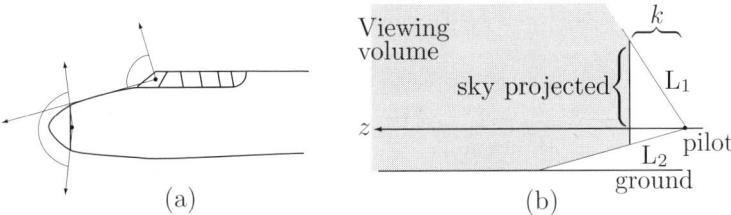

Figure 3.40: (a) Two Fields of View. (b) A Viewing Volume.

Figure 3.40a shows part of a typical World War II bomber. It is obvious that the field of view of the pilots in the cockpit is restricted. They see a lot of sky and part of the airplane, but only distant parts of the ground in front and on the sides. The bombardier, however, has almost a 180° field of view and can see all the way from 6 o'clock (the ground below their feet) to 12 o'clock (straight up).

Figure 3.40b is a schematic diagram showing the viewing volume of the pilot (ignoring the curvature of the Earth). We assume that the flight simulator has to display the pilot's view on a screen placed k units in front of the pilot. It is obvious that the view depends on the precise shape of the aircraft. (This determines the orientation of lines L_1 and L_2.) Most of the screen in the figure is a projection of the sky and only a small part shows a projection of the ground in front of the aircraft. It is also trivial to use similar triangles to obtain the basic perspective expression [Equation (3.1)]

$$\frac{z+k}{y} = \frac{k}{y^*} \quad \text{or} \quad y^* = \frac{ky}{z+k} = \frac{y}{z/k+1}.$$

3.11 Stereoscopic Images

We now turn to an important application of transformations and perspective projection, namely stereoscopic view. This section explains the principles and theory of stereoscopic images, how to create them, and how to view them.

Stereo (from the Greek στερεοσ)—solid, three-dimensional.

It is generally agreed that the concepts of stereoscopy were discovered in 1833 by Charles Wheatstone, who is mostly known for the Wheatstone bridge (an electrical circuit for the precise comparison of resistances). His 1833 lecture to the Royal Society in London on his discoveries has been published and became the first milestone in the history of this topic. In this lecture, he describes how his discovery came about as a result of his acoustical experiments. Wheatstone also developed the first stereoscopic viewer, which worked with mirrors. Initially, a pair of stereo pictures had to be drawn by hand, but with the invention of photography by Louis Daguerre in 1839, it became possible to generate precise pairs of stereo pictures and watch them as a single stereoscopic image.

The ideal graphics output device should be three-dimensional. Unfortunately, only a few such devices are available today and they are expensive and cumbersome. This is why stereo pictures, displayed on a two-dimensional screen or printed on paper, are interesting and have important applications. The reason we see real-life objects in three dimensions is that our eyes are separated (by about 60–70 mm) and hence look at the same object from slightly different positions (Figure 3.41a). They see slightly different images, which are "fused" by the brain to create the three-dimensional image.

The principle of stereo images is therefore to create and display two slightly different images of the same object and to make sure that each eye sees just one image. This may be achieved by displaying the two images in two different colors and watching them through special glasses that allow each color to reach just one eye. Other methods for viewing such a pair of images in stereo are discussed in Section 3.13.

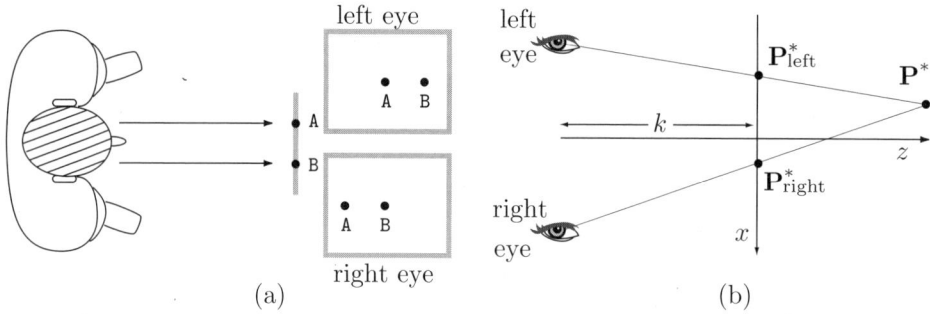

Figure 3.41: Principle of Stereo Images.

The simplest way to calculate the two stereo images is to use translation and perspective projection. This is what makes stereoscopy a useful application of the concepts described earlier. Figure 3.41b shows each eye as a viewer. The left eye is located at $(-e, 0, -k)$ and the right one at $(e, 0, -k)$. To create the image seen by the left eye (the projection $\mathbf{P}^*_{\text{left}}$ of point $\mathbf{P}$), we first have to translate the eye to the origin and then follow with a standard perspective projection. The transformations are

$$\begin{pmatrix} 1 & 0 & 0 & 0 \\ 0 & 1 & 0 & 0 \\ 0 & 0 & 1 & 0 \\ e & 0 & 0 & 1 \end{pmatrix} \begin{pmatrix} 1 & 0 & 0 & 0 \\ 0 & 1 & 0 & 0 \\ 0 & 0 & 0 & r \\ 0 & 0 & 0 & 1 \end{pmatrix} = \begin{pmatrix} 1 & 0 & 0 & 0 \\ 0 & 1 & 0 & 0 \\ 0 & 0 & 0 & r \\ e & 0 & 0 & 1 \end{pmatrix} = \mathbf{T}_{\text{left}}.$$

The transformation for the right eye is similarly

$$\begin{pmatrix} 1 & 0 & 0 & 0 \\ 0 & 1 & 0 & 0 \\ 0 & 0 & 0 & r \\ -e & 0 & 0 & 1 \end{pmatrix} = \mathbf{T}_{\text{right}}.$$

It projects $\mathbf{P}$ to $\mathbf{P}^*_{\text{right}}$.

The stereo pair is created by transforming each point $\mathbf{P}$ on the original image twice, to the two points $\mathbf{P}_{\text{left}} = \mathbf{P}\,\mathbf{T}_{\text{left}}$ and $\mathbf{P}_{\text{right}} = \mathbf{P}\,\mathbf{T}_{\text{right}}$. The value selected for e depends on how the picture is to be viewed. For the dual-color method mentioned earlier, $2e$ should equal the distance between the eyes (about 60–70 mm). This is a small value, so there is not much difference between $\mathbf{P}_{\text{right}}$ and $\mathbf{P}_{\text{left}}$. The two images highly overlap.

For a general point $\mathbf{P} = (x, y, z)$, the projections for both eyes are

$$\mathbf{P}_{\text{left}} = (x, y, z, 1)\mathbf{T}_{\text{left}} = (x + e, y, 0, zr + 1) \rightarrow \left(\frac{x + e}{zr + 1}, \frac{y}{zr + 1} \right),$$

$$\mathbf{P}_{\text{right}} = (x, y, z, 1)\mathbf{T}_{\text{right}} = (x - e, y, 0, zr + 1) \rightarrow \left(\frac{x - e}{zr + 1}, \frac{y}{zr + 1} \right).$$

This means that the smaller z is (i.e., the closer the point is to the viewer), the greater the difference between what the two eyes see. A good way to visualize this is to imagine an object sliding past the viewer. The front of the object slides faster than the back, an effect known as *parallax*.

As an example, consider the two points $\mathbf{P} = (5, 0, 1)$ and $\mathbf{Q} = (5, 0, 2)$. They differ only in their z coordinate. Assuming that $e = 2$ and $r = 3$, their projections are

$$\mathbf{P}_{\text{left}} = \left(\frac{5 + 2}{3 + 1}, 0 \right) = \left(\frac{7}{4}, 0 \right), \quad \mathbf{P}_{\text{right}} = \left(\frac{5 - 2}{3 + 1}, 0 \right) = \left(\frac{3}{4}, 0 \right),$$

$$\mathbf{Q}_{\text{left}} = \left(\frac{5 + 2}{2 \cdot 3 + 1}, 0 \right) = \left(\frac{7}{7}, 0 \right), \quad \mathbf{Q}_{\text{right}} = \left(\frac{5 - 2}{2 \cdot 3 + 1}, 0 \right) = \left(\frac{3}{7}, 0 \right).$$

The difference between $\mathbf{P}_{\text{left}}$ and $\mathbf{P}_{\text{right}}$ is $7/4 - 3/4 = 1$, whereas the difference between $\mathbf{Q}_{\text{left}}$ and $\mathbf{Q}_{\text{right}}$ is only $7/7 - 3/7 = 4/7$.

Figure 3.42 is an example of a stereo pair of a polyline connecting the eight corners of a cube. The *Mathematica* code that did the computations is also listed. Figure 3.43 shows the complete cubes.

A more sophisticated approach to generating a stereo image is shown in Figure 3.44a. The two eyes are located at $(e, 0, -k)$ and $(-e, 0, -k)$, and they view the general point $\mathbf{P} = (x, y, z)$ from different directions. Point $\mathbf{P}$ is projected twice on the projection plane, at points $\mathbf{P}_L$ and $\mathbf{P}_R$, using the general rule for perspective projections. Assuming that the distance between the eyes is $2e$, Figure 3.44c,d shows how to calculate the x coordinates of points $\mathbf{P}_L$ and $\mathbf{P}_R$, respectively. Using similar triangles,

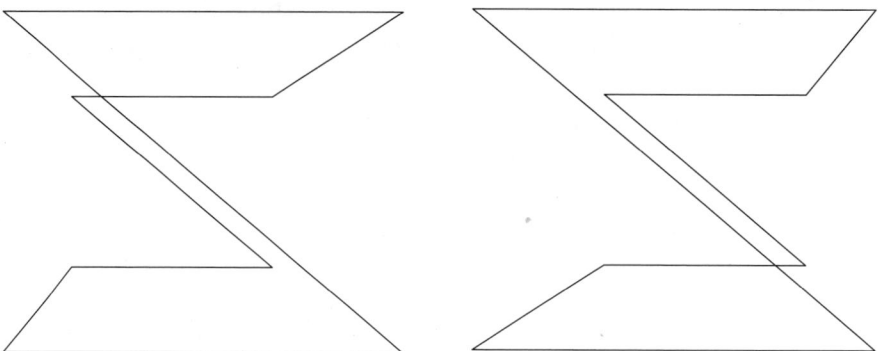

Figure 3.42: Example of a Stereo Image Pair.

```
(* display two cubes as a stereo pair *)
Clear[Trg, Tlf, pt, e, r, qt];
Tlf={{1,0,0,0},{0,1,0,0},{0,0,0,r},{e,0,0,1}};
Trg={{1,0,0,0},{0,1,0,0},{0,0,0,r},{-e,0,0,1}};
pt={{1,1,1,1},{-1,1,1,1},{1,-1,1,1},{-1,-1,1,1},
  {1,1,-1,1},{-1,1,-1,1},{1,-1,-1,1},
  {-1,-1,-1,1},{1,1,1,1}}; e=.1; r=3;
qt=Table[0, {i,9},{j,4}];
Do[qt[[i]]=pt[[i]].Tlf, {i,1,9}]; (* use Tlf for other image *)
Do[qt[[i,1]]=qt[[i,1]]/qt[[i,4]], {i,1,9}];
Do[qt[[i,2]]=qt[[i,2]]/qt[[i,4]], {i,1,9}];
ListPlot[Table[{qt[[i,1]], qt[[i,2]]},{i,1,9}],
  PlotJoined->True, Axes->False]
```

Code for Figure 3.42.

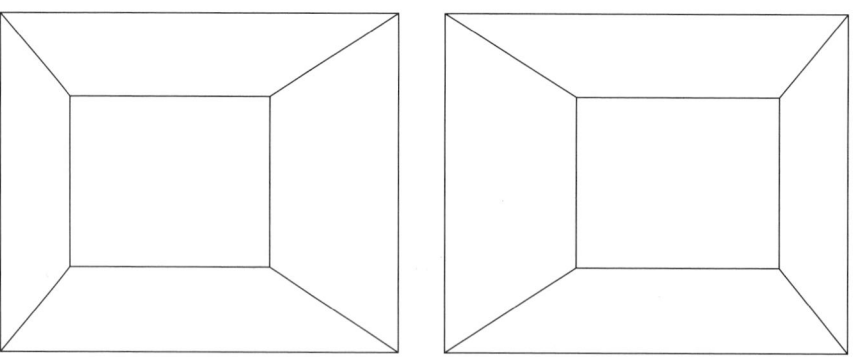

Figure 3.43: Stereo Pair Shown as Complete Cubes.

Figure 3.44c yields

$$\frac{x-e}{k+z} = \frac{x_L - w}{k} \quad \text{or} \quad x_L = \frac{x-e}{1+z/k} + e = \frac{x+ez/k}{1+z/k},$$

and, similarly, from Figure 3.44d we get

$$\frac{x+e}{k+z} = \frac{x_R + w}{k} \quad \text{or} \quad x_R = \frac{x+e}{1+z/k} - e = \frac{x - ez/k}{1+z/k}.$$

Since both eyes are at $y = 0$, the y^* coordinates of both $\mathbf{P}_L$ and $\mathbf{P}_R$ are given by

$$y^* = \frac{y}{1+z/k}.$$

We thus obtain the transformation matrices $\mathbf{T}_L$ and $\mathbf{T}_R$ that transform $\mathbf{P}$ to $\mathbf{P}_L$ and $\mathbf{P}_R$,

$$\mathbf{T}_L = \begin{pmatrix} 1 & 0 & 0 & 0 \\ 0 & 1 & 0 & 0 \\ e/k & 0 & 0 & 1/k \\ 0 & 0 & 0 & 1 \end{pmatrix}, \quad \mathbf{T}_R = \begin{pmatrix} 1 & 0 & 0 & 0 \\ 0 & 1 & 0 & 0 \\ -e/k & 0 & 0 & 1/k \\ 0 & 0 & 0 & 1 \end{pmatrix}. \tag{3.19}$$

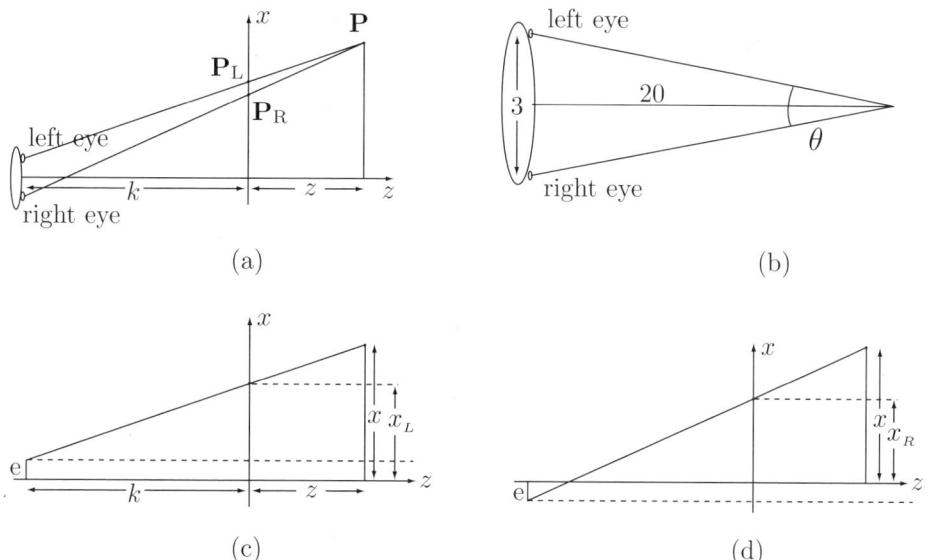

(a)

(b)

(c)

(d)

Figure 3.44: Perspective Projection of a Stereo Pair.

Figure 3.44b shows how to select reasonable values for e and k. We first assume that the distance between the eyes is about 75 mm (about 3 in). Normal reading distance is about 20 in. Using the values 3 and 20, we get $\tan\theta/2 = 1.5/20$, yielding $\theta/2 = 4.29°$ or $\theta = 8.58°$. This is the average stereo angle between the eyes. To get a stereo pair that will look natural and will be free of distortions, we should select values for e and k that should maintain this angle. A natural value for k is 4 in, since this is the focal length of the lenses used by most commercial stereoscopes. If we reduce k from 20 to 4 (a factor of 5), we should reduce e from 3 to $3/5 = 0.6$ to maintain the same stereo angle.

A stereo pair is therefore calculated by substituting $e = 0.6$, $k = 4$ in Equation (3.19) and computing $\mathbf{P}_L = \mathbf{T}_L \cdot \mathbf{P}$ and $\mathbf{P}_R = \mathbf{T}_R \cdot \mathbf{P}$ for every point $\mathbf{P}$ of the object.

⋄ **Exercise 3.30:** What would be good values for e and k assuming a distance of 2.5 in between the eyes?

3.12 Creating a Stereoscopic Image

The discussion in Section 3.11 suggests that the simplest way to obtain a left-eye, right-eye pair of stereoscopic images is to select a camera, choose a good subject, take a picture and then shift the camera along the baseline (normally about 65 mm) to the right and take another picture. This pair of two-dimensional images can then be watched as a single three-dimensional (stereoscopic) image with the methods discussed in Section 3.13. (Actually, what will be seen in three dimensions are those parts that are common to both pictures. Any objects that appear only in one picture because they are near an edge will disappear or will confuse the brain, depending on how the pictures are watched.) Here we show several simple ways to photograph such a pair, and we start with the basic rules for obtaining good stereoscopic images.

The first rule is to take sharp pictures. All the objects in the photograph should be in focus. The professional term for this is a large depth of field. Photographers sometimes take pictures where certain elements, normally in the background, are blurred, while the main subject is sharp. Such a picture may have artistic value, but it does not translate well to three dimensions.

The second rule is to select an appropriate subject. The aim is to produce a stereoscopic, three-dimensional image, preferably also interesting and in color. Thus, the subject must be in color and must have depth. Professional photographers and artists recommend selecting a subject that has three main elements, one near the camera, one far away, and the third in between. A simple example is a nearby gray rock, a green/brown tree in the background, and a white fence running between them. A similar example is a nearby statue in the Palais de Chaillot, the Eiffel Tower in the background, and the Pont d'Iena in between. Once such an image is converted to stereoscopic, the viewer can easily see the relative positions of the three elements. In addition, the subject should have other background elements, because a picture with only three items looks empty and disappointing. Experience shows that the best results are achieved if the distance of the nearest picture element from the camera is 30 times the baseline. For the normal baseline of 6.5 cm, this translates to a distance of 195 cm or about 6.4 ft. However, many stereo enthusiasts have discovered that the baseline does not have to be 6.5 cm as long as a ratio of 30 is obtained. Thus, if the nearest object is 300 cm from the camera, then a baseline of 10 cm will produce a realistic-looking stereoscopic image.

The third rule is to maintain precise vertical alignment of the two pictures. Every picture element must appear at the same height in the two pictures. Thus, the camera should not be tilted, raised, or lowered between the two exposures. It should only be shifted horizontally.

Rule 4 is to avoid having many red and blue (or red and green) objects in the picture. Section 3.13 shows that a stereoscopic image generated as a color anaglyph looks bad if it uses these colors extensively.

Also, make sure the camera is held vertically and is not tilted up or down, as this may cause unwanted converging lines and extra vanishing points, features that tend to confuse the viewer. Only static images can be photographed (images with moving elements, such as clouds, flags, or vehicles, can be photographed with a pair of cameras; see below). Finally, remember which image is for the left eye and which is for the right eye. Switching these two results in a nonworking stereo image.

We now turn to techniques for taking a pair of stereo pictures with a camera.

Perhaps the simplest (and cheapest) technique is to use a small, 6 ft (2 m) ladder. Place the camera on several steps of the ladder until you find the ideal height for your subject. Take a picture, move the camera horizontally about 6.5 cm, and make the second exposure. A ruler or a straight piece of wood makes it easier to slide the camera without tilting or rotating it.

If you own a tripod, you can get better results. The simplest way to use a tripod is to take one picture, lift the tripod, move it to the left or right, and take the second picture. Before you start, draw a straight line on the ground, perpendicular to the line of sight of the camera, and position the tripod such that two of its legs are on the line.

Much more accurate results can be yours if you build a simple jig like the one illustrated in Figure 3.45.

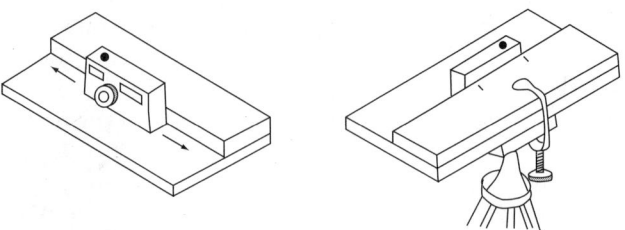

Figure 3.45: A Jig to Photograph a Stereo Pair.

The jig consists of two pieces of wood or plywood that are attached to the tripod with a clamp. The camera is placed on the wider piece, while the smaller piece serves as a support and guide. The dimensions of the two pieces depend on the size of the camera and the length of the required baseline. Before taking any pictures, mark two points on the guide, about 6.5 cm apart, to serve as marks for the standard baseline. With a bit of experience, this primitive device produces very accurate stereo pairs. It is easy to come up with variations on this simple design.

Good-quality kitchen cabinets often have drawers mounted on special ball-bearing metal slides. Anyone planning to take many stereo pictures might want to attach a camera to such a slide and attach the slide to a heavy-duty tripod. Such an arrangement is accurate, easy to use, and lasts a long time. Detailed instructions are available at [berezin 06]. Similar devices are sold commercially by [photo3d 06] and others.

A completely different approach to taking such pairs of pictures is to make or obtain a pair of cameras whose lenses are placed the right distance apart and are operated together with a common shutter release cable (Figure 3.46). Such a device can produce stereo pairs of scenes that change rapidly, such as flocks of birds or racing cars. The two cameras can be placed either side by side [part (b)] or one above the other, as long as the distance between the centers of their lenses is the right one. If one camera is placed on top of the other [part (a)], it is important to leave enough room between them for the shutter release cable. If the cameras are mounted bottom to bottom, care should be taken to align their lenses vertically [part (c)]. In the latter case, the user should verify that the two pictures are vertically aligned (rule 3 above). Every point should be at the same height in the two pictures.

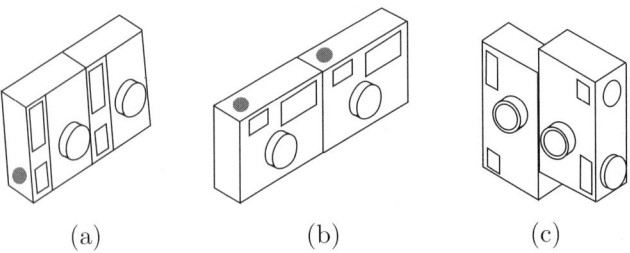

(a)   (b)   (c)

Figure 3.46: A Pair of Cameras.

Such double cameras are available commercially and can also be homemade. Figure 3.47 shows one made in 1998 by Andreas Petersik [Petersik 05] from two Nikon cameras. Yet another solution is to construct a camera with a sliding lens. The lens is shifted to the left and a picture is taken. The lens is then shifted to the right and another picture is taken. If film is used, the two pictures are taken on two adjacent frames of the roll of film. In a digital camera of this type, the CCD sensor slides with the lens and both pictures are captured by the same sensor and stored in different memory areas.

Figure 3.47: A Homemade Double Nikon Camera (Courtesy of Andreas Petersik).

Note. When a pair of stereo pictures is taken, a flash should be avoided because it normally casts shadows. The subject being photographed is shifted in the two pictures because the camera has moved, but any shadows cast on a wall behind the subject are shifted twice because the camera has moved and because the light from the flash is coming from a different direction. Thus, shadows would be placed incorrectly in the two pictures and would interfere with the correct visualization of the brain.

> Our range of expression is small, so that a smile in genuine pleasure photographs indistinguishably from a grimace of pain; they are the same unless we know their history and their nature.
>
> —C. P. Snow, *Strangers and Brothers (George Passant)* (1948)

⋄ **Exercise 3.31:** The two pictures of a stereo pair differ by a horizontal shift, which suggests the following idea. Instead of taking two pictures, take just one, copy it, shift the copy horizontally, and use it as the second picture. What's wrong with this method?

3.13 Viewing a Stereoscopic Image

A stereoscopic image consists of a pair (right-eye, left-eye) of images. To see this pair in three dimensions, we have to view it in a special way. The guiding principle is that our brain must receive from our eyes the same signals it receives when we watch a real three-dimensional image. Given a stereoscopic pair of images on paper or on a screen, the most common techniques to view it are as follows:

1. View it through a stereoscope. This is a simple device that can easily be built at home.

2. The cross-eye technique. The two images of a pair are laid side by side and the viewer has to cross his eyes in order to slide the images and see them fused into a single image.

3. The parallel-view technique. This is similar to the cross-view method but is appropriate for small images.

4. The anaglyph method. The two original images are combined into a single image where they are painted different colors. Special glasses are used to make sure each eye sees only one color.

5. Page-flipped techniques, where the left and right pictures are continually flipped on the screen.

6. Line alternate methods, where the left-eye and right-eye pictures are interleaved on the screen. These are popular with head-mounted displays.

There are other, more sophisticated techniques, such as the Pulfrich effect and dot stereograms. We follow with detailed descriptions of the most common methods.

Stereoscope

A stereoscope (Figure 3.48) is a simple device for viewing a stereo pair. It can easily be made at home from cardboard, wood, and two lenses. In a piece of cardboard, cut two circular holes with a diameter of about 1.5 in each and with about 6.5 cm separation between their centers. Place a lens with a focal length of 4 in in each hole. Look at a stereo pair located about 4 in away through the lenses, using another piece of cardboard to make sure each eye sees only one image. More sophisticated devices are available from several sources, such as [StereoGraphics 05] and [Edmund Scientific 05].

The Cross-Eye View Technique

Note. If you wear glasses, keep them on when trying this method.

The right and left images should be displayed side by side, with the right image on the left and the left image on the right, as illustrated here:

Start by staring at the center point between the two images. Slowly cross your eyes and watch the two images slide closer. With a little patience and practice, you should be able to make the two images overlap. You will then see three images, as illustrated here:

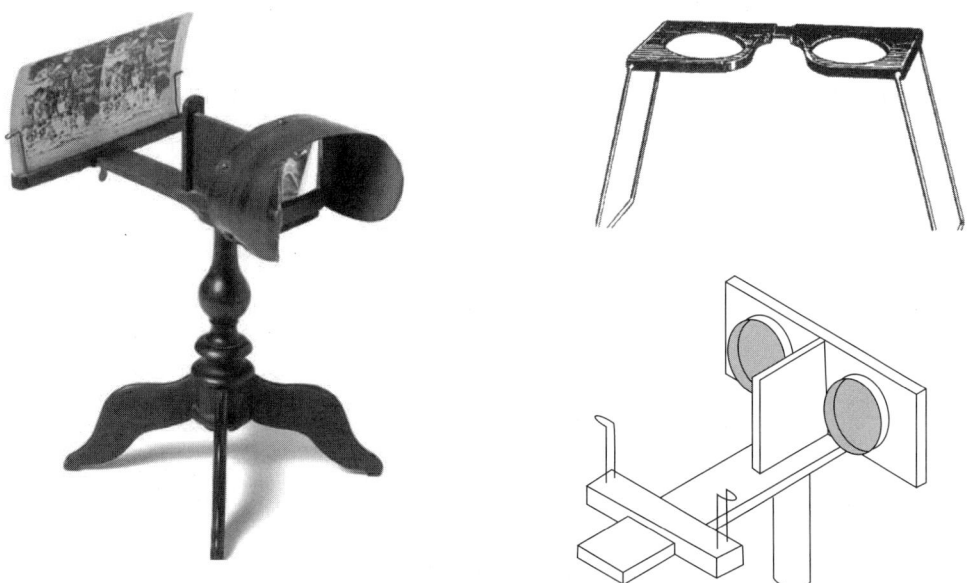

Figure 3.48: Stereoscopes.

R R L

At this point, your right eye sees the right image and your left eye sees the left image. Try to ignore them and concentrate on the central image. When you are successful, the center image will feature depth; it will be stereoscopic.

If you are one of those who find this technique difficult in practice, try the following aid.

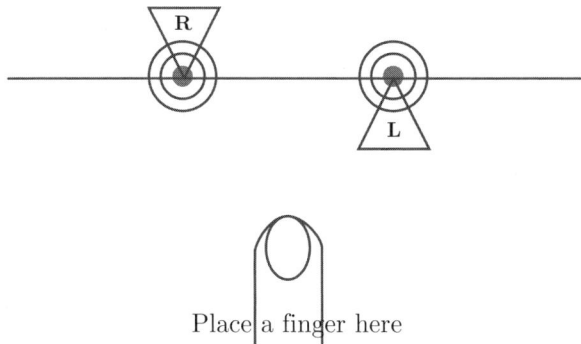

Place a finger here

1. Observe the image above with the R and L targets.
2. Place a finger on the paper, right under the two targets, as indicated.
3. Stare at your fingertip and, while still looking at it, slowly move your finger away from the image pair and toward your eyes.

If you relax, practice this method several times, and do it slowly, you should be able to slide the two targets and align them perfectly. You may need to tilt your head slightly left or right to align the targets vertically.

4. When the two targets fuse, move your eyes slowly from your fingertip to the fused image on the page. Don't forget to keep your eyes crossed during this step.

If this "trick" is successful, apply it to a pair of stereo images such as Figure 3.49. Experience indicates that most people get used to this way of viewing stereoscopic images and don't find it tiring or uncomfortable. However, if you feel discomfort or if your eyes get tired, don't try this method again! There are other ways to enjoy stereoscopic images.

Figure 3.49: A Stereo Pair (color version on page 235).

The Parallel View Technique

This technique (also referred to as relaxed viewing) is appropriate for small images where each image of a stereoscopic pair fits between the eyes. The pair is displayed with the right-eye image on the right and the left-eye image on the left, as illustrated here:

The following steps show how such an image can be viewed stereoscopically without any tools or instruments. Those who wear glasses may get better results trying these steps without their glasses.

1. Watch the image pair from close range so that each eye is over one of the images. This is possible if the images are small enough.

2. Stare straight ahead and try to gaze through the images to infinity. The stereo images will look blurred.

3. Slowly pull your head away from the page while maintaining the same gaze. The two images will turn into four images. Continue to move away while gazing to infinity.

4. At a certain point, the four images will merge into three. Concentrate on the central image and you will suddenly see it in three dimensions. The effect is more noticeable if the original image pair is in vivid colors.

You can try this technique on the image pair of Figure 3.49, but avoid prolonged viewing and concentration, which may lead to eye fatigue.

The Anaglyph Method

This approach to stereoscopic viewing combines the right-eye and left-eye images (partly overlapping) in one image but in different colors, normally red and blue (or cyan) but sometimes red and green. The method requires the use of special glasses with different color filters for the two eyes, as illustrated here. (See also [kspark 05] for several well-known Escher drawings that have been converted to three dimensions, mostly as anaglyphs.)

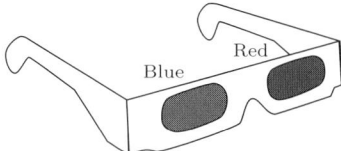

The red filter on the left eye looks red because it reflects red light. Any other colors are partly absorbed and partly transmitted through the filter. Thus, this filter lets the blue parts of the image through to the left eye. Similarly, the blue filter on the right lets only the red parts of the image to the right eye.

Warning. Some people may be sensitive to these glasses. If you feel discomfort or if you get tired very quickly, take off the glasses and take a break. In any case, try to use these glasses for short periods and only to view an anaglyph image. They are not intended for normal use!

> From the Dictionary
>
> Anaglyph. From Late Latin anaglyphus, carved in low relief. Also from Greek anagluphos (to carve).

An anaglyph image is encoded in one of three ways as follows:

Color. The left-eye image is left mostly in its original colors, but certain crucial parts, such as edges, curves, lines, and points, are painted blue (or cyan or green). The right-eye image is treated similarly with red. Thus, a color anaglyph (Figure 3.50a) preserves much of the original colors of the image, but its red/blue (or red/green) stereo information is diluted throughout the image. The result is that many images lose their depth information in this format and don't look three-dimensional. This is especially true if the original image has vivid red or blue colors. However, if an image does look good in a color anaglyph, it looks real and vivid.

Gray. The two original images are converted to grayscale and the same crucial elements are painted red and blue. A gray anaglyph image (Figure 3.50b) is therefore seen in grayscale, but its depth information is normally easy to perceive.

Pure. The right-eye image is entirely converted to shades of red. The left-eye image is treated as in the color anaglyph method. The combined image looks reddish on paper (Figure 3.50c) but has much depth information when seen through the glasses. Some of the original color information naturally is lost.

An interesting difference between the three anaglyphs of Figure 3.50 is the person on the left-hand side (he is clearly seen in Figure 3.49) who completely disappears in the pure version.

Experienced users recommend creating all three anaglyphs of a given image, trying the color, gray, and pure versions (in this order), and selecting the one judged best.

There are many sources of software (much of it free) to generate anaglyphs. Those too lazy to search can check the list at [anabuilder 05].

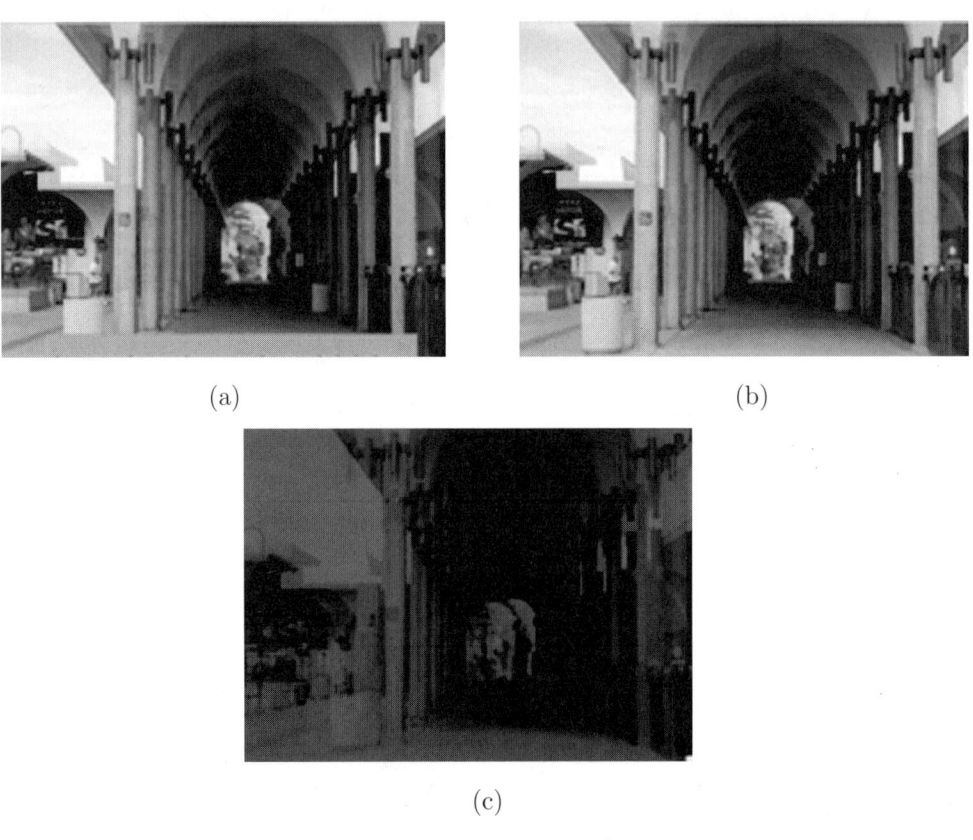

(a)   (b)

(c)

Figure 3.50: Three Anaglyph Encodings (color version on page 235).

Pick a good-quality anaglyph and examine it carefully. You will notice that each crucial picture element **P** is shown twice in the anaglyph, in red and blue. The relative positions of these two color elements is interpreted by the brain as the depth of element **P**. Let's assume that the left-eye view becomes the red parts and the right-eye view becomes the blue parts. If the red and blue parts of **P** overlap, the brain considers **P** to be on the image plane (i.e., the paper or screen on which the anaglyph image is printed or displayed). Such picture elements are said to be at the stereo window and are always comfortable for the eye to watch, regardless of where they are located in the image.

If the red part of a picture element **P** is placed on the anaglyph to the right of the corresponding blue part, the brain perceives **P** as being located in front of the stereo window. Figure 3.51 shows that this effect requires a large separation of the red and blue parts. If the red part of a picture element **P** is placed on the anaglyph to the left of the corresponding blue part, the brain visualizes **P** as located behind the stereo window. Figure 3.51 shows that this can be achieved with only a small separation of the red and blue parts.

If **P** is seen in only one of the two eye views (because it is close to a border of the image), then it is translated to one color only and is not seen in stereo. It may even confuse the brain if the viewer concentrates on **P**.

Figure 3.51 also illustrates the effect of moving the viewer closer to and away from the anaglyph. As we watch an anaglyph from close by, we see the entire image bigger but with less depth. As we move our head away from the anaglyph, the stereo image becomes smaller, but the difference between points A and B increases; the image acquires more depth.

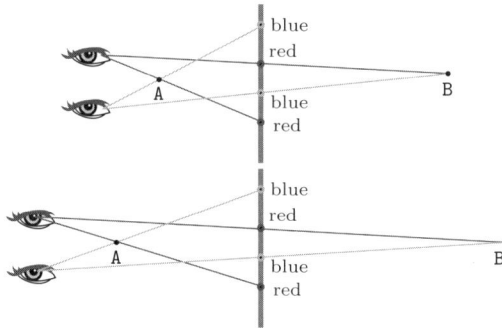

Figure 3.51: Relative Positions of the Red and Blue Parts (color on page 242).

Page-Flipped Techniques

These techniques require a special monitor screen and special shutter glasses. The screen switches rapidly between the left-eye and right-eye images. The glasses are triggered by the monitor hardware to block the right lens when the left-eye image is displayed and block the left lens when the right-eye image is displayed (Figure 3.52). Thus, the brain receives the correct image from each eye, and if the images are sent to the brain at a fast rate, the brain fuses them as usual into a single three-dimensional image. If the switching rate is low, the brain interprets the signals as a flickering image (still three-dimensional). The shutters in the glasses are electronic and are normally made from liquid crystals. The glasses themselves are connected to the monitor (actually, to the video card) through a special cable or through one of the input/output ports (serial or parallel). New types of shutter glasses are wireless.

Left-eye image ➡ Time ➡ Right-eye image

Figure 3.52: Page-Flip Monitor and Shutter Glasses.

This method generates high-quality, high-resolution color stereoscopic images but requires special hardware, so it is not as common as the previous methods.

Line-Alternate Techniques

In the past, most monitors used with computers were cathode-ray tubes (CRTs). Recently, liquid crystal display (LCD) monitors have become popular. Both types of monitors operate as *raster scan* displays and generate an image in the *interleaved* mode. The term "raster scan" means that the image is displayed on the monitor screen row by row, from top to bottom, and each row of pixels is generated from left to right. A complete scan of the screen is known as a refresh. In a CRT, this is achieved by sweeping the electron beam over the screen row by row from the top left corner to the bottom right corner. In an LCD monitor, the individual LCDs are scanned in this order and turned on or off as needed. The term "interleaved" means that each refresh of the screen is done in two parts. The first part refreshes the display of the odd-numbered screen rows, and the second part refreshes the even-numbered rows.

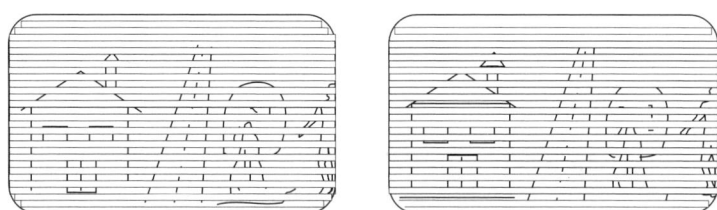

Figure 3.53: Line-Alternate Techniques.

A line-alternate technique for stereoscopic images displays the right-eye image on the odd-numbered rows and the left-eye image on the even-numbered rows or vice versa (Figure 3.53). Special shutter glasses block the left eye from seeing the display during the first part of the refresh (i.e., when the odd-numbered rows are scanned and refreshed) and blocks the right eye during the second part. Such techniques are popular in small, head-mounted displays.

A three-dimensional image created by the various line-alternate techniques is stable and doesn't suffer from flickers because the screen is normally scanned at a flicker-free speed. In addition, there is no loss of color. The downside is loss of resolution because each image is displayed on half the rows of the display.

A variation of the line-alternate technique is a lenticular lens. Figure 3.55 shows the principles of this technique. Each of the two stereo eye images is cut into narrow strips that are then interleaved and viewed through a special lens made of many small half-circular elements (placed at about 100 elements per inch). Each lens element sends one image strip to the left eye and one strip to the right eye.

The Pulfrich Effect

The Pulfrich effect, described by Carl Pulfrich in 1922, is best explained as an optical illusion (but then one might argue that any stereoscopic image is an optical illusion). Imagine an object moving in the plane perpendicular to our line of sight. If we look at the object with both eyes and dim the light reaching one eye, the object seems to move out of this plane and to either approach us or recede from us. The simplest way to observe this effect is to use one sunglass lens, but most pieces of dark glass or plastic, as well as many optical filters, work fine.

It is easy to demonstrate the Pulfrich effect with a swinging pendulum. When viewed normally with both eyes, we can verify that the pendulum swings in a plane back and forth. When a dark lens or filter is placed in front of one eye, the pendulum suddenly seems to be swinging in an ellipse parallel to the ground. The light has to be dimmed to one eye only. Dimming the light equally to both eyes results in a dim pendulum seen swinging in a plane.

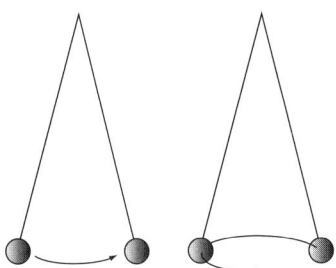

Figure 3.54: The Pulfrich Effect.

There are several Web sites with Java applets that illustrate this effect very convincingly through animation. One such site is [Newbold 05], but a Web search for `pulfrich`, `java`, and `animation` yields many more.

I have never been able to observe these effects myself, for I have been blind in the left eye for 16 years as a result of a traumatic (blutigen) injury of the eye suffered when I was young.

—Carl Pulfrich (1922)

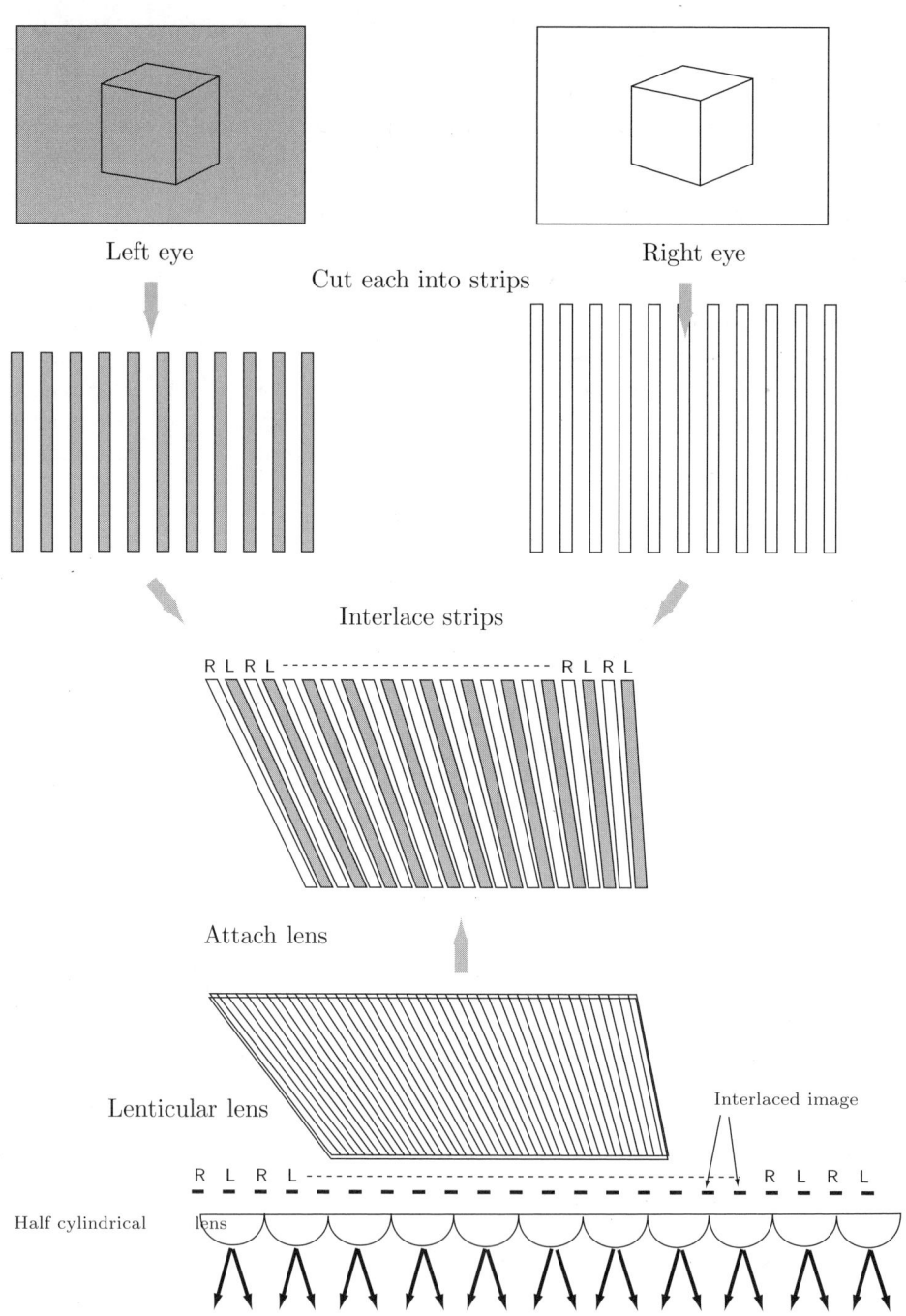

Left eye

Cut each into strips

Right eye

Interlace strips

Attach lens

Lenticular lens

Interlaced image

Half cylindrical lens

Figure 3.55: Lenticular Lens Principles.

Dot Stereograms

Figure 3.56 illustrates the principle of this interesting method (for a complete description, see [Thimbleby et al. 94]). A three-dimensional scene is projected on a screen and a point P_1 is selected at random. The two eyes of the viewer see point P_1 projected at points Q_1 and Q_2. We now select another point P_2 such that its projections for the two eyes are Q_2 and Q_3. Thus, point Q_2 is both the left-eye projection of P_1 and the right-eye projection of P_2. (P_2 must be at the same height as P_1, which is not obvious in our two-dimensional figure.) A little thinking shows that most points on the screen do similar "double duty." The exceptions are points close to the edges of the screen, or points whose P_1 or P_2 are hidden by other parts of the scene. Since Q_2 is common to P_1 and P_2, we face the question of what color to paint it. In fact, Q_1, Q_2, and Q_3 have to be painted the same color.

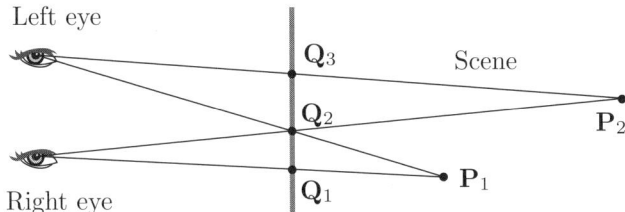

Figure 3.56: Dot Stereograms: The Principle.

The algorithm described in [Thimbleby et al. 94] has to decide what color to paint each point (dot) on the screen and also to determine the two parents, P_1 and P_2, of each point on the screen.

The result of this algorithm is a stereogram that consists of dots and can be watched in three dimensions by crossing the eyes, without the need for special glasses or any other device. There are three types of dot stereograms, as we now discuss.

SIRDS (Single Image Random Dot Stereograms). This is the oldest type. It goes back to the pioneering work of Béla Julesz in the 1960s. Such a stereogram consists of a random pattern of dots, each representing two pixels of the object. Figure 3.57 is an example of this type of stereogram.

SIS (Single Image Stereograms). This is currently the most common type. The picture consists of (slightly modified) tiles. This type of dot stereogram is somewhat more complex to generate, but the basic algorithm is the same.

SIRTS (Single Image Random Text Stereograms). This type is identical to SIRDS but uses ASCII characters instead of dots. The resulting stereogram has low resolution.

A dot stereogram is easier to perceive in three dimensions if it is printed on paper rather than displayed on a screen. Here are two simple methods for viewing this interesting type of stereoscopic image.

In the *pull-back method*, hold the picture close to and in front of your face. Imagine that you are looking straight ahead, right through the picture. When your eyes relax and

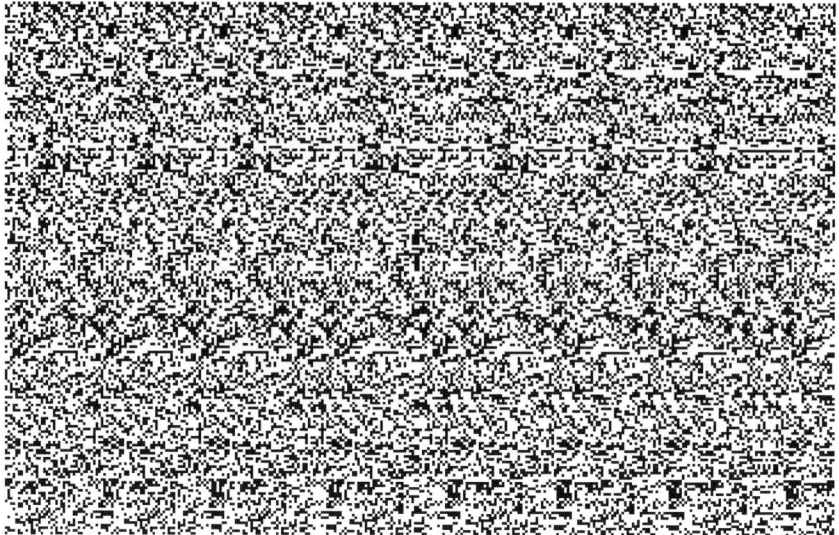

Figure 3.57: Example of an SIRDS Dot Stereogram.

are no longer focused on any point, start moving the picture away from you slowly. When you reach your normal reading distance, you should perceive the three-dimensional image. It's important not to focus on the image.

The reflection method works for stereograms that are printed on reflecting paper. Turn and tilt the paper until it reflects light into your eyes. Focus on the reflection and wait. After a few seconds, you should see the three-dimensional image.

3.14 Autostereoscopic Displays

The autostereoscopic display presents a completely different approach to the problem of creating and viewing three-dimensional images. Such a display generates a three-dimensional image without the need for special glasses, any headgear, or any other auxiliary device. The price for this is a limited field of view. A correct, lifelike three-dimensional image can be seen only from certain points. A viewer positioned elsewhere sees either a confusing image or nothing at all.

The original idea of the autostereoscopic display is due to Adrian Travis [Travis 90] who patented it in 1992 [Travis 92]. Practical autostereoscopic displays are currently being developed by DeepLight, Inc., of Westlake Village, California [deeplight 06].

Imagine two cameras L and R separated by the correct distance for stereoscopic viewing (about 6–7 cm), sending images to a computer (Figure 3.58a). The computer displays the two images alternately at high speed (we say that the images are time multiplexed). It sends image L, followed by image R, followed again by image L, and so on, to a monitor screen. A person is sitting in front of the screen, watching the images. The two time-multiplexed images are synchronized with a fast shutter device such that when image L is displayed, only the left eye of the viewer sees the display and when image R is displayed, only the right eye sees the screen.

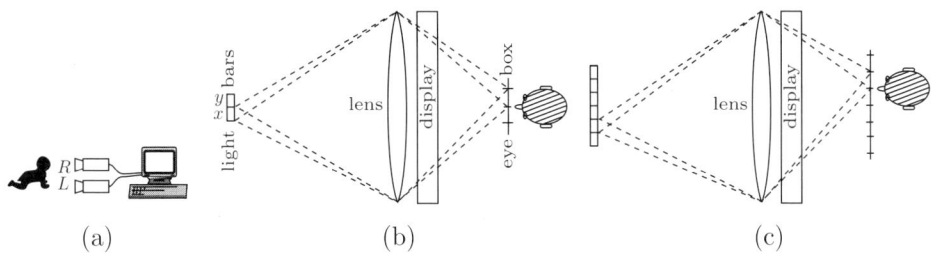

Figure 3.58: Autostereoscopic Display With Light Bars.

The ideal way to achieve such optical synchronization is to use a liquid crystal (LCD) display. This type of display does not generate light and has to be illuminated from behind. Two special light sources and a Fresnel lens are now placed behind the display. Each light source is a narrow vertical rectangle (a light bar) that illuminates the display from a different direction, thereby causing the light from the display to be sent in a different direction. Figure 3.58b illustrates this configuration as seen from above. The viewer has to be positioned at the center of the eye box. (The eye box is simply a region in space, not a screen or a device.). When light from bar x reaches the lens and the display, the image from the display is seen only by the viewer's right eye. A little later, light bar x is turned off and light bar y is turned on, causing the image from the display to shift to the left (in the figure, it is shifted down) and be seen only by the viewer's left eye.

Figure 3.58c shows how this idea can be extended to more than two images. Imagine six cameras positioned in front of a scene. The cameras are set precisely at the same height, they are parallel, and are separated horizontally by seven cm. Six images are sent to the computer and are time-multiplexed by it to the display. Six light bars are synchronized with the images, such that each image is directed by the display to a different area in the eye box. The viewer can now shift his head left and right from area to area within the eye box and can see the scene in three dimensions from five positions with the correct parallax.

Unfortunately, this ideal arrangement is currently impractical because of the following reasons:

1. The images must be sent to the LCD display at a high rate in order to create the illusion of a single, three-dimensional image. In a system with six images, if we want to send each image to the display 60 times a second, we need a refresh rate of $6 \times 60 = 360$ Hz. Unfortunately, the refresh rate of current LCD displays is low. A practical autostereoscopic display must therefore use a high-speed CRT.

2. It is difficult to arrange six cameras at the same height while also keeping them parallel and separated by the right distance. The autostereoscopic display that is currently developed by Deeplight uses two cameras and a special, proprietary algorithm

to generate four additional stereo images, for a total of six images that are then time-multiplexed and sent to the display one by one.

Because of these reasons, autostereoscopic displays currently available use a different arrangement that is illustrated in Figure 3.59. (The figure shows the main components from above.) Light from a high-speed CRT is sent through a lens to form an image. An array of fast LCD shutters "looks" at the image. Each shutter is a narrow rectangle through which the entire image is sent to the Fresnel lens. At any time, only one shutter is open, allowing the image to pass through the shutter and be focused by the Fresnel lens in one area of the eye box.

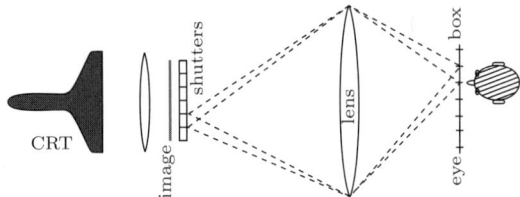

Figure 3.59: Autostereoscopic Display With LCD Shutters.

The shutters are switched rapidly, in synchronization with the image displayed by the CRT, so a viewer looking in two adjacent areas in the eye box sees two (time-multiplexed) stereoscopic images.

Applications of autostereoscopic displays are currently limited by the high cost of the hardware. There is also the fact that only one viewer can see the image at any time, and only a limited number of views is available in the eye box. Current applications include laparoscopic surgery and geologic displays employed in searching for oil and gas deposits. Once costs start coming down, future applications may include three-dimensional graphics design and game playing on personal computers.

Perspective is the rein and rudder of painting.
—Leonardo da Vinci

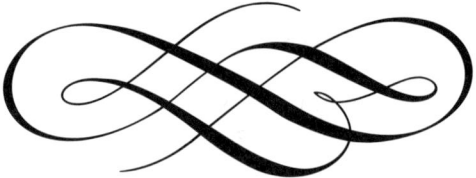

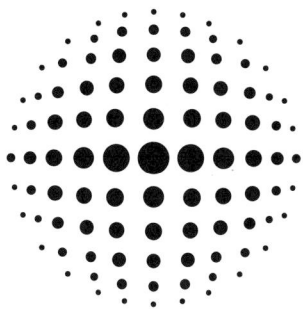

4
Nonlinear Projections

In addition to the parallel and perspective projections, other projections may be developed that are useful for special applications or that create ornamental or artistic effects. Such projections are termed nonlinear because they cannot be expressed by linear transformations such as $x^* = ax + cy + m$ and $y^* = bx + dy + n$. It seems that the number of possible nonlinear projections is vast and is limited only by the imaginations of those who try to develop new ones. This chapter discusses some of the more common nonlinear projections, including the false perspective, the fisheye projection, several 360° panoramic projections, the telescopic and microscopic projections, sphere projections, and circle inversion (a special projection from two dimensions to two dimensions). These projections create aesthetically pleasing (and sometimes confusing) effects and are mathematically simple and easy to derive. However, since they are nonlinear, they generally cannot be represented by means of transformation matrices. [Recall that multiplying a point (x, y, z) by a matrix results in a linear expression such as $ax + by + cz$, but never in nonlinear constructs such as ax^2.]

> Back in the corridor of the building, posters of computer-generated fractal images depicting the "arithmetic limits of iterative nonlinear equations" line the walls.
> —Douglas Rushkoff, *Cyberia: Life in the Trenches of Hyperspace* (1994)

4.1 False Perspective

Equation (3.1) is the main expression for the linear perspective projection; it is duplicated here:

$$x^* = \frac{x}{1 + (z/k)}, \quad y^* = \frac{y}{1 + (z/k)}. \tag{3.1}$$

It shows that the (two-dimensional) coordinates of the projected point $\mathbf{P}^*$ are obtained by dividing by the z coordinate (the depth) of the original point $\mathbf{P}$. False perspective (or pseudoperspective) is a technique to artificially add depth and introduce perspective (or an effect similar to perspective) into a two-dimensional image, thereby making it appear three-dimensional. Points in a two-dimensional image have just x and y coordinates, which makes it natural to modify Equation (3.1) to

$$x^* = \frac{x}{1 + f(x, y)}, \quad y^* = \frac{y}{1 + f(x, y)}, \tag{4.1}$$

where $f(x, y)$ is a function chosen by the user according to the desired effect. For example, the function

$$f(x, y) = -\frac{1}{2} e^{-ax^2 - by^2},$$

where a and b are real constants, returns the value -0.5 for $x = y = 0$ (the origin) and values that approach zero for very large x or y coordinates (positive or negative). Points (x, y) near the origin are therefore projected to $(2x, 2y)$, while points on the edges of the image are hardly affected by this projection. This has the effect of magnifying the center of the image, thereby making it appear closer. Other functions may create different effects. Figure 4.1 shows an example of a 5×5 grid of points moved in such a way.

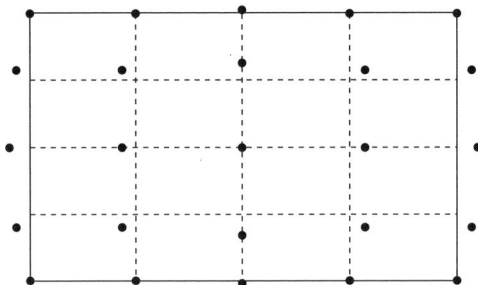

Figure 4.1: Moving Points in False Perspective.

Psychedelics and VR are both ways of creating a new, nonlinear reality, where self-expression is a community event.

If you realize that the world is nonlinear and random, then it means that you can be completely annihilated by chaos for no particular reason at all. These things happen. There's no cosmic justice. And that's a disquieting thing to have to face. It's damaging to people's self-esteem.

—Douglas Rushkoff,

Cyberia: Life in the Trenches of Hyperspace (1994)

4.2 Fisheye Projection

This type of projection is named after the fisheye camera lenses that many photography enthusiasts like to use. The name "fisheye" reflects the shape of such a lens, which resembles the protruding eye of a fish. Such lenses are also used in peepholes installed in doors. The basic idea in this type of projection is to take the half-sphere of space (with infinite radius) located in front of the viewer and project it into a flat circle. The half-sphere is infinite, whereas the circle is finite and may be quite small. Thus, the projected image must be distorted. Just shrinking the image uniformly will make most of its details too small to see. A better idea is to implement nonlinear shrinking that should get more pronounced as we move from the center toward the periphery of the image. Objects close to the center of the image are more visible to a viewer and should therefore

be shrunk only a little. The shrinking should increase for objects located away from the center. In principle, the scale factor should vary from 1 (no shrinking) at the center to 0 (shrinking all the way to zero) for image points on the periphery (i.e., at 180° to the line of sight of the viewer).

> He sat by Chrystal's side, red-complexioned, opulent, with protruding eyes that glanced round whenever he spoke to make sure that all were listening.
> —C. P. Snow, *The Light and the Dark* (1947)

Hemispherical Fisheye Projection

We start with a simple variant that can be called *hemispherical fisheye*. This variant is easy to understand but requires the computations of both the tangent and arctangent for each point being projected. The projection of points in this variant is derived in two steps. In the first step, illustrated in Figure 4.2a, all the points in the hemisphere where z is nonnegative are projected into an infinitely large circle on the xy plane, centered on the origin. In the second step, all the points on this circle are moved closer to the center and end up on the radius-k circle centered at the origin (Figure 4.2b).

The first step employs parallel projection to project points onto a plane. Figure 4.2a shows how the parallel projection of a point simply amounts to clearing its z coordinate. The three-dimensional point (x, y, z) is projected to $(x, y, 0)$ on the infinite circle on the xy plane.

The second step compresses the infinite circle to a radius-k circle nonlinearly. The user selects a positive value k and each point on the xy plane is moved toward the origin by halving its angle of view θ as seen from the standard position $(0, 0, -k)$. (See page 88 for a definition of the standard position.) Figure 4.2b shows a point $\mathbf{P}$ on the xy plane where the angle between the z axis and line VP is θ. The point is moved closer to the origin along the segment PO and becomes $\mathbf{P}^*$ with a view angle of $\theta/2$. Since both $\mathbf{P}$ and $\mathbf{P}^*$ are on the xy plane, we can consider this transformation scaling in two dimensions. The transformed point $\mathbf{P}^*$ equals $s\mathbf{P}$, where the scale factor s is less than one (i.e., shrinking). However, it is easy to see intuitively that points located away from the origin will be scaled more than points closer to the origin. The scale factor s is

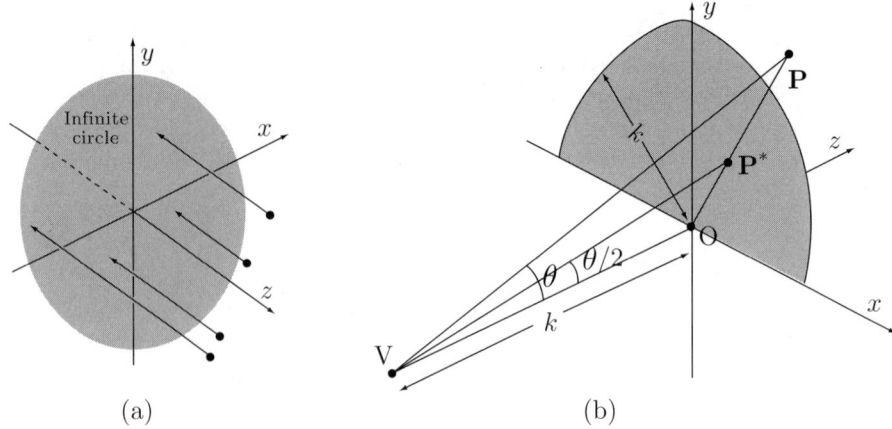

Figure 4.2: Hemispherical Fisheye Projection.

therefore a variable; it depends on $\mathbf{P}$, which is why this type of projection justifies the name *nonlinear*.

The derivation of s starts with Figure 4.2b, which shows that $\tan\theta = |\mathbf{P}|/k$, implying $\theta = \arctan[|\mathbf{P}|/k]$. Similarly, the transformed point satisfies $\tan(\theta/2) = |\mathbf{P}^*|/k$, which yields the scaling factor

$$s = \frac{|\mathbf{P}^*|}{|\mathbf{P}|} = \frac{k\tan(\theta/2)}{|\mathbf{P}|} = \frac{k\tan\big((\arctan[|\mathbf{P}|/k])/2\big)}{|\mathbf{P}|}. \tag{4.2}$$

◇ **Exercise 4.1:** Use mathematical software to compute the scale factors for several $|\mathbf{P}|$ values from 1 to 10,000.

If the programming language or mathematical software being used cannot compute the arctan to the desired accuracy, the following expressions (where h stands for $|\mathbf{P}|$) are equivalent and employ only sines and cosines. From $h/k = \tan\theta$ and $sh/k = \tan(\theta/2)$, we obtain

$$s = \frac{k}{h}\tan(\theta/2) = \frac{\tan(\theta/2)}{\tan\theta} = \frac{1-\cos\theta}{\sin\theta} \bigg/ \frac{\sin\theta}{\cos\theta} = \frac{\cos\theta(1-\cos\theta)}{\sin^2\theta},$$

or equivalently

$$sh = k\tan(\theta/2) = k\frac{1-\cos\theta}{\sin\theta}.$$

Notice that points that are the farthest from the origin on the xy plane have an angle θ in Figure 4.2b close to $90°$. Thus, their projections have an angle close to $45°$. A view angle of $45°$ implies that the distance of such a projected point from the origin equals the distance k of the standard position from the origin. The result is that all the points on the (infinitely large) xy plane are moved by the hemispherical fisheye projection onto the radius-k circle located in the xy plane and centered on the origin.

Figure 4.3 illustrates this process with 50 points. It is easy to see how the distance of a point from the center of the circle affects the amount by which it is moved toward the center. (The code that generated this figure is kept simple. It generates 50 points with random coordinates in the interval $[-10, 10]$, which is why some points are located outside the radius-10 circle.)

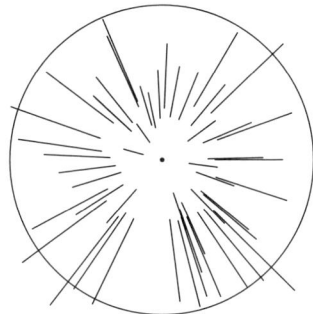

Figure 4.3: Moving Points in Hemispherical Fisheye Projection.

```
(* hemispherical fisheye projection *)
Clear[k, n, P, Q, L]
k=10; n=50;
scal[q_]:=(k Tan[ArcTan[q/k]/2])/q;
P=Table[{Random[Real,{-10.,10.}], Random[Real,{-10., 10.}]},{n}];
Q=Table[Sqrt[P[[i]].P[[i]]], {i, n}];
L=Table[Line[{P[[i]], scal[Q[[i]]] P[[i]]}], {i, n}];
Show[Graphics[L], Graphics[Circle[{0, 0}, 10]],
  Graphics[Point[{0, 0}]],  AspectRatio -> 1]
```

Code for Figure 4.3

It is possible to extend this variant of the fisheye projection to cover more than 180° of space. Figure 4.4a shows how a coverage of up to about 220° can be achieved by bending the xy plane "backward" (i.e., toward the negative z axis) and projecting all the three-dimensional points that are located to the "right" of this bent plane. Once this is done, the points are scaled as before into the radius-k circle.

Figure 4.4b is an example of the type of distortion typical of the hemispherical fisheye projection. The figure shows the old executive office building in Washington, D.C., and it is easy to see that both the vertical lines (the tree in the foreground) and horizontal lines (the fence) are curved and that image elements in the center are more detailed than those near the periphery.

◇ **Exercise 4.2:** Explain why we expect vertical and horizontal straight lines to become curved in a fisheye projection.

Well-known examples of the hemispherical fisheye projection are *Hand with Reflecting Sphere* and *Circle Limit IV (Heaven and Hell)* by M. C. Escher [Ernst 76].

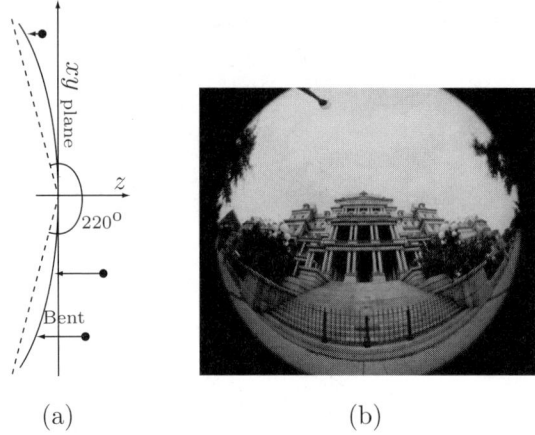

(a)   (b)

Figure 4.4: (a) Extended Hemispherical Fisheye Projection. (b) Example.

Approximate Hemispherical Fisheye Projection

The downside of the hemispherical fisheye projection is the extensive computations required by the tangent and arctangent functions. The method described here uses approximations to simplify the computations. The tradeoff is loss of accuracy, but since the fisheye projection introduces distortions anyway, many viewers may not be able to tell accurate results from approximate ones.

Figure 4.2b illustrates the principle. Each point $\mathbf{P}$ on the infinitely large circle corresponds to an angle θ and is moved toward the origin such that its new angle is $\theta/2$. Thus, we can compute the radii of several concentric circles that correspond to, say, $\theta = 22.5°$, $45°$, $67.5°$, and $89°$. Similarly, we can compute the radii of the corresponding circles (the circles for $\theta/2$ values) on the radius-k circle. Figure 4.5 shows an example.

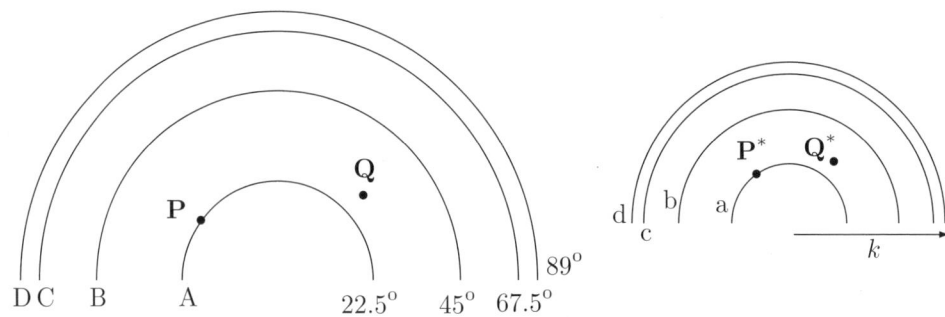

Figure 4.5: Approximate Fisheye Projection.

If a point $\mathbf{P}$ happens to be located on circle A, it is scaled by moving it to the corresponding circle a on the radius-k circle. Its scale factor is the ratio r_a/r_A of the radii of the two circles. If a point $\mathbf{Q}$ happens to be located 30% of the distance between circles A and B, then it is moved 30% of the distance between circles a and b. Its scale factor is $[(1-0.3)r_a+0.3r_b]/[(1-0.3)r_A+0.3r_B]$. This simplifies the computations but introduces inaccuracies because the interpolation between circles is linear. However, the inaccuracies can be reduced as much as desired by precomputing the radii of more circles.

The circle that corresponds to $89°$ is large and the circle for $90°$ has infinite radius. Points whose θ is between $89°$ and $90°$ will be moved to the radius-k circle and placed in the narrow region between the $(89/2)°$ circle and the outer edge. Such points increase the inaccuracies of this method, but this may be acceptable because this region suffers from maximum distortion anyway.

Table 4.6 lists large and small radii for five angles and for $k = 10$. The code that performs the computations is also listed.

n	$\theta°$	R_n	r_n
0	0	0	0
1	22.5	4.142	1.989
2	45	10	4.142
3	67.5	24.142	6.682
4	89	572.9	9.827

Table 4.6

```
k = 10;
angl = {22.5, 45., 67.5, 89.};
k Tan[angl Degree]
k Tan[angl/2 Degree]
```

Code for Table 4.6

A given point (x, y) is at a distance $d = \sqrt{x^2 + y^2}$ from the origin. This distance is compared with all the radii in the table. If d equals an R_n, then the point is multiplied by the scale factor r_n/R_n. Otherwise, we find the smallest R_n such that $R_n < d < R_{n+1}$. The relative distance of the point from R_n is $(d - R_n)/(R_{n+1} - R_n)$. As an example (recall that our table is based on $k = 10$), consider the point $(15, 10)$. Its distance from the origin is $\sqrt{15^2 + 10^2} \approx 18$. Thus, it is between $R_2 = 10$ and $R_3 = 24$. We compute $(18 - 10)/(24 - 10) \approx 0.57$, which tells us that the point is located 57% of the distance from R_2 to R_3.

The scale factor of the point is given by $[(1-0.57)4.142+0.57\cdot6.682]/[(1-0.57)10+ 0.57 \cdot 24] = 5.59/18 = 0.31$, so it has to be moved to $0.31(15, 10) = (4.66, 3.11)$ on the radius 10 circle, where its new distance from the origin is 5.6, or 57% of the distance from $r_2 = 4.142$ to $r_3 = 6.682$.

> A story of particular facts is a mirror which obscures and distorts that which should be beautiful; poetry is a mirror which makes beautiful that which it distorts.
> —Percy Bysshe Shelley, *A Defence of Poetry.*

Angular Fisheye Projection

The hemispherical fisheye projection assigns more importance to those image parts located near the line of sight of the viewer. These parts are displayed in detail, while image elements close to the periphery are displayed in compressed form near the edges of the projection. In contrast, the angular fisheye projection described here assigns the same importance to all the image parts. Each is compressed by the same amount. Perhaps a better name for this method would be "linear fisheye," but the term "linear" seems a misnomer because even this projection introduces distortions and is therefore nonlinear. An important feature of the angular fisheye projection is that it can easily be extended to viewing angles of more than 180° and can even encompass the entire 360° space surrounding the viewer. Figure 4.7 illustrates the principle. The (infinite) sphere of space surrounding the viewer is divided into eight vertical slices of equal viewing angle, each of which is projected into a ring in the final circular projection. We actually see only seven of the eight slices because we are looking at the sphere from an angle. Six points a–f are shown on the sphere with their approximate projections on the circle. Notice that point "d" (shown in gray in slice 5) is supposed to be on the side of the sphere away from us, which is why it is projected on the right-hand side of ring 5.

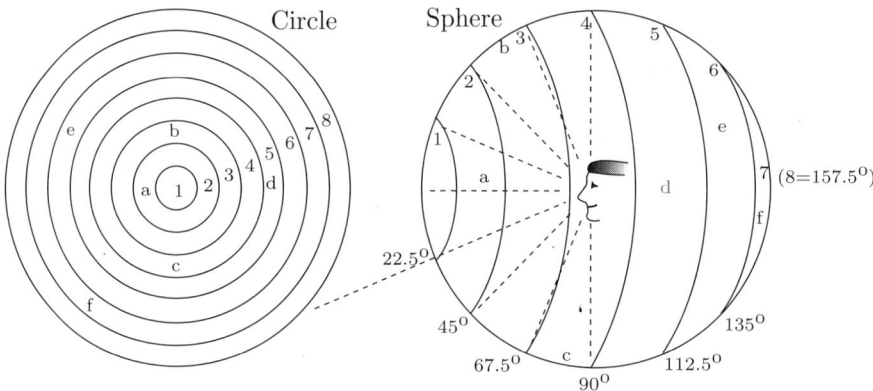

Figure 4.7: Angular Fisheye Projection.

The mathematical analysis of this method is a bit tedious but requires only basic geometry and trigonometry. To start, notice that there is one long dashed line in Figure 4.7. A little thinking should convince the reader that all the points in space along this line are projected to the same point on the radius-k circle. Thus, generating a 360° angular fisheye projection is done by scanning the entire space around the viewer and, for each direction in space, selecting that point on the scene that is the closest to the viewer. This point should be projected to the surface of the sphere and the scan continued to the next direction. Once all the directions have been examined, the surface of the radius-k sphere around the viewer is full of points. The next step is to divide the sphere into slices and project each slice on the radius-k circle. As a result, we can consider a radius-k sphere centered on the viewer and figure out how to scan it and project any point on this sphere to the radius-k circle.

Figure 4.8a shows the half-circle of radius k in the xz plane. Those familiar with the parametric representations of curves and surfaces know that the parametric representation of this half-circle is $k(\cos u, 0, \sin u)$ for $0 \leq u \leq 180°$. Those unfamiliar with parametric methods should either notice that $\cos^2 u + \sin^2 u = 1$ or should refer to [Salomon 05]. A complete sphere of radius k is created when this half-circle is rotated $360°$ about the x axis. The parametric equation of the sphere is therefore the product of the half-circle with the matrix that rotates about the x axis,

$$k(\cos u, 0, \sin u) \begin{pmatrix} 1 & 0 & 0 \\ 0 & \cos w & -\sin w \\ 0 & \sin w & \cos w \end{pmatrix} = k(\cos u, \sin u \sin w, \sin u \cos w), \qquad (4.3)$$

for $0 \le u \le 180°$ and $0 \le w \le 360°$.

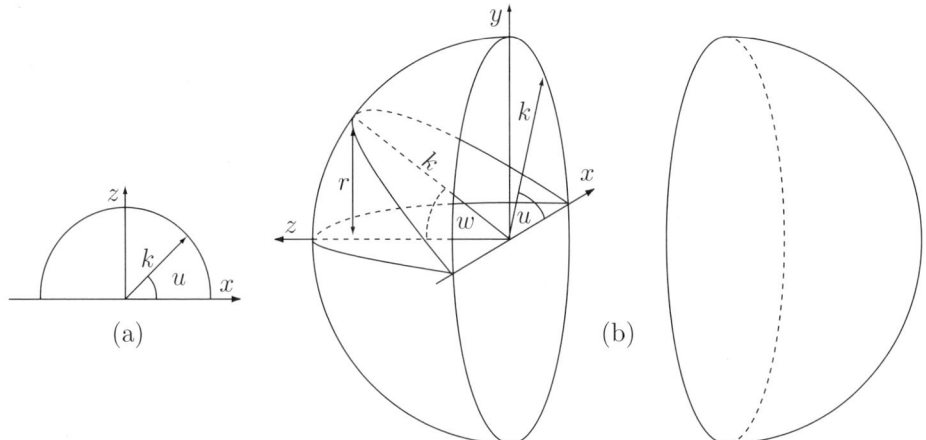

Figure 4.8: Analysis of the Angular Fisheye Projection.

The word *barycentric* is derived from *barycenter*, meaning "center of gravity," because such weights are used to calculate the center of gravity of an object. Barycentric weights have many uses in geometry in general and in curve and surface design in particular.

Figure 4.8b shows the half-circle in the xz plane and how it is rotated. It is clear that the angle w of a point $\mathbf{P}$ on the sphere is one of the parameters of the projected point $\mathbf{P}^*$. This angle determines the distance r of $\mathbf{P}^*$ from the center of the radius-k circle. In the figure, r equals $k \sin w$, but the point is that for $w = 0$ we want $r = 0$, while for $w = 90°$ we want $r = k/2$ and not $r = k$. This is because r values from $k/2$ to k correspond to w values in the "right" hemisphere (i.e., from $90°$ to $270°$). Thus, for w values in the interval $[0, 90]$, we write $r = \frac{k}{2} \sin w$, and Table 4.9 lists the expressions of r for the remaining three intervals of w.

Once we have r, we still need to decide where in the radius-k circle to place $\mathbf{P}^*$, and this is determined by u. This angle varies in the interval $[0, 180°]$, and $\mathbf{P}^*$ has to be placed either in the "top" half (if $0 \le w \le 180°$) or the "bottom" half (if $180° \le w \le 360°$) of the circle, as indicated by Table 4.9.

w	r	r interval	u	$\sin w$
$0 \to 90$	$\frac{k}{2} \sin w$	$[0, k/2]$	top	$0 \to 1$
$90 \to 180$	$(1 - \frac{\sin w}{2})k$	$[k/2, k]$	top	$1 \to 0$
$180 \to 270$	$(1 + \frac{\sin w}{2})k$	$[k, k/2]$	bottom	$0 \to -1$
$270 \to 360$	$-\frac{k}{2} \sin w$	$[k/2, 0]$	bottom	$-1 \to 0$

Table 4.9: Four Cases of w, r, and u.

The complete mapping of the radius-k sphere to the radius-k circle is done in a double loop, where w varies from 0 to 360° in the outer loop and u varies from 0 to 180° in the inner loop. For each pair (u, w), the point of the three-dimensional scene nearest the viewer (who is located at the origin) is determined and is projected by computing its r value from the table and using the pair (r, u), as well as information about "up" or "down" from the table, as the polar coordinates of $\mathbf{P}^*$.

⋄ **Exercise 4.3:** Rewrite Table 4.9 for a 180° angular fisheye projection.

The point directly behind the observer presents a special case. This point is reached when $w = 180°$ (implying $r = k$), in which case any value of u will select this point. This special point is therefore mapped to every point on the circle $r = k$.

⋄ **Exercise 4.4:** Explain the special case of the point directly in front of the viewer.

Often, a three-dimensional scene occupies every direction in space. The scene may consist of several objects with patches of ground, water, and sky filling up every other point. In such cases, every direction (u, w) will correspond to at least one point of the scene. Sometimes, a scene consists of just objects, with no background. In such cases, many pairs (u, w) will not correspond to any point of the scene. For such a pair, its projection on the radius-k circle can be painted white or any other background color.

When the entire space around the viewer is projected into a circle, the angular fisheye projection becomes one of many ways to map a sphere on a plane. Sphere projections are the topic of Section 4.14. Every projection of a sphere into a plane introduces distortions, and the two main distortions of the angular fisheye projection are that (1) straight lines are mapped into curves and that (2) the hemisphere in front of the viewer is projected into the inner half of the circle and can, with some practice, be perceived and understood, but the hemisphere behind the viewer is projected into the outer half of the circle, which is a ring, and this makes it unintuitive to perceive its details.

Figure 4.10 shows two 180° examples (in grayscale and color; see page 236) of the angular fisheye projection. It is possible to see that the distortion is uniform over the entire picture. Also, the many straight lines are curved, but it is obvious that the curvature diminishes in lines that are close to the center of the figure. The figure on the left (courtesy of Joseph Bly [joebly 06]) is a lawn in New York's Central Park. It is obvious that both the vertical lines (the tree in the foreground) and horizontal lines (the horizon and the seats) are curved and that image details in the center are larger than those near the periphery. The figure on the right (courtesy of Dick Termes [termespheres 05]) is

Figure 4.10: Two Angular Fisheye Examples. (The one on the left is courtesy of Joseph Bly [joebly 06]. The one on the right is courtesy of Dick Termes.)

titled *Food for Thought, 2004* and features the La Plazula restaurant at the La Fonda Hotel in Santa Fe, New Mexico.

◇ **Exercise 4.5:** Show why most straight lines are mapped to curves under the angular fisheye projection.

Another point worth mentioning is that the sphere is larger than the circle. Even if u and w are varied in large steps, there may be more directions to scan than there are pixels in the radius-k circle. This suggests another approach to the angular fisheye projection. Instead of scanning the 360° sphere in many directions, scan the radius-k circle pixel by pixel, compute the polar coordinates (r, u) of each pixel, and use them to determine the corresponding direction (u, w) in space. If a point of the scene is found in that direction, it is projected to the pixel without any additional calculations.

Here is a summary of this derivation. (Actual C code can be found in [Bourke 05].) We assume that the circle is embedded in a rectangular bitmap of height H pixels and width W pixels. We scan this rectangle row by row. If the current pixel has coordinates (a, b), we first convert them to normalized coordinates (x, y) in the interval $[-k, +k]$ by

$$x = \left(\frac{2a}{W} - 1\right) k \text{ and } y = \left(\frac{2b}{H} - 1\right) k.$$

The distance of the pixel from the center of the rectangular image is $r = \sqrt{x^2 + y^2}$. If r is greater than k, the pixel is outside the radius-k circle and is ignored. Otherwise, angle u is computed by

$$u = \begin{cases} 0, & r = 0, \\ \pi - \arcsin(y/r), & x < 0, \\ \arcsin(y/r), & x \geq 0. \end{cases}$$

Angle w equals $r/2$, so it is in the interval $[0, k/2]$, and the direction vector is

$$k(\cos u, \sin u \sin w, \sin u \cos w).$$

The distortion introduced by the fisheye projection can be used to convert it to a spherical panoramic projection (which is discussed in more detail in Section 4.6). Imagine a radius-k circle on which a $180°$ fisheye projection is displayed. We scan the circle pixel by pixel and translate the Cartesian coordinates (a, b) of a pixel to polar coordinates $r = \sqrt{a^2 + b^2}$ and $u = \arctan(b/a)$ (if $a = 0$, then $u = 0$ or $u = 180°$, depending on b). Once r is known, we can use the relations $r = \pm k \sin w$ to compute angle w. Once u and w have been computed, we know that pixel (a, b) is the projection of a point $\mathbf{P}$ located in direction $(\cos u, \sin u \sin w, \sin u \cos w)$ on the radius-k hemisphere centered on the viewer. Thus, in principle it is possible to map each pixel in the fisheye projection to a three-dimensional point $\mathbf{P}$ on this hemisphere. We don't know how far from the viewer the original point was because this information was lost when the fisheye projection was prepared, but we know that of all the three-dimensional points in direction (u, w) in the scene, point $\mathbf{P}$ was the nearest to the viewer, blocking all the points directly behind it. In practice, however, this technique is not that simple to implement because the number of pixels in the circle is much smaller than the number of pixels in the hemisphere.

In this month's *Hemispheres Magazine*, the magazine of United Airlines, you'll find my article about exploring the chocolate shops of Paris. I talk about many of my favorite places, why I like them... and what I recommend you get while you're there!
    —David Lebovitz in [davidlebovitz 05], October 2005.

Off-Axis Fisheye Projection

The discussion of both the hemispherical and angular fisheye projections assumes that the viewer is looking at a radius-k circle on which an infinite hemisphere is projected. Figures 4.2b and 4.7 further imply that the line of sight of the viewer passes through the center of the circle. We can say that the viewer is located on the axis of the circle and we can ask what the viewer will see when he moves away from the axis, still looking in the same direction. This is not just a theoretical problem. Many planetariums use a fisheye lens to project an image on a hemispherical dome, where some (or even many) viewers sit away from the center. Those viewers see a twice-distorted image, once because it is a fisheye projection and again because they observe it off-axis.

The mathematics of an off-axis fisheye projection is illustrated in Figure 4.11. We start with four points, depicted as circles and labeled 1 through 4. In part (a) of the figure, the viewer is assumed to be on the axis and the points are shifted toward the viewer by halving their view angles. The shifted points are depicted as small squares. In part (b), the viewer is assumed to be located off-axis, and the four points are shifted toward the viewer by halving their new view angles. The new points are depicted as triangles. It is obvious that points 1 and 2 are shifted more in part (a) than in part (b). Thus, those parts of the image are more distorted when the viewer is on-axis. In

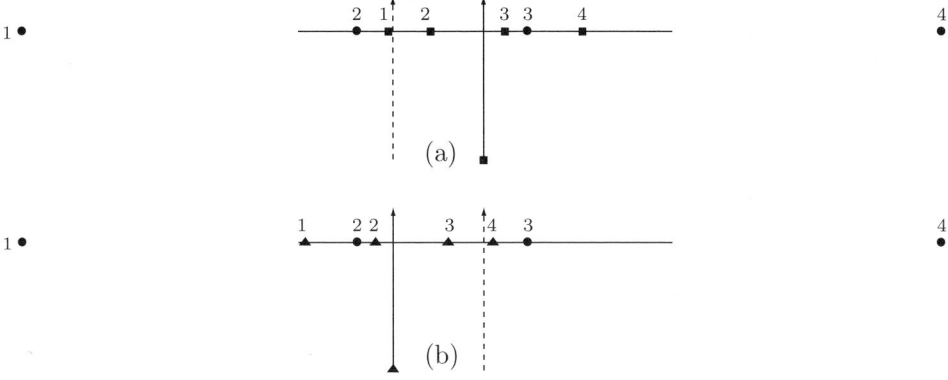

Figure 4.11: Off-Axis Fisheye Projection.

contrast, points 3 and 4 are shifted more when the viewer is off-axis, thereby distorting those parts of the image on the "right" side.

Figure 4.12 illustrates the overall effect of an off-axis projection. It shows 50 points moved toward an off-axis viewer. In the three parts of the figure, from left to right, the viewer is located at $(10, 0)$, $(-5, 5)$, and $(0, 5)$. This figure illustrates the effects of the viewer being off-axis and ignores the distortions (such as straight lines transformed into curves) introduced by the fisheye projection itself.

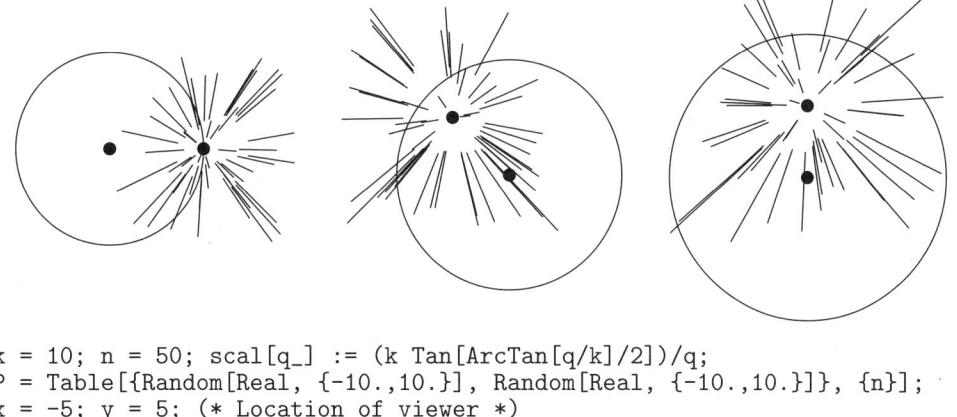

```
k = 10; n = 50; scal[q_] := (k Tan[ArcTan[q/k]/2])/q;
P = Table[{Random[Real, {-10.,10.}], Random[Real, {-10.,10.}]}, {n}];
x = -5; y = 5; (* Location of viewer *)
Pt = P - Table[{x, y}, {n}];
Q = Table[Sqrt[Pt[[i]].Pt[[i]]], {i, n}];
L = Table[Line[{P[[i]]+{x, y}, (scal[Q[[i]]] P[[i]])+{x, y}}], {i, n}];
Show[Graphics[L], Graphics[Circle[{0, 0}, k]],
 Graphics[{AbsolutePointSize[5], Point[{0, 0}]}],
 Graphics[{AbsolutePointSize[5], Point[{x, y}]}],
 AspectRatio -> Automatic, PlotRange -> All]
```

Figure 4.12: Off-Axis Fisheye Projection and Code.

Those who took the trouble to read Chapter 1 know how to compute the off-axis fisheye projection. Figure out how to translate the viewer on the xy plane to the on-axis position, and then use the translation vector (a, b) to translate each point with $(-a, -b)$, project it according to Equation (4.2), then translate the result back with (a, b). If the last translation brings the point outside the radius-k circle, the point is ignored because the off-axis viewer cannot see it.

Rectangular Fisheye Projection

The hemispherical fisheye projection projects the entire 180° space located in front of the viewer, an infinitely large image, into a finite-sized circle, and it does this by distorting the image, especially in areas away from its center. The rectangular fisheye projection discussed here is a compromise on this technique. It creates less distortion but can project only part of the space in front of the viewer. Those parts that are too high above the viewer or too low are not included in this type of projection. Figure 4.13a shows the principle. We imagine a rectangle of infinite width and a finite height h centered on the xy plane. A three-dimensional point (x, y, z) is projected on the rectangle in parallel into the point $(x, y, 0)$, but only if the y coordinate is in the interval $[-h/2, +h/2]$. (The figure shows one point that's too high.) Points above or below the rectangle are not included in the projection.

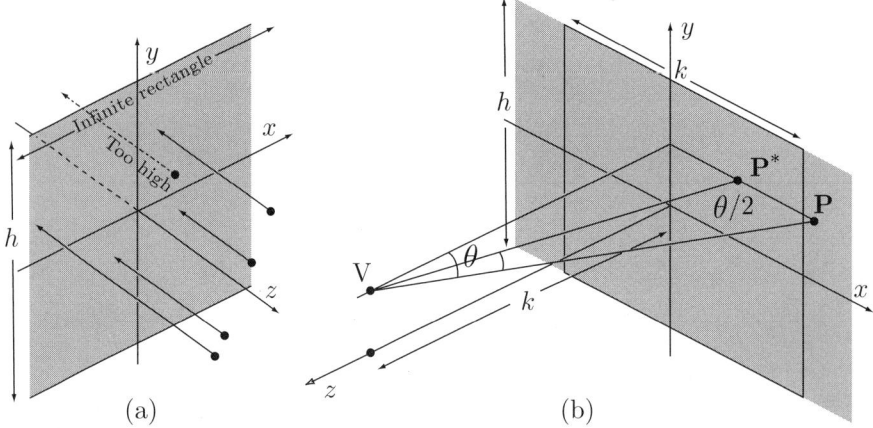

Figure 4.13: Rectangular Fisheye Projection.

Once a point has been projected on the rectangle, it is shifted in the x direction to bring it into the rectangle of width k. This is done by halving its view angle θ, as in the hemispherical fisheye projection, but only in the x direction (Figure 4.13b). The final projection is distorted only in the x direction; all the y dimensions are preserved. The final result is that point (x, y, z) is projected into $(s \cdot x, y, 0)$, where the scale factor s is given by [compare with Equation (4.2)]

$$s = \frac{k \tan\big((\arctan[|x|/k])/2\big)}{|x|}.$$

This variant of the fisheye projection is a relative of the semicylindrical fisheye projection. We start with half a cylinder, on which three-dimensional points are projected in parallel. The semicylinder is then unrolled and viewed as a flat rectangle. Notice that points "d" and "e" in Figure 4.14 are close in three-dimensional space, but their projections on the cylinder are separated. This type of projection magnifies details close to the vertical edges of the final projection, which is the opposite of the other fisheye variants.

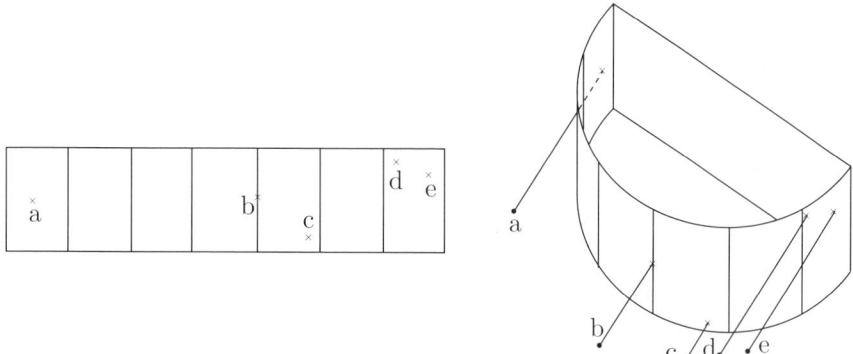

Figure 4.14: Semicylindrical Fisheye Projection.

4.2.1 Fisheye Menus

The topic of this section is not a projection from three dimensions to two dimensions, but it is included here because it is a useful and interesting application of the fisheye principle: the technique of local magnification combined with global shrinking. Often, a computer program has to display a long, dynamic menu of items. An address book has to display the list of addresses, an Internet browser must display a list of URLs, and a commercial Web site should display a list of items described on the site or offered for sale. The user watches such a menu—normally with other menus, text, images, and miscellaneous items—on the computer monitor, where "real estate" (i.e., space) is limited.

Software designers have been aware of this problem for a long time and have come up with various solutions. Perhaps the simplest solution is to shrink the size of individual menu items as more items are added to the menu and it gets taller than the screen. This solution can only go so far because text under a certain size (typically 5 printer's points) is impossible to read on a screen, where the pixel resolution is typically 72 dots per inch (dpi). A slightly better solution is to scroll the screen. Once the menu is taller than the screen (or taller than the window assigned to the menu), a scroll bar appears on the side, so the user can scroll the menu up or down. Sometimes arrows at the top and bottom are used instead of a scroll bar. This is a simple, effective, and very common solution. Its only downside is that only part of the menu is displayed at any given time, but if the menu items are sorted in some way, which they often are, this may not present a serious problem.

Another common technique is to use hierarchical "cascading" menus, where the main menu is kept small, but any items in it can have a submenu. Selecting an item, normally with a mouse, opens (after a short delay, allowing the mouse to slide to another item) its submenu and lists its items, which may have subsubmenus. This allows for very large menus, but again only a small part of the menu is displayed at any time. Another disadvantage of this type of menu is the time it takes to open a submenu, examine it, and, if it is the wrong one, slide to another submenu.

A more sophisticated solution is the fisheye menu. In such a menu, all the items are displayed simultaneously on the screen or in the window. If there are many items, most are shrunk to small sizes or even very small sizes, where it is impossible to read or perceive an item. Sliding the cursor along such a list magnifies the items closest to the cursor, so they can be read or observed at their full size. Items slightly away from the cursor are displayed at somewhat smaller sizes, and items far from the cursor are displayed at very small sizes. Figure 4.15 shows two examples of fisheye menus. One is a long list of text items (country names from [fisheyemenu 05]) sorted alphabetically. It is obvious that sliding the cursor along such a list is a fast and easy way to select any desired name, even though at any given time most of the list is too small to read. The other example is the Macintosh dock, a feature familiar to Macintosh users since the introduction of OS X in 2001. The dock is a graphical menu with icons of files, folders, and applications that are commonly used. A dock item is selected by sliding the cursor along the dock. The icon sizes vary from small to medium to large and back to small in real time, making it easy for the user to locate any desired item. Once an item is found, merely selecting it also launches it.

When a menu is short, all its items can be displayed in full size and the entire menu fits comfortably on the screen. When items are added to the menu, it gets taller until the time comes to shrink items. The algorithm for that must consider three features:

1. The total height of the menu must equal the height of the screen regardless of the number of items. The only exception is a menu that's too short even when all its items are displayed at maximum size.

2. The maximum font size (or size of the graphical icon) must be specified by the user, with a reasonable default value. Some fisheye menus require a large maximum size, while others can be used with a fairly small maximum size.

3. The item at the cursor location is displayed at the maximum size, and all the items within a distance of $f/2$ items above it and $f/2$ items below it must be displayed at a size that will make it possible for the user to read and identify them. The sizes of the remaining items are selected such that the entire menu will fill up the screen. This creates a dynamic bubble of f readable items around the main item, which enables the user to identify items adjacent to the main item and select any of them with ease. The parameter f is referred to as the focus length of the fisheye menu and should be specified by the user, with a reasonable default value. Notice that large-sized items require larger spacing between them, while the spacing between the smaller items can be shrunk accordingly.

A large focus length, such as 10 or 20, will cause the peripheral items to be very small, while a small focus length, such as 2 or 3, will force the user to slide the cursor

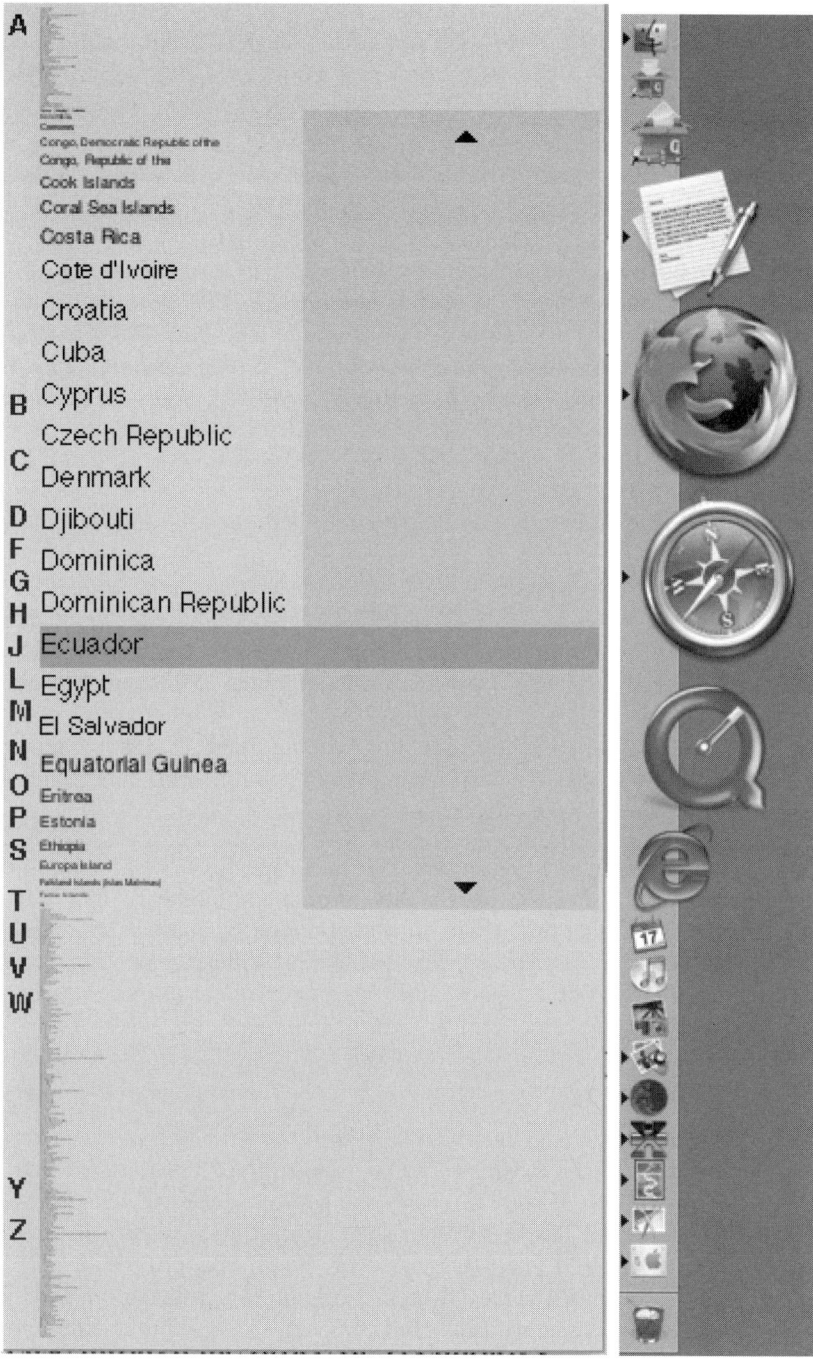

Figure 4.15: Fisheye Menus (color version on page 237).

slowly in order to be able to read the current two or three large items. Thus, the choice of focus length is a compromise between fast selection and ease of reading.

When a menu becomes very large, most of its items are shrunk to the size of a dot. In such a case, it helps to embed index items in the menu. These items are always kept at a readable size and are used to locate the start of any desired region in the menu. This idea is illustrated in the left part of Figure 4.15, where the index items are the single letters "A" through "Z." A user looking for an item that starts with "Q" can quickly slide the cursor to the index "Q," where the first few relevant items will immediately be readable.

A fast implementation of fisheye menus is a must and is based on arrays or other data structures, each of which contains relevant data at a certain size. If the menu items are text, then fonts at several sizes must be available. If the items are icons, then each new icon added to the menu must be immediately prepared at several sizes and added to the appropriate data structures.

For more information on fisheye menus, see [fisheyemenu 05].

4.3 Circle Inversion

This projection is an exception, perhaps the only one, to the material in this chapter. It projects (or rather transforms) a two-dimensional image to another two-dimensional image. In spite of this, it is included here because of its simplicity and mathematical elegance. Circle inversion was the brainchild, around 1830, of Jakob Steiner. It has been researched and studied extensively since its first publication, and much is known about it (as is shown by a simple Internet search).

He [Steiner] is a middle-aged man, of pretty stout proportions, has a long intellectual face, with beard and moustache and a fine prominent forehead, hair dark rather inclining to turn grey. The first thing that strikes you on his face is a dash of care and anxiety, almost pain, as if arising from physical suffering—he has rheumatism. He never prepares his lectures beforehand. He thus often stumbles or fails to prove what he wishes at the moment, and at every such failure he is sure to make some characteristic remark.

—Thomas Hirst, *Diary* (1852)

Figure 4.16 illustrates the principle. The figure shows the unit circle centered on the origin and an arbitrary point $\mathbf{P}$ with polar coordinates (r, θ). Circle inversion projects $\mathbf{P}$ to $\mathbf{P}^* = (1/r, \theta)$. Both $\mathbf{P}$ and $\mathbf{P}^*$ have the same angle θ, which places them on the same straight line that passes through the origin. If $r > 1$, then $\mathbf{P}$ is outside the unit circle and $\mathbf{P}^*$ is inside it (because $1/r < 1$). Thus, this projection inverts points with respect to the unit circle centered on the origin. It is easy to see that points on the circumference of the circle are projected to themselves and that circle inversion is undefined for the origin, where $r = 0$. (Although we can say that the origin is projected to the *point at infinity*, but this claim is not very useful and may cause confusion with parallel lines, which are also sometimes said to meet at infinity.) Since $\mathbf{P}$ is moved to $\mathbf{P}^*$

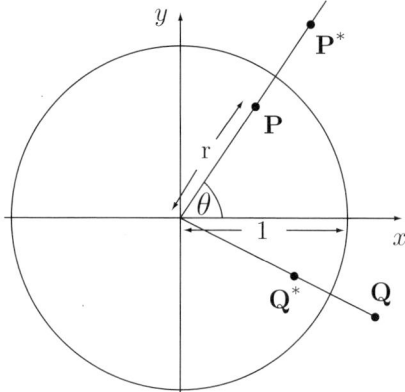

Figure 4.16: Circle Inversion.

along the line that connects **P** to the origin, we can think of this projection as scaling. From $\mathbf{P}^* = (1/r, \theta)$, we obtain $x^{*2} + y^{*2} = 1/r^2$ and this implies

$$\mathbf{P}^* = (x^*, y^*) = \frac{(x, y)}{x^2 + y^2} = \frac{\mathbf{P}}{x^2 + y^2} = s\mathbf{P}$$

because this relation means that

$$x^{*2} + y^{*2} = \frac{x^2}{(x^2 + y^2)^2} + \frac{y^2}{(x^2 + y^2)^2} = \frac{1}{(x^2 + y^2)} = 1/r^2.$$

Notice that the scale factor s depends on **P**, showing that this type of projection is nonlinear.

Currently, there are several applets on the Internet that make it easy to explore the properties of circle inversion. This projection has a number of interesting features, the most important of which are the following:

1. Any circle that intersects the unit circle at right angles is projected to itself.

2. The angle between two projected lines is preserved. Thus, circle inversion is a *conformal* projection.

3. Circles that do not pass through the origin are projected into circles (that do not pass through the origin and generally have a different radius).

4. Similarly, lines that do not pass through the origin are projected into circles that *do* pass through the origin (Figure 4.17).

5. A circle centered on the origin is projected to another circle similarly centered.

6. Lines through the origin are projected to themselves (except that the projection of the origin is undefined).

7. The inverse of an inverse is the original point. Thus, $(\mathbf{P}^*)^* = \mathbf{P}$. (This is trivial.)

Curves that are their own inverse are called anallagmatic.

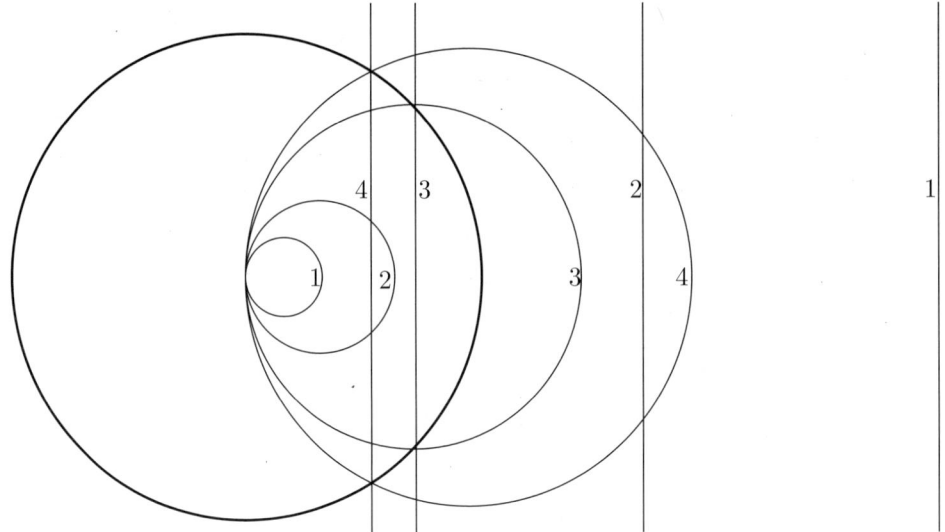

Figure 4.17: Four Circles and Lines.

◇ **Exercise 4.6:** Search the mathematical literature or the Internet (or just think about this) to find another anallagmatic curve.

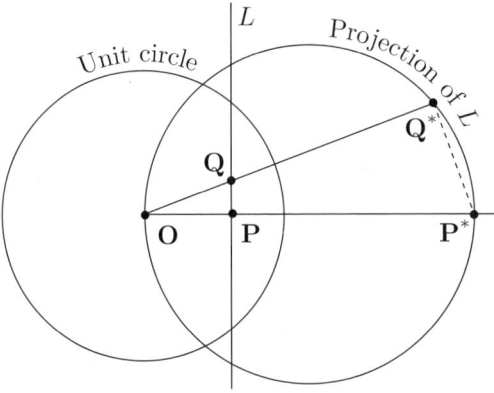

Figure 4.18: Circular Inversion of a Line.

Here is a proof of feature 4. Figure 4.18 shows a line L that does not pass through the origin. Consequently, there must be a perpendicular to L from the origin. The point where this perpendicular meets L is denoted $\mathbf{P}$ and its projection is denoted $\mathbf{P}^*$. We now select another arbitrary point $\mathbf{Q}$ on L and denote its projection $\mathbf{Q}^*$. It is obvious that $OP \cdot OP^* = 1$ and $OQ \cdot OQ^* = 1$, so we conclude that $OP/OQ^* = OQ/OP^*$. This shows that triangles OPQ and OP^*Q^* are similar (notice that they have a common angle), which, in turn, implies that angles OPQ and OQ^*P^* are equal. Since the former is a right angle, the latter must be also. However, point $\mathbf{Q}$ is an arbitrary point on L, so angle OQ^*P^* equals 90° for any point $\mathbf{Q}$ on L, showing that the projection $\mathbf{Q}^*$ lies on a circle that passes through the origin O and has a diameter OP^*. The projection of $\mathbf{P}$ is $\mathbf{P}^*$, and the projection of the origin is the point (or points) at infinity. Line L of Figure 4.18 passes inside the unit circle. For lines outside this circle, the diagram looks different but the proof is identical.

⋄ **Exercise 4.7:** Use similar arguments to prove feature 3.

⋄ **Exercise 4.8:** The discussion so far has assumed inversion with respect to the unit circle. Given a circle C of radius R about the origin, show how to project a point $\mathbf{P}$ with respect to it.

Figure 4.19 shows a simple geometric construction of the inverse of a point $\mathbf{P}$. In part (a) of the figure, $\mathbf{P}$ is inside the circle. Line L_1 is constructed from the center through $\mathbf{P}$ and continues outside the circle. Line L_2 is then constructed perpendicular to L_1. Point $\mathbf{A}$ is the intersection of L_2 with the circle. A tangent L_3 to the circle is constructed at $\mathbf{A}$, and $\mathbf{P}^*$ is placed at the intersection of the tangent and L_1. Part (b) shows the similar construction when $\mathbf{P}$ is outside the circle.

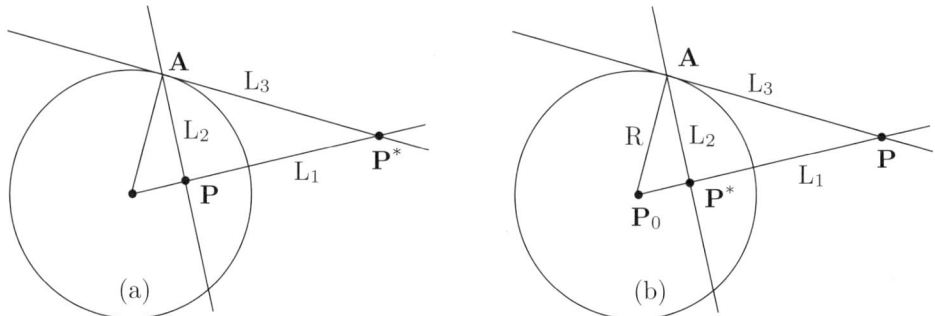

Figure 4.19: Construction of Circle Inversion.

Figure 4.19b illustrates another feature of circle inversion. Up to now, we assumed that the inversion is about a unit circle centered on the origin. Given a circle of radius R, the two triangles $\mathbf{PP_0A}$ and $\mathbf{P^*P_0A}$ are similar, implying that $\mathbf{P_0P}/R = R/\mathbf{P_0P^*}$ or $R^2 = \mathbf{P_0P} \times \mathbf{P_0P^*}$. The quantity R^2 is termed the *circle power*. The inverse $\mathbf{P}^*$ of a point $\mathbf{P}$ with respect to an inversion circle of radius R centered at $\mathbf{P_0}$ is given by

$$\mathbf{P}^* = \mathbf{P}_0 + R^2 \frac{\mathbf{P} - \mathbf{P}_0}{|\mathbf{P} - \mathbf{P}_0|^2}.$$

As is common with nonlinear projections, it is possible to come up with many variants of circle inversions. For example, project point (r, θ) to $(1/r, 180° + \theta)$. An obvious (but perhaps not very useful) extension of circle inversion is sphere inversion, where the spaces inside and outside a sphere are swapped. Reference [Coxeter 69] presents the complete theory of circle inversions. A more general treatment of inversive geometry can be found in [Stothers 05].

Figure 4.20 (after [Gardner 84]) shows the circle inversion of a chessboard.

 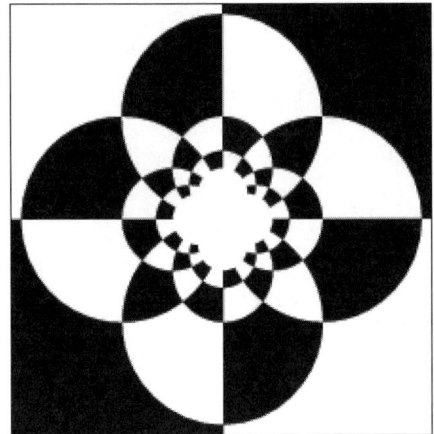

Figure 4.20: The Circle Inversion of a Chessboard.

4.4 Panoramic Projections

Visitors to an exceptionally lovely spot sometimes wish they could see the view behind them as well as in front of them simultaneously. This kind of effect is generated by the various *panoramic projections*. A panorama is defined as an unbroken view of an entire surrounding area, and panoramas have always been a favorite with artists, painters, and photographers. The insert below discusses the Mesdag panorama, one of the few surviving large panoramas painted in the 18th and 19th centuries. When cameras came into general use in the early 20th century, inventors started developing panoramic cameras (Section 4.10). With the advent of fast, inexpensive personal computers and digital cameras in the 1980s, it became possible, even easy, to take a sequence of (partially overlapping) photographs with any camera and stitch them by software into a single picture that depicts a large area, sometimes an entire 360° view around a point, including parts that are very high or very low and cannot normally be included in a single picture. The price for including so much visual information in one picture is distortion. Any method for projecting a three-dimensional scene into a panoramic picture introduces some distortion. Straight lines become curved and familiar shapes may look funny or become completely unrecognizable.

The main types of panoramic projections described here are the cylindrical, spherical, and cubic. All three are based on the same principle, but only the first is popular because it manages to squeeze the most visual data into a flat image with the minimum of distortion. Section 4.8 presents a different approach to panoramic projections, where they are considered variants of the linear perspective projection but with several vanishing points (up to six) placed at certain strategic locations in the projection. Section 4.9 mentions other techniques for panoramic projections.

The Mesdag Panorama

The Mesdag Panorama is a painting depicting a 360° panoramic view of the surroundings of Scheveningen, a fishing port northwest of The Hague, as seen by the painter in 1881.

The painting is huge, measuring 120×14 meters (390×45 feet) for an area of about 17,000 square feet. It is folded into a cylinder and several observers can enter from below and stand at the center, turning, watching, and admiring.

The Mesdag panorama was painted by the 19th-century Dutch painter Hendrik Willem Mesdag, with the help of S. Mesdag-van Houten, Theophile de Bock, B.J. Blommers, G.H. Breitner, and A. Nijberck.

Similar panoramas were exhibited throughout Europe and America during the 19th century (they were sometimes called *cycloramas*). The Mesdag panorama is one of the last panorama paintings still in existence. It can be viewed at the Museum Panorama Mesdag in The Hague, The Netherlands.

See [Mesdag Documentation Society 98] for more information.

4.5 Cylindrical Panoramic Projection

Imagine a rectangle made of transparent material being rolled into a cylinder and placed around an observer (Figure 4.21a). The observer is located at the origin, which is also the center of the cylinder, and is looking at the view outside through the transparent surface of the cylinder. The observer now starts turning around. We imagine that everything that the observer sees is magically fused into the cylinder material. (In the absence of magic, the observer may simply use a paintbrush or a magic marker to paint what he sees through the cylinder.) As an example, point $\mathbf{P}$ in Figure 4.21a is projected to point $\mathbf{P}^*$ by connecting $\mathbf{P}$ to the observer as in linear perspective. After the observer has turned a full circle, the surface of the cylinder is entirely covered with images. The cylinder is now unrolled and is hung flat on a wall, to be viewed as a rectangular picture. The image shown in such a picture is a 360° cylindrical panorama (or a cylindrical projection) of the view seen by the observer. Notice that certain details seen by the observer are too high or too low to be seen through the cylinder. Point $\mathbf{Q}$ in Figure 4.21a is such an example. Thus, the unrolled cylinder does not contain the entire scene surrounding the observer. The top and bottom parts are missing, and the sizes of the missing parts depend on the height of the cylinder.

Figure 4.21a shows a cylinder centered about the origin. It is easy to see how a three-dimensional point $\mathbf{P}$ is projected to a point $\mathbf{P}^*$ on the cylinder. Figure 4.21b shows the cylinder unrolled. Point **P** is located in the same place in space, but its projection has moved with the opening of the cylinder.

Figure 4.21c shows the geometry of the problem. We assume that the dimensions of the original rectangle are $2Y \times 2Z$. When rolled into a cylinder of radius R, the perimeter of the cylinder satisfies $2\pi R = 2Y$, so $R = Y/\pi$. Consider an arbitrary three-dimensional point $\mathbf{P} = (x, y, z)$ viewed by the observer. When the cylinder is eventually unrolled, **P** will be projected to a point $\mathbf{P}^* = (x^*, y^*, z^*)$ and our problem is to determine the coordinates of $\mathbf{P}^*$ as functions of x, y, z, Y, and Z.

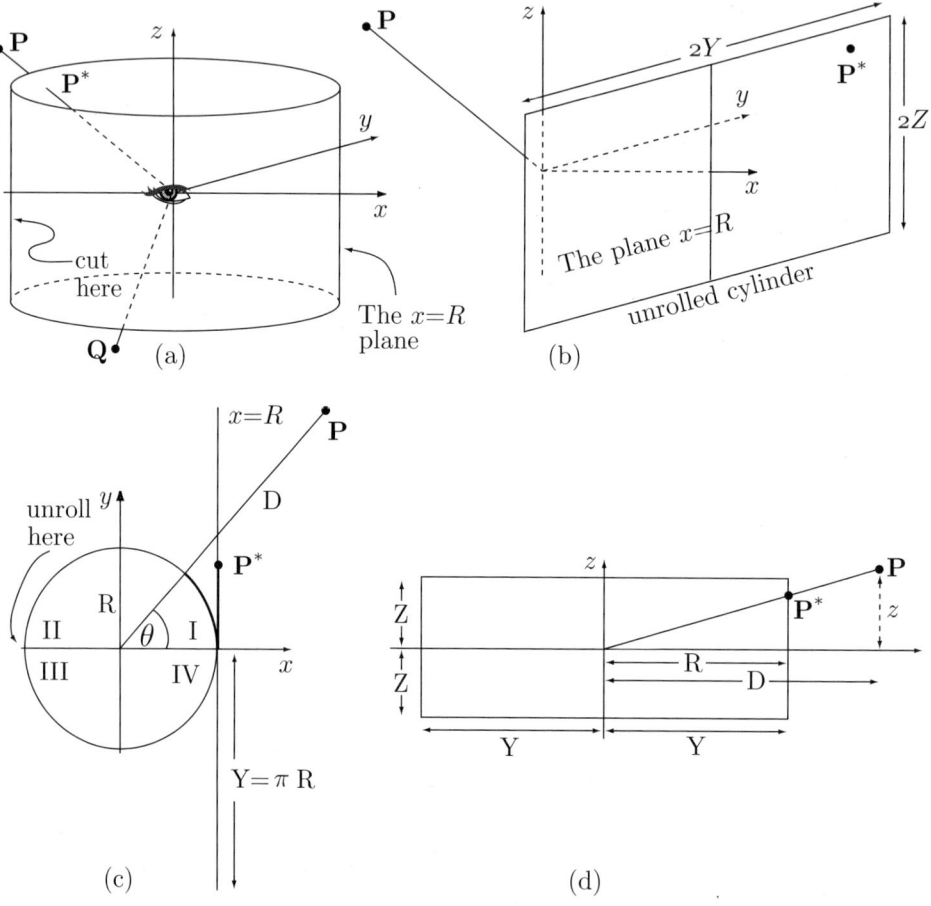

Figure 4.21: A 360° Panoramic Projection.

The x^* coordinate is trivial to determine. The figure shows that all the points on the unrolled cylinder have the same x coordinate. We can set it to R or, even simpler, to zero. The y^* coordinate should equal the length of the arc subtended by θ, which is $R\theta$. Angle θ depends on the x and y coordinates of **P** but not on its z coordinate. The relation is $(x, y) = D(\cos\theta, \sin\theta)$, where D is the distance (projected on the xy plane) of $\mathbf{P}$ from the origin. This distance is $\sqrt{x^2 + y^2}$. From this we get

$$\frac{(x, y)}{\sqrt{x^2 + y^2}} = (\cos\theta, \sin\theta),$$

or

$$\theta = \arcsin\frac{y}{\sqrt{x^2 + y^2}} = \arccos\frac{x}{\sqrt{x^2 + y^2}} = \arctan\left(\frac{y}{x}\right).$$

Notice that the signs of x and y determine the quadrant number. If θ is in quadrant III or IV, then y^* should be negative.

The z^* coordinate is determined by perspective projection. Figure 4.21d shows how this is done with similar triangles:

$$\frac{z}{D} = \frac{z^*}{R} \rightarrow z^* = \frac{z\,R}{D} = \frac{z\,Y}{\pi\sqrt{x^2 + y^2}}.$$

◇ **Exercise 4.9:** It seems that the projected point $\mathbf{P}^*$ is given by

$$(x^*, y^*, z^*) = \left(0, \pm R\theta, \frac{z\,Y}{\pi\sqrt{x^2 + y^2}}\right),$$

so its coordinates depend on x, y, z, and Y, but not on Z. What's the explanation?

The panoramic projection leads naturally to the concept of *curved perspective* (see also Section 4.8). This concept comes up when we consider the panoramic projection of a straight line. Figure 4.22a shows a cylinder and a line A in space. Several projection lines are shown going from A to the center of the cylinder. These lines are contained in a plane L, and we know from elementary geometry that the intersection of a cylinder and a plane is, in general, an ellipse (Figure 4.22b). The projection of A on the cylinder is therefore an elliptical arc. When the cylinder is unrolled, this arc turns into a sinusoidal curve (Figure 4.22c).

◇ **Exercise 4.10:** Prove this claim!

This behavior means that the panoramic projection converts straight lines into curves, resulting in what can be termed *curved perspective*. Two special cases should be considered. One is when the plane is perpendicular to the cylinder (corresponding to an angle $\theta = 0°$ in Figure Ans.15, page 271), and the other occurs when it is parallel to the axis of the cylinder (corresponding to an angle $\theta = 90°$ in Figure Ans.15). In the former case, the intersection is a circle and the sinusoidal curve has zero amplitude (i.e., it degenerates into a straight segment). In the latter case, the intersection is an infinite ellipse and the sinusoidal curve has infinite amplitude; it degenerates into three lines.

Figure 4.22d shows an observer positioned at the center of a cylinder and looking to the north. Three horizontal infinitely long lines are shown. The projections of lines 1 and 3 are ellipses and become the sinusoids shown in Figure 4.22e. The projection

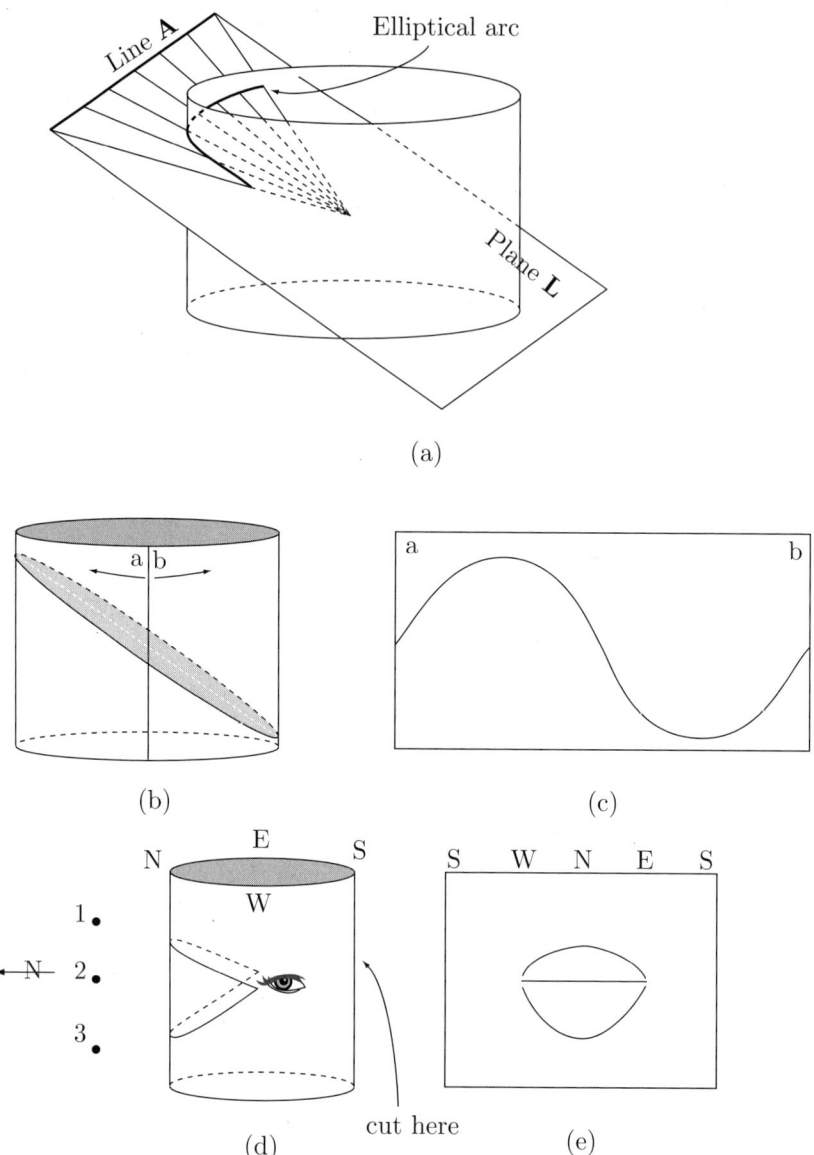

(a)

(b)   (c)

(d)   cut here   (e)

Figure 4.22: Projections of Straight Segments.

of line 2 is a half-circle (not shown) that becomes a straight line when the cylinder is unrolled. This shows how horizontal straight lines are projected by curved perspective into either horizontal segments or curves. The three segments are projected into the cylinder in the region bounded by the W and E directions. Two segments become curves (whose curvature depends on the height of the projected segment), and the central one remains straight. Vertical lines are always projected into vertical straight segments.

Figure 4.23 is an extension of Figure 4.22e. It illustrates the 360° cylindrical projection of horizontal straight segments in four directions. Part (a) of the figure shows four segments and their directions. Part (b) shows how each segment becomes a curve on the unrolled cylinder. Segment 1, to the north, is projected into a curve between W and E (several curves are shown, which are the projections of segments at various heights). Segment 4, to the south, is projected from E to W through S, so it is displayed in two halves. Segment 2, to the west, is projected from S through W to N, and segment 3 is projected from N through E to S. Some straight vertical segments are also shown. Such a grid corresponds to the continuous four-point perspective of Section 4.8.

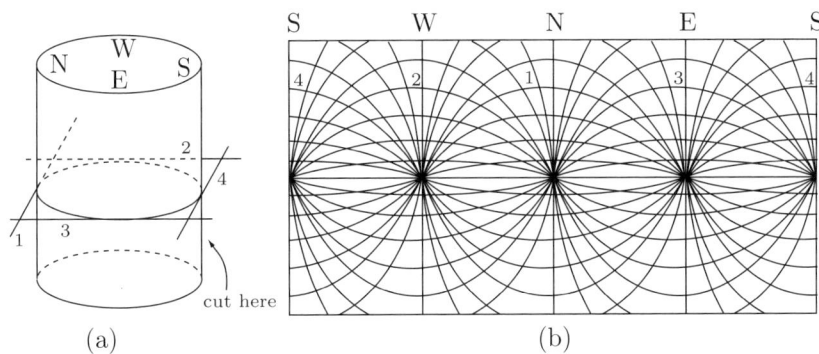

Figure 4.23: (a) Four segments. (b) Cylindrical Projections of Horizontal Segments.

Such a grid is handy when we want to compute or paint the cylindrical projection of a three-dimensional scene on a rectangular canvas. This can be done either manually or by special software. Any point in the space around the cylinder of Figure 4.23a is projected onto the surface of the cylinder by moving it to the surface along the segment that connects it to the center of the cylinder. Once a point is on the surface of the cylinder, it is easy to tell where it should go on the grid of Figure 4.23b.

> Art, like morality, consists in drawing the line somewhere.
> —G. K. Chesterton
> A great artist is always before his time or behind it.
> —George Moore

Figure 4.24 (courtesy of Dick Termes) is an example of such a drawing. It depicts a familiar scene, so there is no need to include the original three-dimensional image or any hints. The reader should especially note how the vertical lines are straight and how horizontal lines are curved mostly around the center of the drawing, as discussed in the answer to Exercise 4.10. This figure is also an example of the four-point continuous perspective discussed in Section 4.8.

Figure 4.24: Cylindrical Panoramic Projection (courtesy of Dick Termes).

Almost everything in Dick Termes' world is round—the sun breaking through morning haze, the tennis ball he batted back and forth before breakfast, and the four geodesic domes in which he lives and works.

For more than 36 years, Termes has eschewed traditional flat canvases to create his art on polycarbonate globes he calls "Termespheres." He came up with the idea while completing his master's degree at the University of Wyoming in the late 1960s, and it has been his passion ever since. Termes estimates he has painted more than 300 major spheres so far—about a third of those by commission—and his work is displayed internationally from North Pole High School in Alaska to the Sphere Museum in Tokyo, Japan.

"In art, the most important thing to find is an original thing to do," he says. "There have been lots of paintings done over thousands of years, most on flat surfaces. The sphere adds a whole new set of geometries that fits with the real world better than a flat surface. Three-dimensional space is what we live in."

—David Eisenhauer, *University of Wyoming Magazine*

Figure 4.25a (courtesy of Ari Salomon [helloari 05]) shows three examples of cylindrical panoramas. Each was made by taking several overlapping photographs and stitching them with appropriate software. Part (a), a bathroom in Paris, France, is vertical. It was made by taking pictures with a 20% overlap and tilting the camera to point higher and higher between images. It is obvious that the vertical lines are curved while the horizontal lines remain straight (but not completely parallel since the camera was held by hand during the shots). Part (b) is a street scene in Tel-Aviv, Israel. After watching this image for a few seconds and trying to "digest" it, it becomes clear that we are looking at three parallel streets (even though they seem to diverge). On the right-hand side, we see cars going toward the center of the image (away from our viewpoint). On

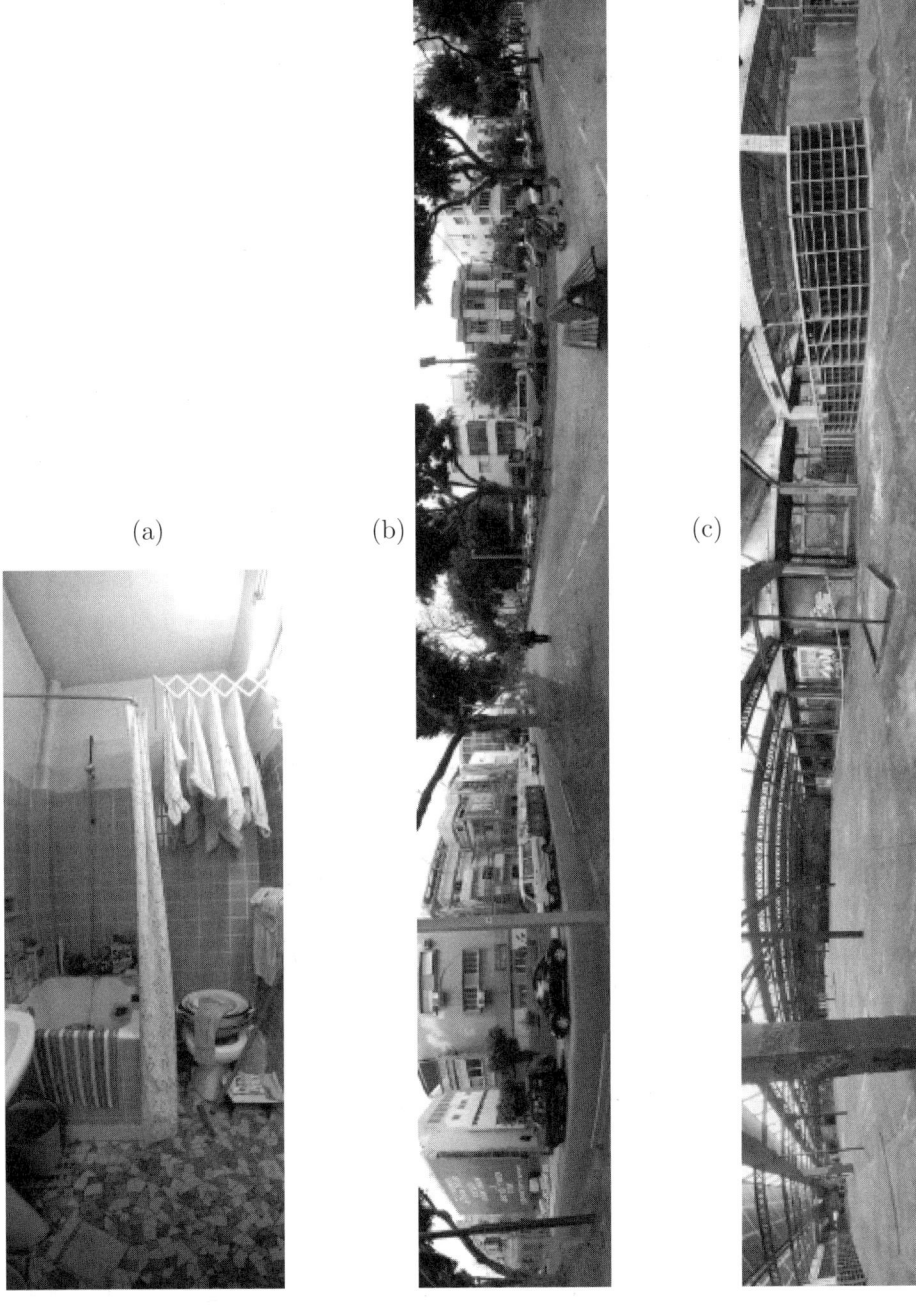

Figure 4.25: (a) Vertical and (b,c) Horizontal Cylindrical Projections (courtesy of Ari Salomon).

the left, cars are parked pointing toward us. (One such car can be seen at the extreme *right* of the image.) These are the two directions of the same street. The center street, where we see a park bench, stroller, and people walking, is a paved walkway sandwiched between the two directions of the street. The implicit assumption behind this image is that viewers' familiarity with street scenes will help them to "straighten out" the distortions in the image and thus to enjoy it. The reader should also notice that vertical lines in this image seem to tilt toward the edges of the image, and this tilting becomes more pronounced for lines close to the edges. This is probably an artifact of the particular software used to create these images. Part (c) of this figure shows a large space serving as artists' studios in Lyon, France. Here we see the four sets of curved horizontal lines that are the hallmark of Figure 4.23b. The vertical lines are also tilted as in part (b).

An intuitive way to understand and accept curved perspective is to print the curved projection of a familiar scene on paper, roll the sheet of paper into a cylinder, go inside into the center, and look around at the scene. (This may be simple if the projection incorporates less than 360°.) When seen this way, any curves on the paper that are the projections of straight lines should look straight. This method also provides a simple test of any software used to compute and render the projection.

Commercial software for creating cylinder-shaped panoramas already exists. Popular examples are the Apple QuickTime VR *Authoring Studio*, *PhotoVista* from Live Picture Inc., and *PhotoStitch*, which comes with every Canon digital camera. A qualitative discussion of curved perspective can be found in [Ernst 76], pp. 102–103. The well-known drawing *High and Low* by M. C. Escher is an example of curved perspective.

4.6 Spherical Panoramic Projection

The following quotation, from [Ernst 76], suggests a way to generalize the cylindrical panoramic projection of the previous section.

> Perhaps it has already struck you that the cylinder perspective used by Escher, leading to curved lines in place of the straight lines prescribed by traditional perspective, could be developed even further. Why not a spherical picture around the eye of the viewer instead of a cylindrical one? A fish-eye objective produces scenes as they would appear on a spherical picture. Escher certainly did give some thought to this, but he did not put the idea into practice, and therefore we will not pursue this further.

The idea raised by Ernst (but not pursued by Escher) is to imagine a transparent sphere placed around the observer, where everything seen by the observer through the sphere is fused (or painted by the observer) onto the sphere's material. The sphere is then somehow flattened, resulting in a full 360° spherical perspective. The trouble with this idea is that a sphere cannot be unrolled into a flat surface without introducing further distortions (see Section 4.14).

We start with what is perhaps the simplest approach to the problem of deforming and flattening a sphere. Once a three-dimensional point $\mathbf{P}$ has been projected onto the surface of the sphere, it becomes a point $\mathbf{P}^*$ with longitude and latitude. We construct a rectangle of width 360 and height 180 units and project $\mathbf{P}^*$ on the rectangle by simply using its longitude and latitude as the x and y coordinates, respectively, on the rectangle. Figure 4.26 illustrates the Earth in this projection, and the deformation is immediately obvious. On the rectangle, the lines of latitude are the same length, so polar latitudes, which on the sphere are short, have to be stretched.

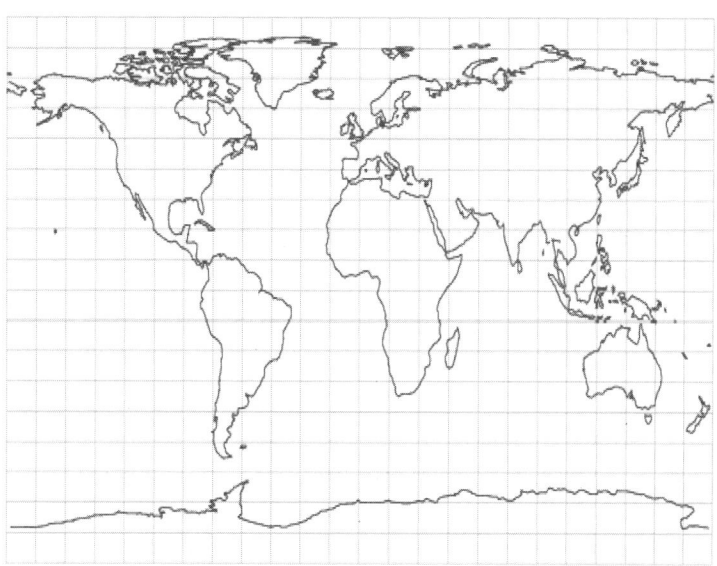

Figure 4.26: Equidirectional Projection of a Sphere.

When the entire $360°$ space around an observer is projected onto the rectangle in this way, the regions directly above and below the observer (which often are less important) are stretched and feature much detail. The regions at the height of the observer (the equator), however, lack detail, but are to scale. This projection is sometimes used in map making and is referred to as equirectangular projection, rectangular projection, plane chart, or plate carre.

The remainder of this section describes another, highly distorted version of spherical panoramic projection. This version is another manifestation of the concept of curved perspective.

> What you see on these screens up here is a fantasy; a computer enhanced hallucination!
> —John Wood (as Stephen Falken) in *WarGames* (1983).

Imagine a transparent sphere of radius R centered on the origin, where an observer is located, looking through the sphere in the z direction. The sphere is now truncated by selecting a value θ in the range $[0, \pi/2]$ and removing the parts of the sphere above and below latitude θ. The remaining part is shaped like a barrel (Figure 4.27a). The

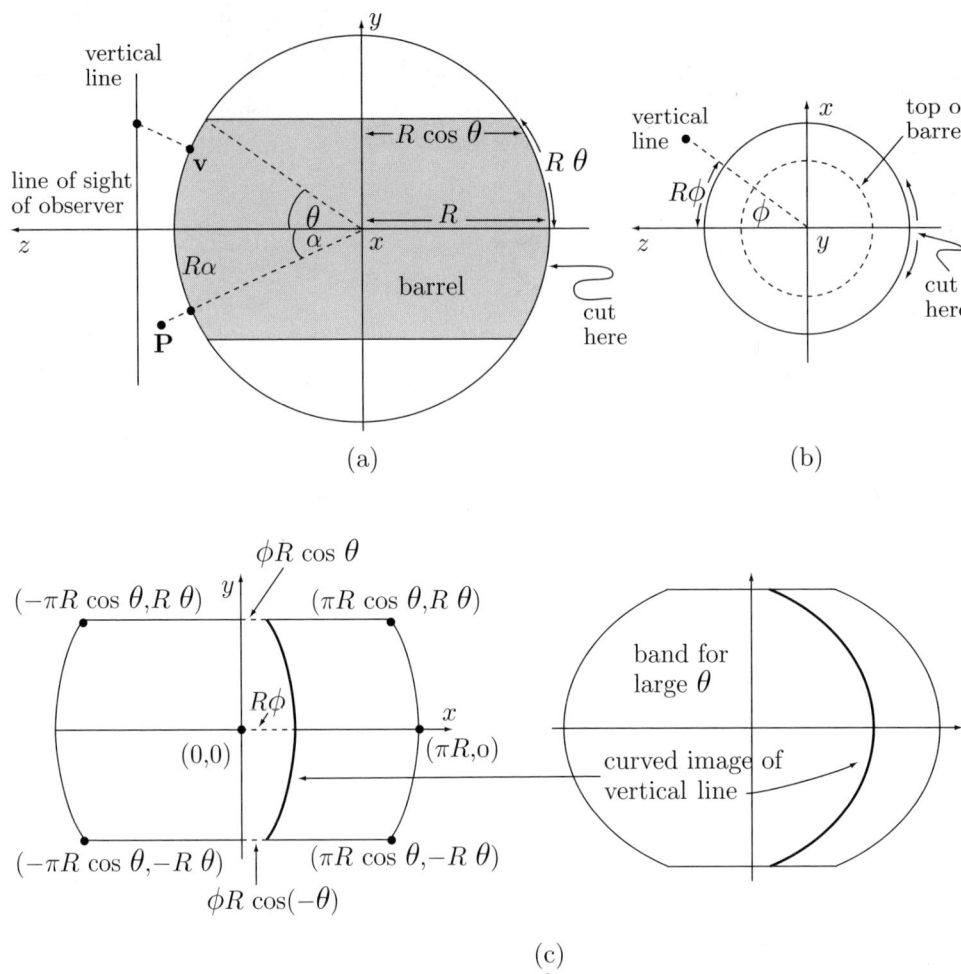

Figure 4.27: Spherical Panoramic Perspective.

barrel is now cut behind the observer and is unrolled into a flat, two-dimensional figure resembling a Band-Aid (Figure 4.27c) that's called a *band* or a *capsule* (see also Figure 4.58). The image seen by the observer through the barrel is displayed on this band, in contrast with the cylindrical panoramic projection, where the projected image is displayed on a rectangle.

At its center, the band has a width of $2\pi R$ (the circumference of the sphere), while at the top and bottom its width equals $2\pi R \cos \theta$. The height of the band is $2R\theta$. Truncating the sphere into a barrel makes it possible to control the amount of distortion in the final projected image. Small values of θ result in a narrow band whose shape is close to a rectangle. Only a small part of the scene around the observer is displayed on this band, but with a minimum of distortion. When θ is set close to $\pi/2$, the band becomes taller and its shape approaches a circle. It includes more of the scene (only those parts located directly above and below the observer are omitted) but with more distortions, especially at the top and bottom.

As in the cylindrical panoramic projection, horizontal lines are projected on the band as sinusoids, but we now show that even vertical lines, which in the cylindrical projection are projected straight, now become curved. Figure 4.27b shows the barrel from above (i.e., looking in the y direction). A long vertical line (parallel to the y axis) is shown, and we assume that a general point on this line is projected to a point $\mathbf{v}$ on the barrel. After the barrel is unrolled, the y coordinate of point $\mathbf{v}$ varies in the range $[-R\theta, +R\theta]$. The x coordinate depends on the y coordinate and equals the radius of the barrel at height y times the angle ϕ. The radius of the barrel at height y is easily seen to be $R\cos(y/R)$, so point $\mathbf{v}$ is located on the band at position $\big(\phi R\cos(y/R), y\big)$, where $-R\theta \leq y \leq +R\theta$. This position varies from $\big(\phi R\cos(-\theta), -R\theta\big)$ to $(\phi R, 0)$ to $\big(\phi R\cos(\theta), R\theta\big)$ when y varies from $-R\theta$ to 0 to $R\theta$. The projection of the vertical line on the band is therefore the thick curve shown in Figure 4.27c. It is easy to see that the closer θ is to $\pi/2$ (or $180°$), the smaller $\cos\theta$ is and the more curved (distorted) the projection.

Given an arbitrary point $\mathbf{P} = (x, y, z)$, it is relatively easy to calculate the xy coordinates of its projection on the band. Figure 4.27b shows the situation on the xz plane and makes it clear that the x coordinate of the projected point on the band is the arc $R\phi$. Since $\tan\phi = x/z$, we get the x coordinate as $R\arctan(x/z)$. Similarly, Figure 4.27a shows that the y coordinate of the projected point on the band is the arc $R\alpha$ or $R\arctan(y/z)$. Thus, the projected point has band coordinates $(R\phi, R\alpha)$ or $\big(R\arctan(x/z), R\arctan(y/z)\big)$. Both ϕ and α can vary in the interval $[-\pi, +\pi]$, so the projected x coordinate varies in $[-\pi R, +\pi R]$. The projected y coordinate varies in the same interval, but it is clear from the figure that any point $\mathbf{P}$ for which $|\alpha|$ is greater than $|\theta|$ is projected outside the barrel (i.e., on one of the sphere parts that have been removed) and should consequently be rejected.

The *IPIX Wizard* software [IPIX 05] can create a spherical panorama from two scanned fisheye photographs.

To some people, spherical panoramas may seem less interesting (and perhaps also less useful) than cylindrical panoramas, as the following 1998 quotation, from David Palermo, a virtual-reality professional, suggests: "Our market is not craving [sphere-shaped panoramas] right now. You can convey a sense of place without looking at the sky or floor."

> For me it remains an open question whether [this work] pertains to the realm of mathematics or to that of art.
>
> —M. C. Escher

4.6.1 Curvilinear Perspective

However, Figure 4.28 (courtesy of Dick Termes) suggests that it is possible to create full spherical panoramas that show everything an observer sees in front of him and behind him, while also maintaining their artistic value in spite of the many vertical and horizontal distortions. The reader should especially note that the few vertical and

Figure 4.28: Spherical Panoramic Perspective (courtesy of Dick Termes).

horizontal lines located close to the center of the picture (noticeable in the upper half) are essentially straight. The five-point grid of Figure 4.34 is an artist's tool that helps draw such pictures. Reference [New Perspective 98] has more on such tools.

This section explains the principles behind the five-point grid. The material presented here is based on the concept of *curvilinear perspective*, developed by Albert Flocon and André Barre [Flocon and Barre 68]. Curvilinear perspective is a two-step spherical panoramic projection whereby points in the 180° space in front of the observer are first projected on a hemisphere and then from the hemisphere onto a flat circle. When this is repeated for the 180° space behind the observer, the result is two circles that contain the entire 360° of space surrounding the observer.

> Their book beckons us to join with the fun and excitement, but it is also a revolutionary manifesto, a call to liberation from dogma. Not "Down with Traditional Perspective!" but "Down with the Tyranny of Official Rules." Not "Learn the Only True Perspective!" but "Let a Hundred Flowers Bloom!"
> —Robert Hansen in [Flocon and Barre 68]

Figure 4.29a illustrates the first step. A point **P** in space is projected to a point **P*** on a hemisphere. The observer is located at the center of the sphere. Part (b) of the figure shows how the hemisphere is projected onto a flat circle. The center of the circle is tangent to point **R** on the sphere (the point right in front of the observer). Given a point **Q** on the sphere, we draw the great-circle arc from **R** to **Q**. Denoting the length of this arc by L, point **Q** is projected to the point at distance L from the center of the circle in the direction from **R** to **Q**. This particular projection of a hemisphere to a circle was proposed in the 16th century by Guillaume Postel and has the useful property that its distortions of angles and distances are minimal. Clearly, the distance between **R** and **Q** on the hemisphere is preserved on the circle, whereas the distance between points **A** and **B** on the hemisphere of Figure 4.29c suffers a minimal distortion. For a 30° angle, the ratio between the arc length **AB** and its projection is only 1.01, and for a 90° angle this ratio is 1.57, much smaller than distance distortions caused by other sphere projections.

◇ **Exercise 4.11:** Show how to determine the distance between points **A** and **B** on the hemisphere of Figure 4.29c and on the circle of the same figure. Compute the ratio of these distances and show that it equals 1.01 for a 30° angle and 1.57 for a 90° angle.

Normally, the radius of the circle is $R(\pi/2)$ because this is the length of the longest radial arc on a hemisphere of radius R. However, it is possible to extend the Postel projection to project an arc of length r on the hemisphere to a segment of length $s\,r$ on the circle, where s is any desired scale factor. The radius of the circle in such a case is $s\,R(\pi/2)$.

When the two steps of curvilinear perspective are performed for a vertical line, it becomes a vertical curve on the circle (Figure 4.29d). This curve is very close to a circular arc and for all practical purposes can be approximated by such an arc. Similarly, a horizontal line in space is projected to a horizontal circular arc on the final circle. Lines that are parallel to the line of sight of the observer are projected on the circle to straight segments that converge at the center. Thus, the five-point grid of Figure 4.34 serves as

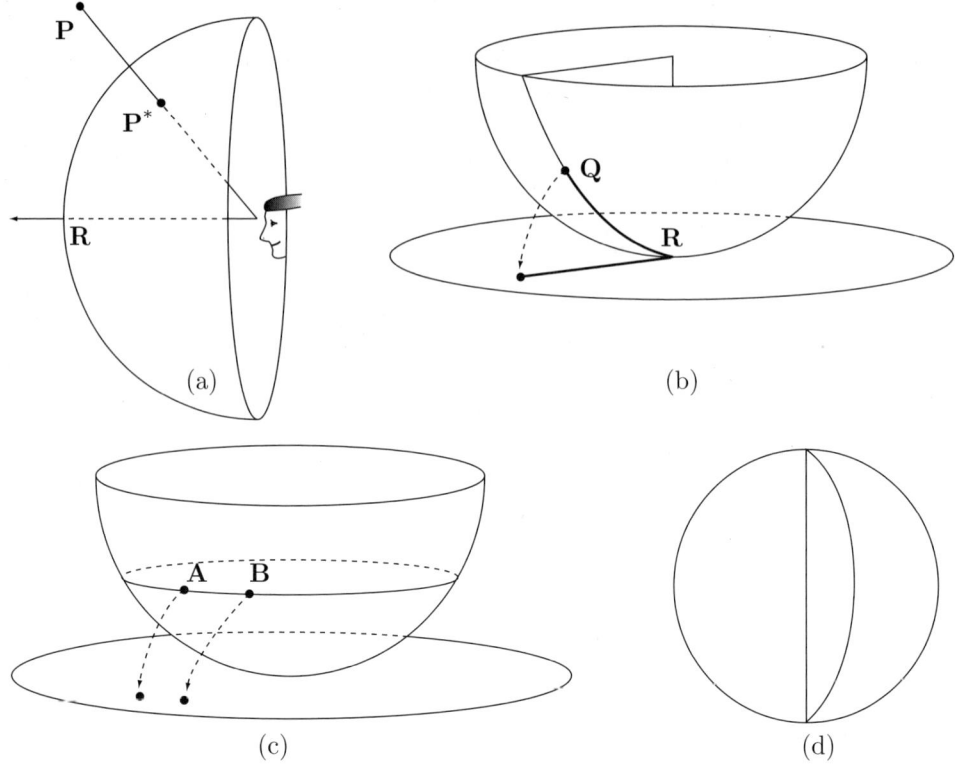

Figure 4.29: Principle of Curvilinear Perspective.

a useful artist's tool to draw the curvilinear perspective projection of any scene on a circle of radius $s\,R(\pi/2)$ in a single step.

4.7 Cubic Panoramic Projection

The principle of the cubic panoramic projection is similar to those of the other panoramic projections. We imagine an observer located at the center of a cube (Figure 4.30a) and looking at the three-dimensional scene outside. Everything the observer sees is etched on the sides of the cube (or is painted there by the observer), and the cube is then flattened into six squares connected as in Figure 4.30b,c. This creates a full 360° panorama in six parts.

The main advantage of the cubic panoramic projection is the absence of distortion. Straight lines are projected into straight lines, and the only deviation from total linearity is discontinuous slopes at the boundaries between the six planes of the cube. This behavior is best illustrated by Figure 4.32 (courtesy of Shinji Araya) but is also demonstrated here rigorously by means of an example.

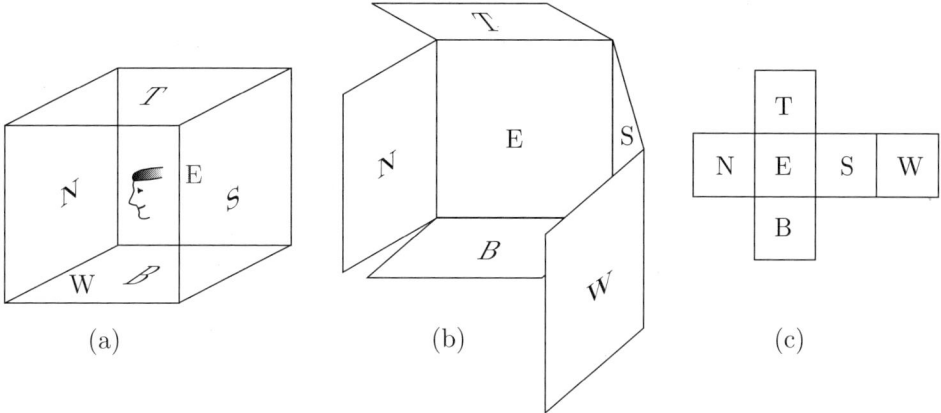

Figure 4.30: Cubic Panoramic Projection.

Figure 4.31a shows two faces (we'll call them panels) of a cube viewed from the positive z direction. Each face of the cube is $2k$ units long, and we see the two panels located at $x = k$ and $y = k$. Figure 4.31b shows the two panels after they have been rotated to stand side by side, and we look at their outside surfaces. To best visualize this, imagine that there are hinges between the two panels, so they look like a folding closet door (notice the direction of the x axis). The figure indicates that the $x = k$ panel is parallel to the yz plane, which is why all points on it have coordinates of the form (k, y, z), while the $y = k$ panel is parallel to the xz plane and all its points are of the form (x, k, z).

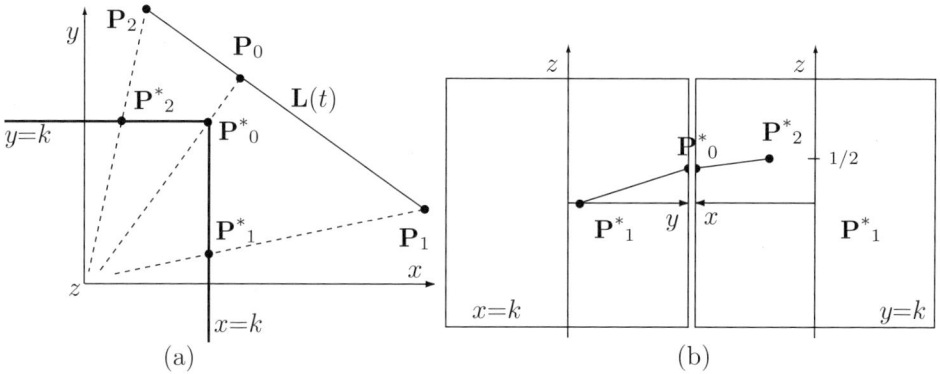

Figure 4.31: Cubic Projection of a Straight Segment.

We arbitrarily select the two points $\mathbf{P}_1 = (4k, k/2, 0)$ and $\mathbf{P}_2 = (k/2, 2k, 1)$. The former is projected to the $x = k$ panel, where points have coordinates (k, y, z), which is why it is projected to $\mathbf{P}_1^* = (k, k/8, 0)$. The latter is projected to the $y = k$ panel, where its y coordinate must be k, so it is projected to $\mathbf{P}_2^* = (k/4, k, 1/2)$. We denote by $\mathbf{L}(t)$ the straight segment connecting $\mathbf{P}_1$ to $\mathbf{P}_2$ and compute it (from Equation (Ans.7)) as the weighted sum $\mathbf{L}(t) = (1-t)\mathbf{P}_1 + t\,\mathbf{P}_2 = (4k - 7tk/2, k/2 + 3tk/2, t)$. Next, we determine the coordinates of point $\mathbf{P}_0$ on this segment. This point will be projected to the cube corner where $x = y = k$, so its x and y coordinates must be equal even before it is projected. Since $\mathbf{P}_0$ is on segment $\mathbf{L}(t)$, it must equal $\mathbf{L}(t_0)$ for some t_0. Thus, we can compute t_0 from the relation $4k - 7t_0 k/2 = k/2 + 3t_0 k/2$, which yields $t_0 = 7/10$. The coordinates of $\mathbf{P}_0$ are therefore $\mathbf{L}(t_0) = (\frac{31}{20}k, \frac{31}{20}k, \frac{7}{10})$, and this is projected to $\mathbf{P}_0^* = (k, k, \frac{7}{10} \times \frac{20}{31}) = (k, k, 14/31)$.

Once the z coordinate of $\mathbf{P}_0^*$ is known, we can compute the slopes of the two segments that constitute the projection of $\mathbf{L}(t)$. On the $y = k$ panel, the slope is

$$\frac{\frac{1}{2} - \frac{14}{31}}{\frac{3k}{4}} = \frac{2}{31k},$$

whereas on the $x = k$ panel it is

$$\frac{\frac{14}{31} - 0}{\frac{7k}{8}} = \frac{16}{31k}.$$

The straight segment connecting $\mathbf{P}_1$ to $\mathbf{P}_2$ has been projected into two segments that are straight but travel with different slopes on the two panels. Because of the symmetry of a cube, there is no difference between horizontal and vertical lines and they all feature the same discontinuity of slope between panels.

$\diamond$ **Exercise 4.12:** In what cases will the slopes be continuous across a panel boundary?

It is clear that a panorama made of six squares doesn't create a satisfying visual sensation, and Figure 4.32 (courtesy of Shinji Araya) proves this claim. The figure shows a beautiful scene, but the projection seems fractionated and unnatural. This lack of artistic value is why the cubic panoramic projection was not seen much in the past. Currently, however, cubic panoramas are very popular because version 5 of the popular *QuickTime* software for the Macintosh computer can create this type of panoramic projection and can also scroll it on the monitor screen such that the viewer can eventually examine a field of view that encompasses 180° vertically and a full 360° horizontally. The main advantage of this scrolling is that it eliminates the discontinuities of the slopes between panels. The image seems to flow smoothly on the screen without any jumps or distortions. Such a panorama cannot be included in a book, but many can be found on the Internet by searching under "cubic panorama."

MakeCubic is a simple OSX-ready app for creating cubic QTVR movies from six faces or from equirectangular (a kind of sphere-to-rectangle projection which is used in some java-based players and other places) images.

—From `http://developer.apple.com/quicktime/quicktimeintro/tools/`

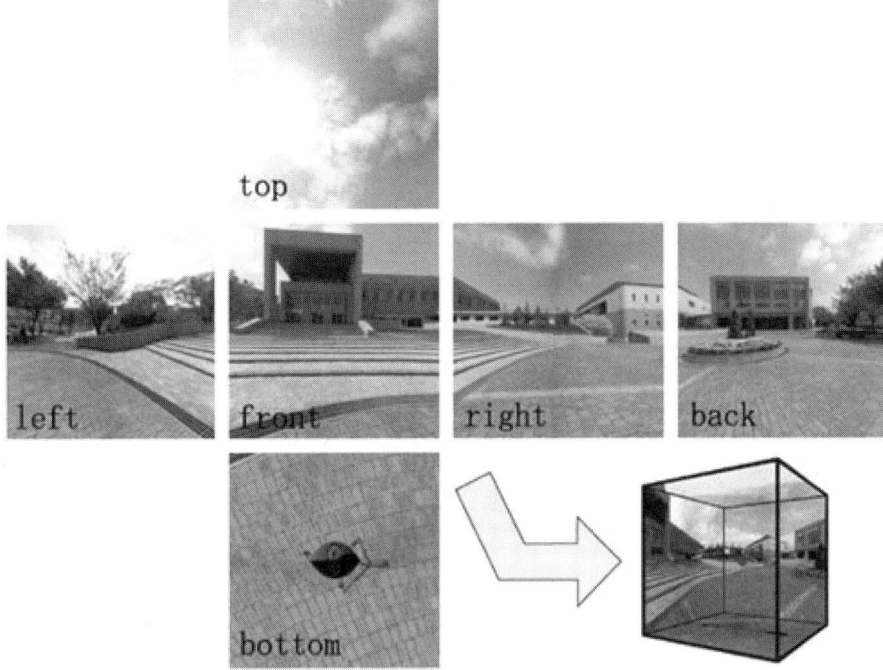

Figure 4.32: Cubic Panoramic Perspective (color version on page 239).

(Courtesy of Professor Shinji Araya, Fukuoka Institute of Technology.)

4.8 Six-Point Perspective

Chapter 3 introduces the concept of n-point perspective, where n can be 1, 2, or 3. This section extends the term "n-point" and discusses n values up to 6. The discussion is based on the work of and terms coined by Dick Termes, who also created the images, art, and grids in this section.

Figure 3.14 shows Alberti's method of traversals in one-point perspective. The important feature of this figure for our present discussion is the converging grid. Certain lines in this grid converge to a vanishing point and thereby turn the grid into an aid to the artist. Such a one-point grid becomes a tool that helps to draw any image in one-point perspective. Section 3.3 discusses perspective in curved objects and employs a similar grid (Figure Ans.6).

Figure 4.33 shows grids for 1, 2, and 3 vanishing points and artistic drawings based on them. It is natural to accept these drawings. They look familiar and don't seem distorted or unusual (although the viewpoint in some of them may be unusual). They are drawn in linear perspective.

In contrast, drawings based on similar grids with more than three vanishing points are distorted. They belong in the realm of nonlinear projections. Figure 4.34 shows grids for four and five vanishing points, and it is immediately clear that they must

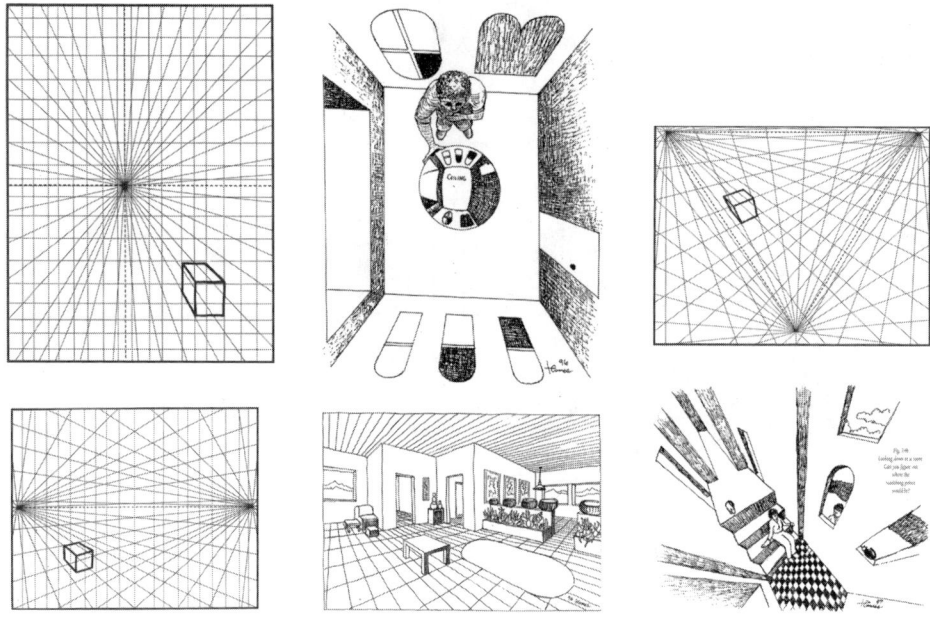

Figure 4.33: Grids and Art for 1, 2, and 3 Vanishing Points (courtesy of Dick Termes).

introduce distortions in any artwork based on them. The former grid shows straight lines bending and converging to four points. The vanishing points on the left and right sides are familiar. They result in the familiar two-point perspective. The extra two vanishing points, at the top and bottom of the grid, force all the vertical lines to bend and introduce distortions in this way. The result is an image (see example to the right of the grid) that becomes more distorted as the eye moves up or down away from the center of the image. This type of distortion has its own artistic value but it is not immediately clear to which of the projections discussed in this chapter it corresponds. A closer look at the four-point grid of Figure 4.34, however, shows its resemblance to Figure 4.23b, which corresponds to the cylindrical panoramic projection. Thus, a complete 360° cylindrical projection, such as the one depicted by Figure 4.25c, can be obtained by placing four four-point grids side by side. This type of grid is referred to by Dick Termes as a continuous four-point perspective.

Initially, the five-point grid of Figure 4.34 looks unfamiliar and strange. It is not trivial to guess the type of distortion that results from bending lines in five different directions, toward the four extreme points on the periphery as well as toward the center. However, a glance at Figure 4.10 should convince the reader that the effect of five-point perspective is similar (perhaps even identical) to the angular fisheye projection (page 151) as well as the spherical panoramic projection of Section 4.6. All the horizontal and vertical lines, except those passing through the middle of the figure, are curved. This drawing shows only half a sphere (180° vertically and horizontally), but it points the way toward depicting a complete sphere on a flat surface. Simply place two five-point

Figure 4.34: Grids and Art for 4 and 5 Vanishing Points (courtesy of Dick Termes).

perspective images side by side or one above the other. The result, which Dick Termes terms six-point perspective (no pun intended), is shown in Figure 4.28. Section 4.6.1 discusses an approach to the construction of the five-point grid that is based on the Postel sphere projection.

Some viewers are impatient with attempts to create panoramic projections on flat surfaces. Such people may like the solution adopted by Dick Termes, namely to actually sit inside a sphere and paint a spherical panoramic projection on its surface. The result, which is naturally termed a *Termesphere* [termespheres 05], is a unique kind of art, but cannot be included in a book. (See Figure C.5 for a rough idea.) A side benefit of this technique is that the finished sphere can easily be converted to two flat disks in six-point perspective [Keith 01]. The original sphere is made of two thin polyethylene hemispheres. Once they are painted with acrylic paint, each hemisphere is heated until the polyethylene melts to become a plastic disk. The painting on the two disks is now in six-point perspective. An added advantage of this process is that such disks can be copied to make more disks that can, in turn, be blown into hemispheres by the same heating process.

> My dad's work is like taking your eyeball out of your head, putting it in a building, and when it spins you can see everything from that one point in space.
>
> —Lang Termes (as a child)

4.9 Other Panoramic Projections

The cylindrical and cubic projections of Sections 4.5 and 4.7 have a common feature that makes them attractive. The cylinder and the cube can be unrolled or opened into a flat surface without additional distortions. Other geometric shapes have the same feature, and this section mentions the most important of them, namely the five Platonic solids (Figure 4.35) and the cone.

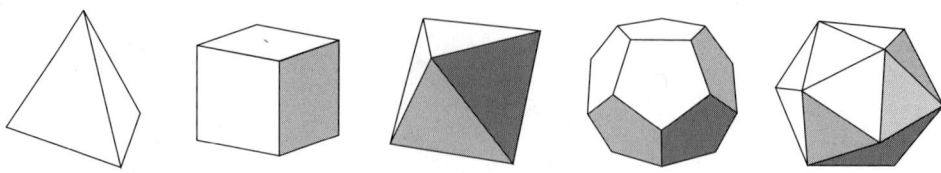

Figure 4.35: The Five Platonic Solids.

A polyhedron whose faces are congruent convex regular polygons is known as a Platonic solid. These figures were known in antiquity, and Euclid has already proved that there are only five of them, the tetrahedron (a pyramid of four triangles), the cube, the octahedron (eight faces, each a triangle), the dodecahedron (12 faces, each a pentagon), and the icosahedron (20 triangles). Many properties and pictures of these solids can be found in [Steinhaus 83].

One of the most original works of art depicting Platonic solids is the wood engraving *Stars* by M. C. Escher. It takes a while to disentangle the many details in this picture and locate the intersecting octahedra, tetrahedra, cubes, and other figures. The only items that stand out immediately are the chameleons, placed by the artist inside the polyhedra to attract nonmathematically-oriented viewers and capture their attention.

The principle of projection is always the same. We imagine an observer located somewhere inside the surface, at the center or at some other preferred point, looking at the three-dimensional scene outside and painting it on the surface. The surface is then opened or unrolled to become a flat panoramic projection. In practice, only the cylinder and the cube are commonly used for panoramic projections. It is rare to find a pyramidal or a conic panoramic projection because opening and flattening such surfaces results in a two-dimensional picture that looks foreign and unfamiliar and is often difficult to visualize, perceive, and enjoy, even though it does not create any distortions.

Figure 4.36 (courtesy of Dick Termes) is a typical example. It shows a panorama of the interior of St. Peter's Basilica in Rome projected on a dodecahedron. It is immediately obvious that in spite of the high precision of the drawing and the many details that are easy to observe, it is difficult, perhaps even impossible, to place the 12 individual pentagons of the projection in the viewer's mind and grasp them as a single coherent work of art. Such a projection is best viewed after it is cut out, folded, and glued together to actually form a dodecahedron (notice the matching tabs designed to help in this process). The details of this process and how such pictures are taken are described on page 191.

Conic Panoramic Projection

Given a cone of height H and radius R, we imagine an observer located at the center of the base of the cone. Such an observer sees the hemisphere of space above him and projects it on the cone, which is later cut and laid flat. It is also possible to place the observer at the center of the cone, where he can see the entire 360° of space around him, but this results in even more distortion because part of the lower hemisphere is seen by the observer through the lower sides of the cone, while the rest of this hemisphere is seen through the flat bottom.

Figure 4.36: A Panoramic Projection on a Dodecahedron (courtesy of Dick Termes).

Reference [lampshade 05] shows how to apply the conic panoramic projection to create original lampshades.

The derivation presented here starts with a point $\mathbf{P} = (x, y, z)$ that is projected onto the surface of the cone. Once the surface is opened, the coordinates of the projected point $\mathbf{P}^*$ are given (Figure 4.37) in terms of the angle θ it makes with the top of the cone, and the distance r from the top. These are polar coordinates on the open cone.

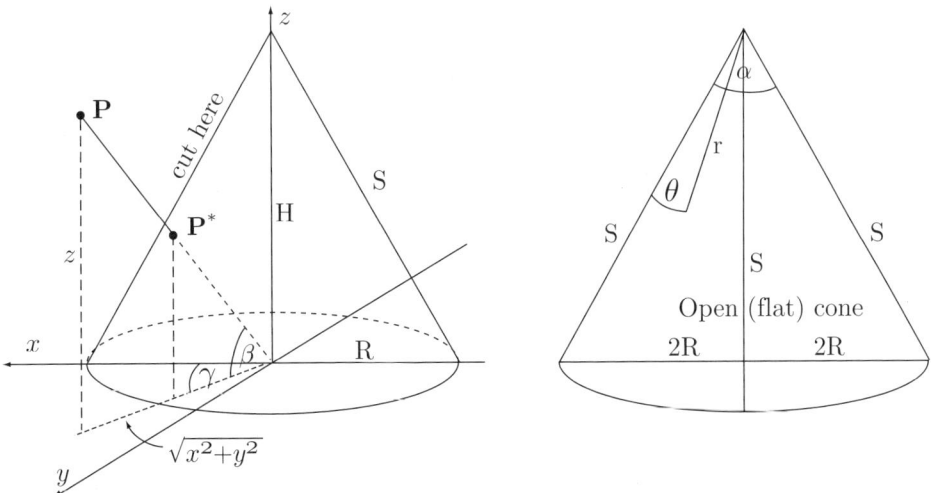

Figure 4.37: Conic Panoramic Projection.

The height S of the open cone is given by $S = \sqrt{H^2 + R^2}$. The vertical angle β between the xy plane and the direction of $\mathbf{P}$ is given by $\tan\beta = z/\sqrt{x^2 + y^2}$. (Notice that β varies from 0 to 90°.) Once β is known, the polar coordinate r is determined by $r = S(1 - \sin\beta)$. It varies between 0 and S. The top angle α of the open cone is computed from $2R/S = \cos(\alpha/2)$, and the polar coordinate θ lies between 0 and α, so it is given by $\theta = \alpha\gamma/(2\pi)$, where γ is determined by the x and y coordinates of $\mathbf{P}$ by means of $\tan\gamma = |y/x|$ (90, 180, or 270 degrees may have to be added to γ depending on the quadrant, see Figure 4.21).

4.10 Panoramic Cameras

A typical dictionary definition of *panorama* is "a picture taken in three-dimensional space and presented on a continuous surface encircling the viewer."

There are a large variety of lenses available for current cameras (both digital and film based), ranging from extreme wide angle to powerful telephoto, but even the widest wide-angle lenses cannot capture an image that spans more than 180°. Most fisheye lenses can capture 180° images, but the result is highly distorted, especially along the edges. Professional as well as amateur photographers like to be able to stand at a given point and capture an image of everything visible from all directions, which explains why panoramic cameras are popular. Inexpensive high-resolution digital cameras have become powerful and popular, and this has encouraged the development of panoramic software. Given a digital camera and a tripod, it is easy to take a series of overlapping photos, input them directly from the camera into the computer, and stitch them by software into a panorama (normally cylindrical). In spite of this, special panoramic cameras, both digital and film-based, the latter of which have been made since the 1840s, are still being made and used.

An important resource for information on all aspects of panoramic cameras is the International Association of Panoramic Photographers [IAPP 05], whose mission is "to educate, promote, exchange artistic and technical ideas, and to expand public awareness regarding panoramic photography." Two important resources maintained by this organization are a list, located at [cameraInproduction 05], of panoramic cameras in production and a timeline of panoramic cameras, located at [cameraTimeline 05].

A fun guide for do-it-yourselfers is [funsci 05]. Information on panoramic cameras and creating panoramic images can be found on many Internet sites. See, for example, [shortcourses 05] and [philohome 05].

A new reference book for this topic is [Jacobs 05].

There are currently three types of cameras that capture panoramic images: a rotating camera, a swing-lens camera, and a camera with a parabolic panoramic lens system. The first two can produce undistorted images, while the third type produces a highly distorted image that has to be "unfolded" by special software to look like other types (normally cylindrical or cubic) of panoramas. A description of all three (followed by a note on pinhole cameras) follows.

A rotating camera, as its name implies, works by rotating on its base, transferring the image to the film while moving the film in the opposite direction, so the film stays stationary relative to the ground. Examples of this type are the Swiss-made RoundShot [roundshot 05], some of whose models are digital, the Globuscope [Globuscope 05], and the Hulcherama camera, invented and built by Charles A. Hulcher [hulchercamera 05]. Following is some information on the latter type.

The Hulcherama is a slit-scanning panoramic camera that works by rotating on its base. An electronically controlled motor is responsible for uniform rotation. (The rate of rotation may be varied from 1 s to 144 s per revolution.) During the rotation, the image passes through the lens and then through an adjustable narrow slit onto the film (Figure 4.38a). The slit masks out most of the image but lets a narrow portion pass through, which is how any optical distortion is minimized. As the camera rotates in one direction, the film moves past the slit in the opposite direction. The camera rotation and film movement are synchronized so that the film is stationary relative to the image being photographed. As the camera makes a complete revolution, 8.9 inches of film pass behind the slit, creating a 360° panoramic image with a height of 2.25 inches. The aspect ratio is therefore a pleasing $2.25 : 8.9 \approx 1 : 4$. It is possible to let the camera rotate more than one revolution (possibly varying the image each time), and a roll of 120-format film is long enough for three revolutions (the Hulcherama uses standard 120 or 220 roll film).

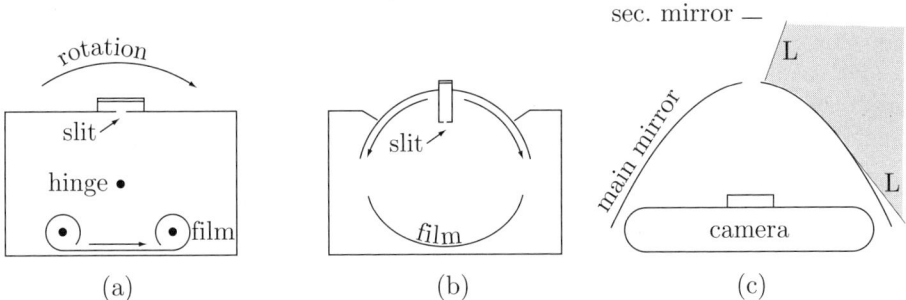

Figure 4.38: Panoramic Cameras.

A swing-lens camera (Figure 4.38b) has a lens that rotates during an exposure, thereby "painting" the image on the film through a narrow, vertical, constant-width slit. In order to keep the same distance between the film and the lens, the film has to be curved. An advantage of this type of camera is that the lens only has to cover the vertical dimension of the film and the width of the slit, so it does not have to be complex. The downside of this type is the limited field of view, which is less than 180°. A complete 360° panorama is created by taking several shots and combining them using special equipment (for a film camera) or special software (for a digital camera). Examples of this type are the Widelux (now discontinued) and the Noblex.

The Noblex [Noblex 05] is a family of cameras that consists of models 135, 150, and 175. Model 135 takes a 136°-wide image and uses standard 35-mm film. The Noblex-150 provides a 146° angle of view, uses 120 film, and produces six 5-inch-wide images on a roll. It can take multiple exposures on the same film.

A panoramic lens system (Figure 4.38c) is somewhat similar to a reflecting telescope. Its main part is a *convex* parabolic mirror (in contrast to the mirrors used in telescopes, which are concave) that captures the entire (or almost the entire) half-sphere of image above it and sends it up, where it is reflected by a small, flat mirror and sent down through a hole in the main mirror to a camera. There are no moving parts, no rotating parts, no need for multiple images, and no need to stitch multiple photos together. The price for all this (aside from the price of the camera and mirror) is image distortion. This lens can, in principle, be used with any camera, digital or film.

Since the mirror captures everything above it and on all sides, the only way for the photographer to stay out of the picture is to crawl under the camera. A panoramic lens system is therefore used while mounted on a tripod or a pole and operated from below.

An example of this type is the Portal S1 panoramic lens system made by the BeHere company [BeHere 05]. It is 12.5 inches in diameter, 13 inches tall, and weighs less than 10 pounds. It has a 35-mm Nikon mount, so any Nikon-compatible camera body, digital or film, can be used with the Portal S1. The depth of field of the Portal is from one inch to infinity. (There is no need to focus the camera.) Its lateral field of view is, of course, 360°, but its vertical field of view is limited to the gray area in the figure and equals 100° (the angle between the two lines marked L). When anything outside this area is reflected in the main mirror, it cannot reach the secondary mirror.

If a film camera is used, the film can later be scanned and then processed with special software provided by the manufacturer. This software flattens the donut-shaped image and can also perform other processing such as evening out the lighting, correcting brightness and contrast, and slightly sharpening the edges. The image can then be saved in one of the popular panoramic formats such as QuickTime VR.

⋄ **Exercise 4.13:** Explain why the image produced by a panoramic lens system is shaped like a donut.

The OmniAlert panoramic video camera system from Remotereality [remotereality 05] also employs a parabolic mirror, but the mirror points down, toward the camera, which results in a circular picture with no hole. This camera has been developed for security and surveillance applications, where a wide field of view is important. The video camera is mounted on a high pole right under the parabolic mirror and uses special software to detect and track moving objects in its field of view and alert operators to any suspicious activities.

The *360 One VR* parabolic mirror system, from Kaidan [Kaidan 05], also uses a down-pointing mirror and can be attached to several different cameras. Special software must be used to convert the highly distorted image to a flat panorama (Flash VR, cylindrical, QuickTime VR cylindrical, spherical, cubic, or QuickTime VR cubic) that can be displayed and printed.

See also [eclipsechaser 05] for astronomical applications of this type of panoramic camera.

Note: The pinhole used to be the first camera of many a poor youngster. This is simply a box with a small hole in front and film or light-sensitive paper loaded in back. The shutter can be as simple as a piece of tape that's removed to expose the film, then reapplied manually, or it can be a purchased, cable-operated shutter assembly. If the hole is small enough, the resulting image is sharp; if the film is wide, this primitive device can produce wide-angle images.

The total photograph. We now turn to a completely different approach to the problem of creating a panorama with a camera. This approach, termed by its inventor *the total photograph*, was developed and patented by Dick Termes in 1980 and is described in [Termes 80]. To understand this technique, consider the cubic panoramic projections of Section 4.7. We imagine an observer located at the center of a cube (Figure 4.30a) and looking at the three-dimensional scene outside. Everything the observer sees is etched on the sides of the cube or is painted there by the observer. Given a three-dimensional scene and a camera, the problem is to generate such a cube. In general, we want a method where we can project a scene on the sides of any of the five regular polyhedra, as discussed in Section 4.9.

The first step is to decide what regular polyhedron we want. For example, we may want to create a panorama on the 20 triangular sides (or faces) of an icosahedron. We use suitable material, such as wood, plastic, or metal, to construct a solid icosahedron and mount it on a good-quality, stable camera tripod. (The tripod may have to be loaded with extra weight to make it extremely stable.) The icosahedron stays fixed while pictures are taken. We drill small holes (labeled #34 in Figure 4.39) in each of the 20 sides of the icosahedron to enable us to quickly attach a special bracket to any side. A camera is mounted on the bracket. (The camera has to have a wide field of view, so pictures taken from adjacent faces of the polyhedron do overlap). We then place the bracket with the camera in one of the 20 sides of the icosahedron and, while holding it stable in our hand, take a snapshot. This guarantees that the center of the camera lens is right over the center of the polygon face. It is also important to make sure that the camera's line of sight is perpendicular to the polygon face. We repeat this for the 19 remaining sides to end up with 20 pictures, each showing what a viewer located on that face of the icosahedron would see.

Figure 4.39 is taken from the patent application. The first five figures show the five Platonic solids, each with two holes on each face, for quick mounting of the bracket. Part 6 of the figure shows the bracket, part 7 shows a camera mounted on the bracket, and part 8 is an exploded view of an icosahedron mounted on a camera tripod and the bracket mounted on one side.

The only problem is that the camera is located outside the icosahedron, not inside. Thus, the camera sees more than an inside observer would see through each face. The 20 photographs therefore partially overlap and we need to identify the overlapping parts and remove them. The result should be 20 triangular pictures, each corresponding to what an observer inside the icosahedron would see through one face. These triangles can then be pasted together to form an actual icosahedron.

Figure 4.40 illustrates this process. Two partially overlapping pictures are placed such that the overlapping parts match precisely (part 9). The centers of the pictures are then identified and connected by a straight segment (#56 in part 10), and another

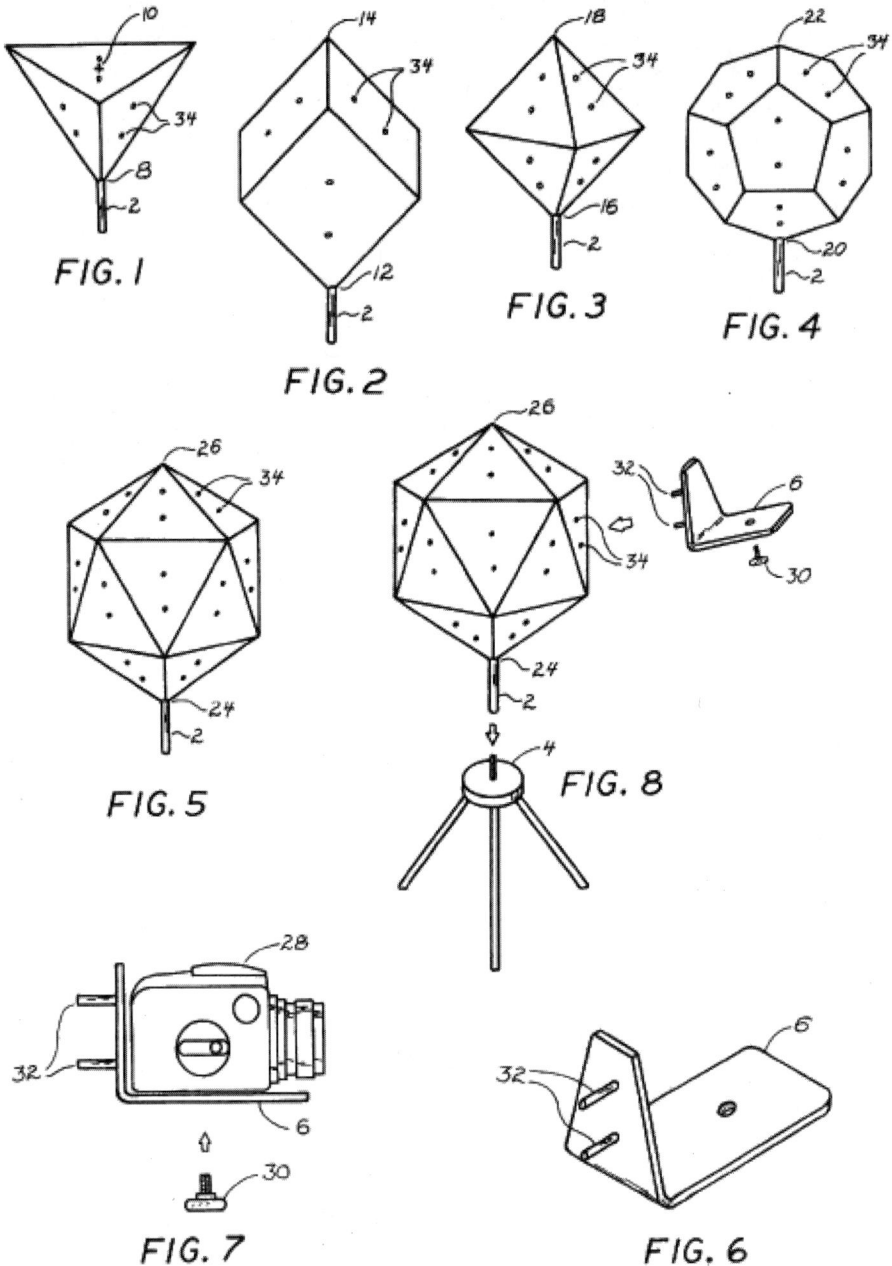

Figure 4.39: Details of Invention (courtesy of Dick Termes).

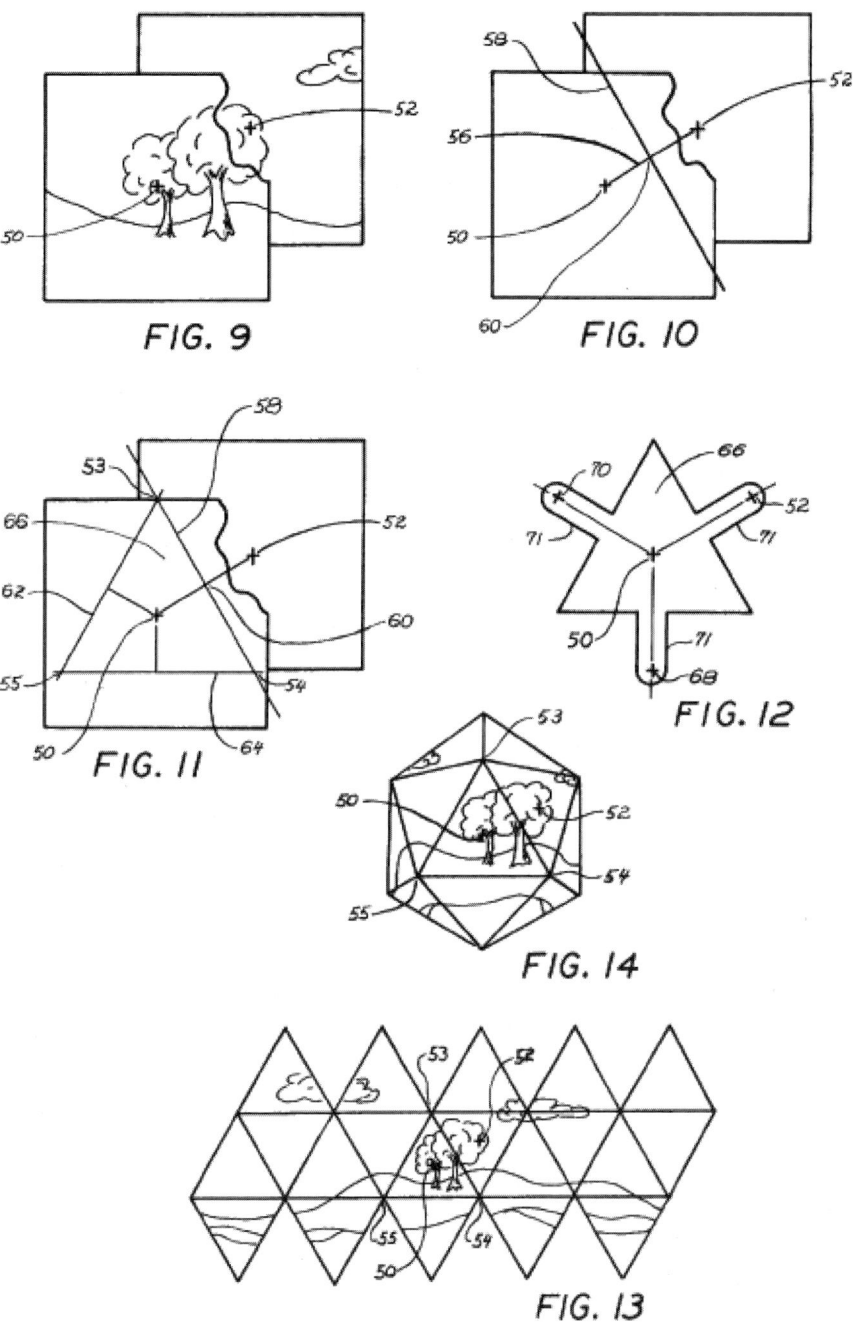

Figure 4.40: Details of Invention (courtesy of Dick Termes).

segment (#58) is drawn, perpendicularly bisecting the first one, as shown in part 10. Once this is done, it is easy to construct the two segments 62 and 64 of part 11 and end up with an equilateral triangle on the picture. The picture is then trimmed as in part 12, with small tabs that are later used to paste this picture to several (up to three) other ones. Part 13 shows how the 20 triangles resulting from this process are mounted in one horizontal strip that can later be converted to an actual icosahedron (part 14).

Each face of a dodecahedron is a pentagon, and each side of a cube is a square, but the details of removing overlapping parts and trimming each picture in these cases are similar to the triangular case. Figure 4.36 is an example of a panorama constructed on a dodecahedron.

From around 1930 on, therefore, the standard photographic image on 35 mm film was 15.6 mm high by 20.8 mm wide, a proportion of roughly four by three. The same proportion of height to width (the aspect ratio) is obtained on the screen when such a frame is projected, and this shape of image (ratio 1:1.33) came to be called the "Academy ratio." But substantial variations are possible even on conventional 35 mm film. Masks or caches can cut the height of each frame, and thus increase the aspect ratio of the projected image: alternatively, special lenses can be used which "squeeze" a wider image on to the film (through a procedure called *anamorphosis*, often used by Renaissance painters) and "unsqueeze" it again when the film is projected. A French optical scientist called Chrétien invented the anamorphic lens and its application to the cinema in the 1920s; Autant-Lara experimented with it in a film version of Jack London's *To Make A Fire*, but the Hypergonar, as Chrétien called his invention, failed to catch on, and development work on it stopped.

—David Bellos, *Jacques Tati* (1999)

4.11 Telescopic Projection

Seen through a microscope, small objects look bigger than they are. The telescope, however, does not enlarge objects; it brings them closer. Objects close to the telescope are brought a little closer, while objects located far away are moved much closer. This short section discusses the mathematics of the telescopic projections, but it should be emphasized that this is not a projection from three dimensions to two dimensions, but rather a three-dimensional transformation. (This is also true for the microscopic projection.) Nevertheless, these topics are discussed here because of their nonlinearity.

The diameter of the moon is 3,476 kilometers (2,160 miles). When we see the moon through a telescope, its diameter seems only a few centimeters or a few meters, much smaller than the real diameter. This shows that the telescope does not increase the size of the object being viewed. Instead, it decreases the apparent distance of the object.

Figure 3.4 is a perspective projection of a long row of telephone poles. The poles, which are the same height and are equally spaced, seem to get smaller and closer together as they get farther from the viewer. This is a common effect of linear perspective.

Looking at the same poles through a telescope brings them closer and makes them look bigger, but not by the same amount. Poles closer to the telescope move just a little closer to the viewer, while poles far away move much closer and also get bigger (although still smaller than nearby poles).

In order to compute such a projection mathematically, we need an expression that will take a quantity z (the distance of a telephone pole) and will shrink it nonlinearly to z^* such that $z = 0$ (a telephone pole at the viewer's position) will result in $z^* = 0$ (no movement) and large values of z will yield z^* values in the interval $[0, k]$ and approaching k slowly. One choice for such an expression is

$$z^* = kz/(z + k), \tag{4.4}$$

where k is a parameter selected by the user. This expression is similar to the *thin lens equation* from optics and also Equation (3.1). The *Mathematica* code

```
k=10.;
Table[k z/(z+k), {z,0,100,5}]
Table[%[[i+1]]-%[[i]], {i,1,20}]
Table[Point[{%%[[i]],0}], {i,1,21}];
Show[Graphics[%]]
```

selects $k = 10$ and 21 z values from 0 to 100 in steps of 5. It produces the 21 numbers 0, 3.33, 5, 6, 6.67, 7.14, 7.5, 7.78, 8, 8.18, 8.33, 8.46, 8.57, 8.67, 8.75, 8.82, 8.89, 8.95, 9, 9.05, and 9.09. They start at zero and approach k (Figure 4.41a). The third line computes the 20 differences between consecutive numbers. It produces 3.33, 1.67, 1, 0.67, 0.47, 0.36, 0.28, 0.22, 0.18, 0.15, 0.13, 0.11, 0.10, 0.083, 0.074, 0.065, 0.058, 0.053, 0.048, and 0.043, and it is obvious that the differences get smaller and smaller, showing that points brought in from infinity converge at distance k from the viewer.

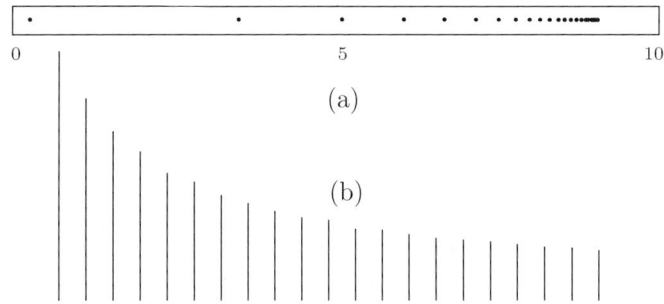

Figure 4.41: (a) Twenty Nonuniformly Spaced Points. (b) Varying Heights.

◇ **Exercise 4.14:** What should be the distance z of a point in order for it to be moved to a distance $z^* = k/2$ by the telescopic transformation?

> In prophetic utterances, time is often telescoped.
> —Anonymous

The heights of the transformed telephone poles can be determined by a similar expression. A pole located right at the viewer's location should maintain its height, while poles that are moved closer should become taller but should remain smaller than the nearest pole. If the nearest pole is l units tall, then the expression

$$l^* = l \left[1 - \frac{z\,r}{z+l} \right]$$

produces l^* values that range from l (for $z = 0$) to $(1 - r)l$ (for very large z). The *Mathematica* code

```
l=20.; r=0.1;
Table[l(1-(z r/(z+l))), {z,0,100,5}]
Table[%[[i]]-%[[i+1]], {i,1,20}]
Table[Line[{{i, 17}, {i, %%[[i]]}}], {i,1,21}]
Show[Graphics[%]]
```

selects $l = 20$ and $r = 0.1$ to obtain l^* values ranging from l to $0.9l = 18$. The results are 20, 19.6, 19.33, 19.14, 19, 18.89, 18.8, 18.72, 18.67, 18.62, 18.57, 18.53, 18.5, 18.47, 18.44, 18.42, 18.40, 18.38, 18.363, 18.35, and 18.33. Figure 4.41b shows the top parts of the poles to illustrate how the differences in height between consecutive poles diminish. The third line of the code yields the 20 differences 0.4, 0.27, 0.19, 0.14, 0.11, 0.09, 0.073, 0.061, 0.051, 0.044, 0.038, 0.033, 0.029, 0.026, 0.0234, 0.021, 0.019, 0.017, 0.016, and 0.014. Thus, the height differences between consecutive telephone poles get smaller and smaller.

After a three-dimensional scene has been telescoped point by point, we can use perspective projection to display it in two dimensions.

> Love looks through a telescope; envy, through a microscope.
> —Josh Billings

4.12 Microscopic Projection

A sample observed through a microscope is normally thin. We can therefore assume that points that go through a microscopic projection have the same (or similar) z coordinates. In contrast to a telescope, which brings points closer to the observer, a microscope "opens up" the points. Figure 4.42 shows how this is done by moving points away from the z axis. If the view angle of a point $\mathbf{P}$ is θ, then the microscope places its projection $\mathbf{P}^*$ such that its view angle is $m\theta$, where m is the magnification power of the microscope. Thus, the projection rule is

$$\frac{x}{z + k} = \tan\theta \quad \text{and} \quad \frac{x^*}{z + k} = \tan(m\theta). \tag{4.5}$$

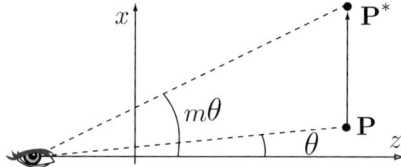

Figure 4.42: Microscopic Projection.

Computing x^* therefore involves the two steps $\theta = \arctan(x/(z+k))$ and $x^* = (z+k)\tan(m\theta)$. For small angles, $\tan\theta$ is close to θ, so we can write as an approximation

$$\frac{x^*}{z+k} = m\frac{x}{z+k} \quad \text{or} \quad x^* = mx.$$

This is a linear scaling transformation where both x and y are scaled by a factor of m, while z is left unchanged. The transformation matrix is

$$\begin{pmatrix} m & 0 & 0 & 0 \\ 0 & m & 0 & 0 \\ 0 & 0 & 1 & 0 \\ 0 & 0 & 0 & 1 \end{pmatrix}.$$

> Nature composes some of her loveliest poems for the microscope and the telescope.
> —Theodore Roszak, *Where the Wasteland Ends* (1972)

4.13 Anamorphosis

An anamorphosis is a distorted image that can be visualized and perceived only when viewed in a special way. The two most common types of anamorphosis are oblique and catoptric. The former type has to be viewed from an unusual angle or from a specific location or distance. The latter has to be seen reflected in a special mirror.

> **Anamorphosis**
>
> A distorted or monstrous projection or representation of an image on a plane or curved surface, which, when viewed from a certain point, or as reflected from a curved mirror or through a polyhedron, appears regular and in proportion; a deformation of an image.
>
> —From *Webster's Dictionary* (1913)

Figure 4.43 illustrates oblique anamorphosis. We imagine the artist painting a subject as if seen through a window. A conventional window is perpendicular to the

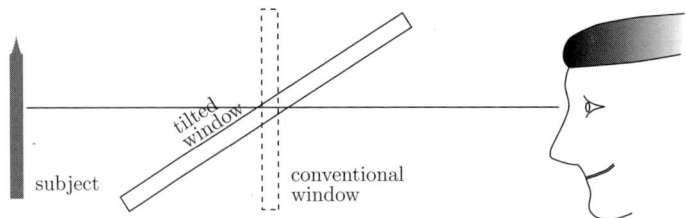

Figure 4.43: An Anamorphosis Window.

line of sight of the artist, whereas an anamorphosis window is tilted at a sharp angle to the line of sight.

The Hungarian artist István Orosz has produced striking examples [Orosz 05] of catoptric anamorphosis. An example of oblique anamorphosis is the well-known painting *The Ambassadors* by Hans Holbein the young [Holbein 05]. It features, in the foreground, a small detail, the distorted image of a skull. In order to actually see the skull, it has to be viewed from a point to the right of the painting and very close to it

A cylindrical anamorphosis is a popular variant of oblique anamorphosis. A cylindrical mirror is placed on a flat plane and a deformed image is drawn on the plane. When viewed in the mirror, the image looks correct.

Web site `www.anamorphosis.com` [anamorphosis 05] is a lively introduction to anamorphosis, with many examples and special software, *Anamorph Me* [Anamorph Me 05], that can input an image in one of several popular formats and prepare an anamorphosis (either oblique or catoptric). The four variations of Figure C.6 were generated by this software.

Figure 4.44 shows how to create an anamorphosis manually. Start with an image, cover it with a regular grid, stretch the grid and distort it, and then copy the details of the image from each original grid box to the corresponding box (which is no longer a rectangle) in the new grid. In order to obtain a cylindrical anamorphosis, the square (or rectangular) grid covering the original image has to be stretched and bent into a circular arc, as depicted in the figure.

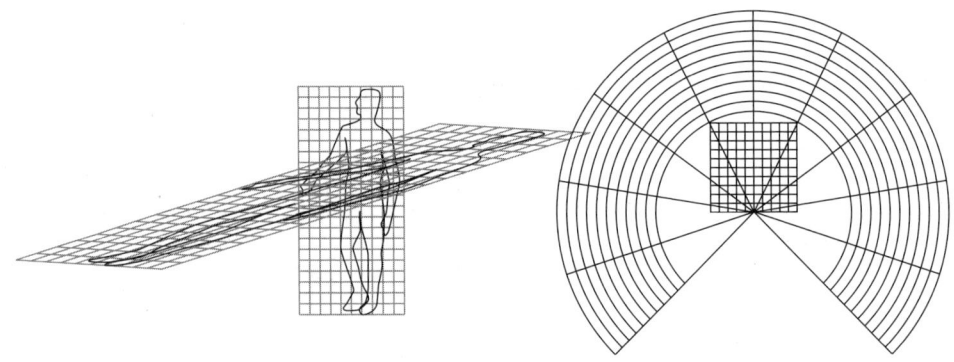

Figure 4.44: Creating an Anamorphosis.

4.14 Map Projections

According to [wikipedia 05], the ancients generally believed that the Earth is flat, but by the time of Pliny the Elder (the first century A.D.) its spherical shape had already been generally acknowledged. Many scientists and cartographers strongly believed in a round Earth, which led Columbus to risk his life, in 1492, trying to reach Japan by going west. Today, most of us believe that the Earth is a sphere (more accurately, a spheroid, since it is slightly flattened at the poles), but there is still a persistent minority that believes otherwise (see [flatearthsociety 05] for an interesting example). Regardless of anyone's beliefs or convictions, our aim in this section is to describe the chief methods for projecting a sphere on a flat plane.

The equation of a sphere of radius R centered on the origin is $x^2 + y^2 + z^2 = R^2$. This is a special case of the ellipsoid

$$\frac{x^2}{a^2} + \frac{y^2}{b^2} + \frac{z^2}{c^2} = 1 \text{ and the spheroid } \frac{x^2 + y^2}{a^2} + \frac{z^2}{c^2} = 1.$$

The sphere may also be described in spherical coordinates as (compare with Equation (4.3))

$$x = R\cos\theta\sin\phi, \quad y = R\sin\theta\sin\phi, \quad z = R\cos\phi,$$

where θ is the longitude (or azimuthal coordinate), which varies from 0 to 2π, and ϕ is the colatitude (or polar coordinate, the latitude measured from the north pole), which varies from 0 to π.

◇ **Exercise 4.15:** Look up (in a dictionary or on the Internet) the definitions of latitude, longitude, antipode, and graticule.

First, let's convince ourselves that projecting a sphere on a plane is a practical, important problem. After all, we have globes of the Earth, so perhaps we don't need maps as well. A globe is a true representation of the Earth's surface because it maintains the true scale of areas and distances and shows the correct shapes of regions and the correct angles between lines. However, its use is limited. Only one half of a globe can be viewed at a time. Normally, the size or scale of a globe is too small to show the details of a small region, such as a town, and large globes are expensive and difficult to handle. Maps, on the other hand, are much more versatile. A flat map is portable because it can be rolled or folded. It is easy to print maps in large quantities and store them digitally in a computer where they can be edited, processed, displayed, and printed.

There is vast literature on map projections, map making, and cartographic technique. Distilling it to just four items yields, in the opinion of this author, [Pearson 90] (very mathematical), [Snyder 87], [Snyder 93], and [Furuti 97].

The main problem with mapping a globe is the fact that a sphere is an undevelopable surface. Any attempt to open, unfold, or unroll a sphere to lie flat results in stretching and deforming it in some way. (This is also mentioned in Section 4.6.) Thus, every projection of a sphere onto a flat plane must introduce distortions, and the problem of mapping a globe is to design and develop sphere projections that eliminate or minimize certain distortions (while perhaps increasing others). Thus, we can say that cartography is the art and science of designing and choosing the least inappropriate projection for a given application. A map that preserves distances may be useful in certain applications even if it corrupts angles. Similarly, a map that minimizes distortions around the equator may be ideal for certain countries, such as Ecuador, even if it deforms the shapes of regions close to the poles.

An important requirement in sphere projection is to preserve spatial relationships. If a region A lies to the north of another region B on the globe, it should also appear to the north of B on the projection (i.e., on the map resulting from the projection).

Other than preserving spatial relationships, any sphere projection is a compromise, displaying some properties accurately while deforming others. Thus, when classifying sphere projections, one attribute that should be considered is the extent to which a projection preserves or distorts certain properties. Following is a list and a short discussion of the most important properties of maps. These properties are identified by answering, for a given map, the following questions:

- Can distances be accurately measured?

- How easy is it to determine the shortest path between two points?

- Are directions between points preserved?

- Are shapes of geographical features preserved?

- Are areas preserved to scale?

- Which regions suffer the most distortion, and what kind of distortion?

These features are discussed here.

Scale. A map has to shrink the globe down to a convenient size that is determined by the scale. In a 1:10,000 scale map, points separated by two units on the map represent geographical locations separated by 20,000 units on the sphere. However, no map satisfies this condition perfectly. Scale on a flat map changes with location on the map and the direction between the points. Measuring arbitrary distances on a map can at best serve as an estimate of the real distances on the sphere. Recall that the shortest distance between two points on a sphere is a great-circle arc, but such an arc is only rarely represented by a straight line on a map.

However, some projections produce maps where certain lines are to scale. Distances measured along those lines are accurate. Such lines are called *standard lines*. In a sinusoidal projection centered on the equator, all latitudes (parallels) are standard lines. In an azimuthal equidistant map, all lines that pass through the central point are standard. In a cylindrical equidistant map, the vertical lines (longitudes) and the equator are standard lines.

A small-scale map portrays a large area and a large-scale map portrays a small area of the Earth. It is intuitively clear that a small region of a sphere is not much different from a flat plane, which is why a large-scale map is not sensitive to the projection algorithm. When mapping a small area of a sphere, practically any projection method will produce a map where distances, areas, and angles are fairly accurate. The problem of distortion arises when a large area of a sphere has to be mapped. In such a case, no projection method will produce ideal results, and the algorithm used has to be selected depending on the application at hand. One projection method may be suitable for navigation, while another may produce maps useful for surveying.

The shortest path between any two points on a sphere is a great-circle arc, also called a geodesic. Thus, a projection where great circles are displayed as straight lines is ideal for measuring shortest paths. No sphere projection can generate such a map, but the stereographic projection comes close to satisfying this requirement because it preserves circles. Any circle on the sphere is mapped by this projection to a circle. In particular, a great circle passing through the center of the projection is mapped to a circle with infinite radius, a straight line. Thus, straight lines through the center of a stereographic projection are great circles and indicate shortest paths. The downside is that this projection can show only one hemisphere, which limits its use in air navigation to short and medium distances. The gnomonic projection maps all great circles, not just those passing through the central point, into straight lines, but this projection projects even less than a hemisphere.

A map prepared especially for determining property taxes should allow for accurate measurements of areas. If the scale of the map is s and if the area of a certain region is A, then the area of the region as measured on the map should be A/s. Such a map is termed equal-area and may distort the shapes of areas and display wrong distances between points.

Even a quick glance at a Mercator map shows a huge Greenland about the same size as all of Africa, obviously not to scale because the ratio of their areas is $1:13.7$. In this projection, areas close to the poles appear bigger than they should. In contrast, the Mollweide projection preserves areas.

It is easy to tell when a familiar shape becomes distorted or deformed. On the other hand, it is not obvious how to measure distortion quantitatively. We are familiar with the shape of the continents on Earth, so when a landmass gets distorted by a projection, we recognize the deformation, but it took cartographers several centuries to come up with a simple measure that shows the amount and direction of the distortion. This measure was introduced by Nicolas Tissot in the 19th century and is known today as Tissot's indicatrix. The idea is simply to add a grid of small circles to the globe area being mapped. The circles are mapped with the other items in the area (land areas, oceans, rivers, etc.), and a quick look at a circle shows the amount and direction of its distortion. A circle may retain its shape and area, it may get scaled but keep its shape, or it may become deformed.

Figure 4.45a shows the Tissot indicatrix for the sinusoidal projection. It is obvious that distortion is minimal around the equator and increases toward the poles. Also, the circles are distorted, but their area is preserved. In contrast, part (b) of the figure indicates that the Mercator projection, which is conformal, does not distort shapes but increases areas as we move away from the equator. (At the poles, the Tissot circles would become infinitely large.)

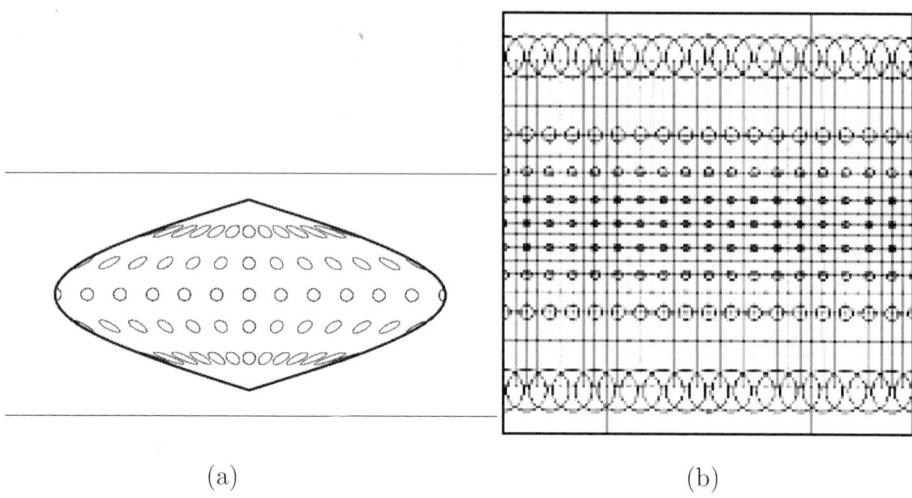

(a)   (b)

Figure 4.45: Tissot Indicatrix for Sinusoidal and Mercator Projections.

A map prepared for determining the routes of new highways should be equidistant; it should preserve distances. If the distance between two points on the sphere is L, then the distance between them on the map should be L/s. In practice, an equidistant map often shows true distances only from one point, the center of projection.

An azimuthal or zenithal projection preserves angles. Ideally, if the angle between three points is α, then the angle between the same points on the map should be the same α. In practice, azimuthal maps maintain true angles only from one central point, and even this property is achieved at the price of great distortions of areas and distances.

A map projection is conformal (also referred to as orthomorphic or equiangular) when (1) all angles at any point are preserved, (2) lines of latitude and longitude intersect at right angles, and (3) the shapes of small areas are preserved. Such a map corrupts the size of large areas.

Table 4.46 lists the pairs of properties that can be combined in a single projection.

Projection	Area	Scale	Angle	Shape
Equal-area	—	no	yes	no
Equidistant	no	—	yes	no
Azimuthal	yes	yes	—	yes
Conformal	no	no	yes	—

Table 4.46: Properties That Can Be Combined.

May I repeat what I told you here: treat nature by means of the cylinder, the sphere, the cone, everything brought into proper perspective so that each side of an object or a plane is directed towards a central point.

—Paul Cézanne to Emile Bernard, 15 April 1904

Developable surfaces. A developable surface is one that can be opened or unrolled to become flat without introducing any distortions or deforming it. A plane is developable, as are the cone and the cylinder. As a result, most methods for projecting a globe start by projecting it on a cone or a cylinder (while introducing distortions) and then unfolding this projection to become flat.

A developable surface is constructed by rolling or twisting a flat sheet of material without stretching or shrinking it. A ruled (or lofted) surface is linear in one direction. The parametric expression of such a surface is of the form $\mathbf{P}(u, w)$ and it is linear either in u or in w. Such surfaces are simple but are not always developable.

⋄ **Exercise 4.16:** Are there any other developable surfaces in addition to the cylinder, cone, and plane?

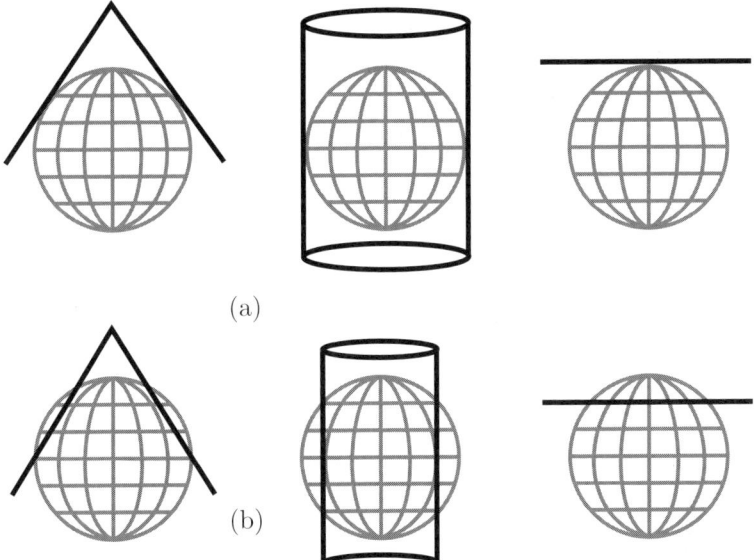

(a)

(b)

Figure 4.47: Principles of Projection.

Figure 4.47 illustrates the principle of employing developable surfaces for sphere projection. The cylinder, cone, or plane can either be tangent to the sphere [part (a) of the figure] or secant to it [part (b)]. In the latter case, the cylinder and cone intersect the sphere in two circles and the plane intersects it in a single circle. The areas of contact between the sphere and the developable surface are called the standard parallel or the standard line. These areas are important because they correspond to the regions of least distortion in the map. The difference between the various projection algorithms is in the precise way they project points on the sphere to the developable surface.

Definitions of secant
Line, ray, or segment that contains a chord of a circle. A line that crosses the circle only twice. A line extending through a circle, connecting two nonadjacent points. A straight line that intersects a curve at two or more points. Ratio of the hypotenuse to the adjacent side of a right-angle triangle.

Figure 4.49 shows how the orientation of the developable surface can vary relative to the poles of the globe. The surface can be polar, equatorial, or oblique. It is obvious that the orientation significantly affects the graticule and thus the appearance of the map. In the oblique projections, the poles are no longer at the top and bottom of the map but have migrated to unexpected places.

Azimuthal projections, also called *planar projections*, are those that project (normally only part of) a sphere directly to a plane, so there is no need to unroll and flatten a developable surface. The plane is tangent to the sphere at a point that becomes the center of projection. If the center is a pole, then lines of latitude become concentric circles on the projection and lines of longitude become straight segments that converge at the center. Figure 4.48 shows that these projections preserve directions from the center but distort distances and areas as well as directions from other points.

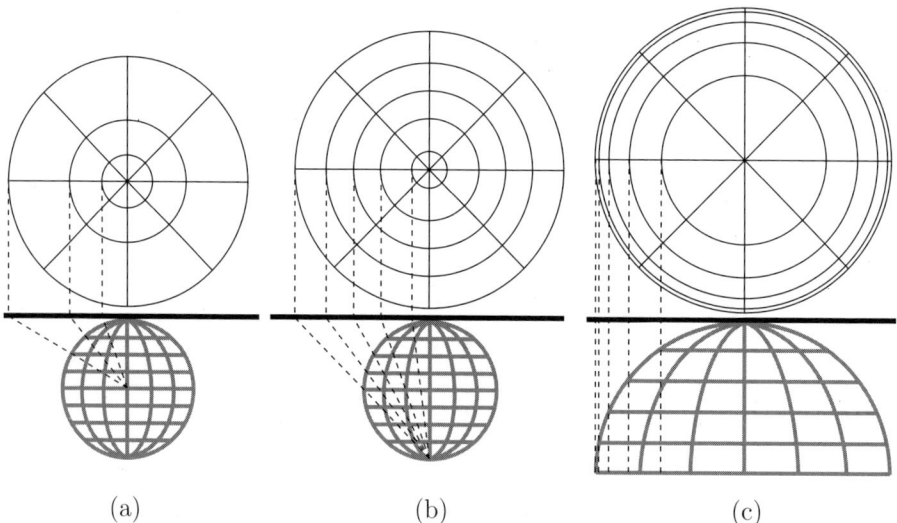

(a)   (b)   (c)

Figure 4.48: Azimuthal Projections.

The case where the center of projection is at the center of the sphere is called a gnomonic projection [part (a) of the figure]. Each line of latitude becomes a circle, but

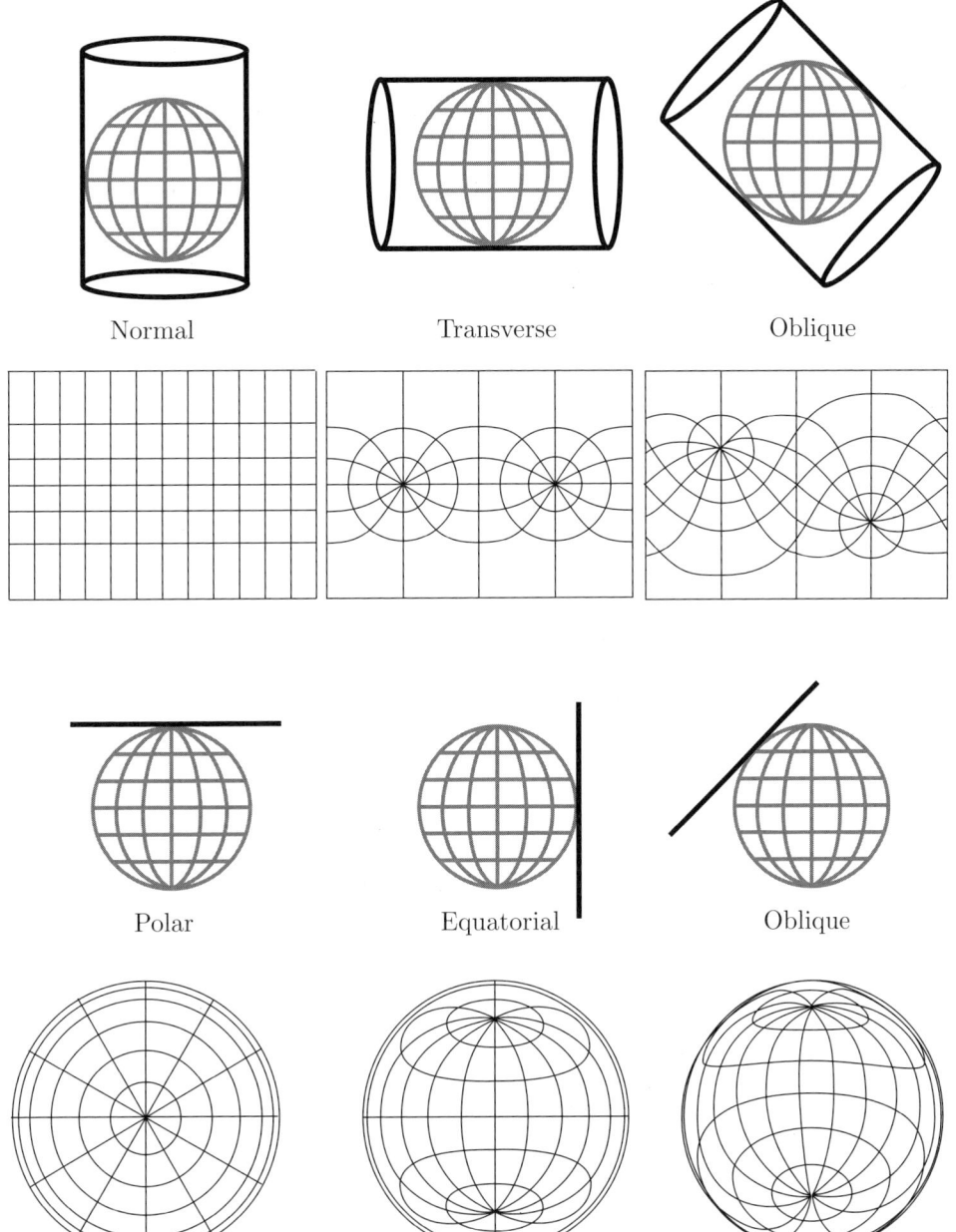

Figure 4.49: Various Orientations.

the distance between consecutive circles shrinks for high latitudes. Thus, equatorial regions are shown in more detail, while polar regions are shrunk in this type of projection. The figure demonstrates that this projection is limited to less than half the sphere; it cannot include the equator. On the other hand, any great circle is displayed in this projection as a straight segment. (A great circle is one whose center is at the center of the sphere.) Great circles are important for navigation because a great circle arc is the shortest distance between two points on the surface of a sphere. This is why the gnomonic projection is commonly used in air navigation. This projection is neither conformal nor equal-area.

Part (b) of the figure shows a stereographic projection. This is the case where the center of projection is at the pole opposite the plane of projection. The circles of latitude are uniformly spaced, which results in uniform distortions throughout.

When the center of projection is at infinity on the side of the sphere opposite that of the projection plane, the lines of projections are parallel and the projection is referred to as orthographic. Part (c) of the figure illustrates this type, and it is obvious that the pole that's tangent to the plane of projection is shown in much detail and little distortion, thereby making this projection ideal for mapping polar regions. Figure 4.50 illustrates the coordinate transformation for the orthographic projection. The polar coordinates of the projected point are $\theta =$ longitude and $r = R \cos (\text{latitude})$.

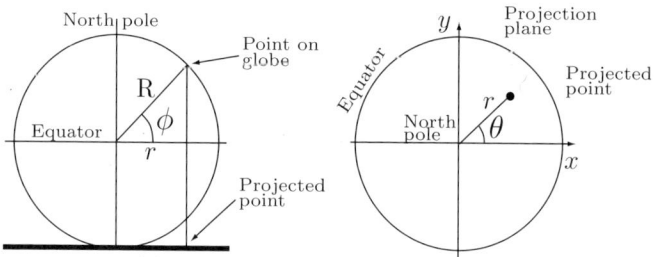

Figure 4.50: Polar Coordinates in Orthographic Projection.

Cylindrical projections. Figure 4.51 shows the three main ways to project a sphere on a cylinder tangent to it. Part (a) of the figure illustrates a perspective projection from the center of the sphere. In part (b), points on the sphere are projected to the cylinder in parallel, while the projection principle in part (c) is to project equal arc lengths on the sphere to equal vertical segments on the cylinder.

In all three types of cylindrical projections, unrolling the cylinder results in equally spaced longitudes on the map. However, in a perspective cylindrical projection, the spaces between consecutive latitudes on the map increase as we move toward the pole and approach infinity at the pole. Thus, it is impractical to extend this projection beyond about latitude 80°. The simple projection depicted in Figure 4.51a projects a point at latitude ϕ and longitude θ on the globe to Cartesian coordinates $x = \theta - \theta_0$ (where θ_0 is the longitude at the center of the map) and $y = R \tan \phi$ on the map. Such a projection stretches the vertical dimensions of any regions between about latitude 30° and the poles, resulting in so much distortion that it is rarely used.

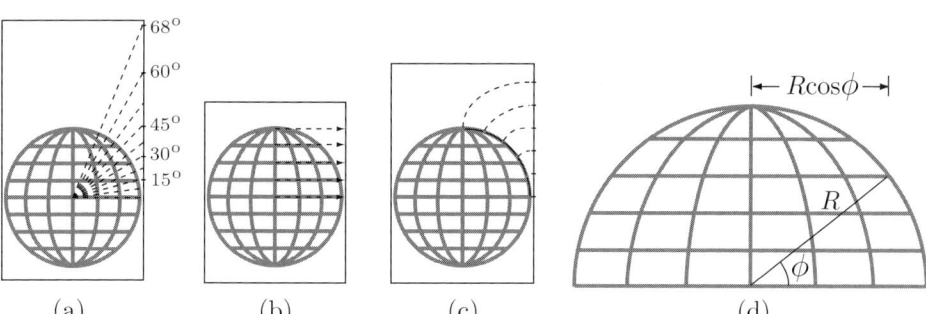

Figure 4.51: Three Types of Cylindrical Projections.

Mercator projection. A common variant of the cylindrical perspective projection is the popular Mercator projection, developed by Gerhardus Mercator in 1569. Its principle is to increase the distance between consecutive latitudes in proportion to the increased distance between meridians. This effect is illustrated in Figure 4.51d. The circumference of a globe of radius R at the equator is $2\pi R$ and at latitude ϕ it is $2\pi R \cos\phi$. Thus, the width of a longitude at latitude ϕ [the distance between longitude $\theta°$ and $(\theta+1)°$] is smaller than the width of a longitude at the equator by a factor of $\cos\phi$. In a cylindrical projection, the longitudes are shown as parallel lines, which means that at latitude ϕ, the width of a longitude in the projection has been artificially increased by a factor of $1/\cos\phi$. The width of a meridian can be considered the horizontal scale, so the principle of the Mercator projection is to also increase the vertical scale by the same factor. In the basic cylindrical projection, the y coordinate depends on the latitude ϕ as $y = R \tan\phi$. Now, we have to change the dependence such that a small change $\Delta\phi$ in ϕ changes y by a factor of $R\Delta\phi/\cos\phi$. The basic equation of y as a function of ϕ is therefore

$$dy = \frac{R\,d\phi}{\cos\phi},$$

which integrates to yield

$$y(\phi) = R \ln\,\tan\left[\frac{\pi}{4} - \frac{\phi}{2}\right].$$

Any integration constant is eliminated if we impose the condition that $\phi = 0$ implies $y = 0$.

Now imagine a small region at latitude ϕ. Both its width and its height have been increased by a factor of $1/\cos\phi$, so its area is increased by a factor of $1/\cos^2\phi$, but its shape hasn't changed. A large region tends to spread beyond a single latitude, so its shape is distorted. Thus, the Mercator projection preserves the shapes of small regions and makes it relatively easy to compute their true areas. Large regions are distorted and also appear very large. Greenland, for example, appears bigger than South America, even though the latter is nine times bigger than Greenland.

"What's the good of Mercator's north poles and equators, tropics, zones, and meridian lines?" so the Bellman would cry: and the crew would reply "They are merely conventional signs!"

—Lewis Carroll, *The Hunting of the Snark* (1876)

Figure 4.54 shows the standard Mercator projection of the Earth (where the cylinder is tangent to the equator), and Figure 4.55 is the oblique 45° Mercator projection introduced by Charles Peirce in 1894.

Cylindrical equal-area projection. When the cylinder is aligned with the rotation axis of the globe, any cylindrical projection results in uniformly spaced, parallel meridians and parallel latitudes. However, the latitudes don't have to be spaced uniformly, and their spacings can be adjusted to preserve areas. There is essentially only one way to design a cylindrical equal-area projection, and it was first described by Johann H. Lambert in 1772.

In a cylindrical projection, the x coordinate for longitude θ on the unrolled cylinder is the length of the arc between θ and θ_0. Thus $x = R(\theta - \theta_0)$. We have to adjust the space between consecutive latitudes such that any area on the cylinder will equal the corresponding area on the sphere, and this is easy to achieve by comparing areas on the sphere and the cylinder. The total surface area of a sphere is $2\pi R^2$, so the area below latitude ϕ is $2\pi R^2 \sin\phi$. The area of a cylinder below a certain height y is $2\pi Ry$, so equating the expressions $2\pi R^2 \sin\phi$ and $2\pi Ry$ results in $y = R\sin\phi$.

Table 4.52 lists y values for $R = 1$ and for latitudes from 0 to 90° and the stretch factor for each. This factor is the extra amount the y coordinate has to be moved relative to its "natural" position. For example, for $\phi = 30°$, the natural position for the y coordinate is 0.3, but it has moved to 0.5, a stretch factor of 1.67. Figure 4.53 illustrates how each latitude is raised (the dashed lines in the Northern Hemisphere) in order to preserve areas. The figure illustrates the fact that such a projection is useful in the equatorial regions but useless in the polar regions, where the small gaps between consecutive latitudes make it impossible to distinguish shapes, borders, and distances.

$\phi°$	y	Stretch	$\phi°$	y	Stretch
0	0	0.00	50	0.77	1.53
10	0.17	1.74	60	0.87	1.44
20	0.34	1.71	70	0.94	1.34
30	0.50	1.67	80	0.98	1.23
40	0.64	1.61	90	1.00	1.11

Table 4.52 Cylindrical Equal-Area Projection.

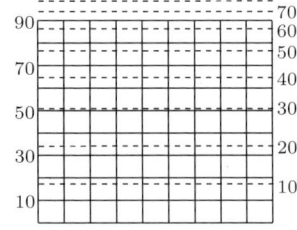

Figure 4.53 Cylindrical Equal-Area Projection.

Lambert's design for an equal-area projection can be varied and is used by several similar equal-area cylindrical projections. These vary the standard parallels, the general map proportions, and the ways of distorting shapes. These projections can be converted back to Lambert's by rescaling both the width and height.

Cylindrical equidistant projection. Perhaps the most familiar feature of the cylindrical projections discussed so far is the straight, parallel, and equidistant meridians.

Gerardus Mercator was a well-known 16th century cartographer who is remembered mostly for the useful projection now named after him. He was Flemish (his birth name was Gerard de Cremer) of German ancestry, and the name "Mercator" means "merchant" or "marketer." Although not a traveler himself, he became interested in geography, maps, and cartography as a young man. His first project, in the mid-1530s, was to construct, with two collaborators, a globe of the Earth. Later, he produced maps of the Holy Land, the world, and Flanders.

After being charged with heresy and spending time in prison, he moved to the town of Duisburg, where he became a professional cartographer and also taught mathematics. In 1564 he reached the peak of his career when he became court cosmographer to Duke Wilhelm of Cleve. His famous projection was conceived a few years later as an aid to sea navigation.

(Continues...)

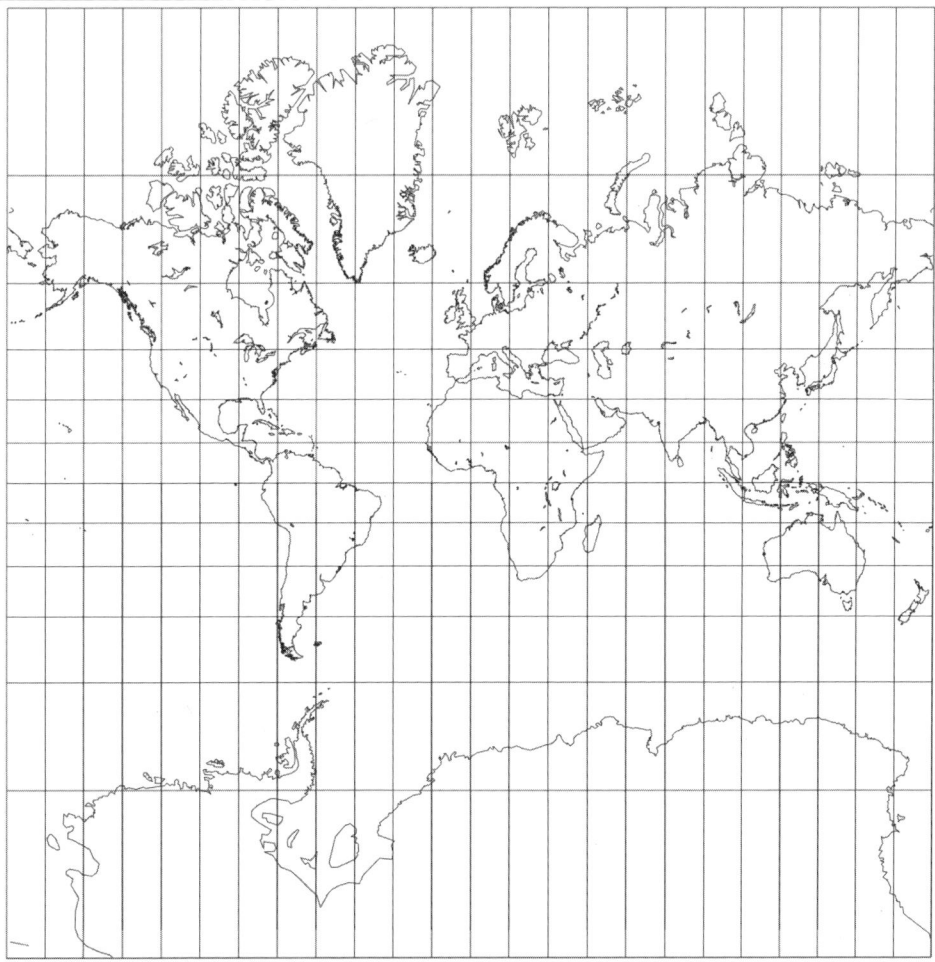

Figure 4.54: Mercator Projection.

(Mercator's life, continued)

In 1552, in Duisburg, he opened a cartographic workshop, where he completed a six-panel map of Europe (in 1554) and produced more maps. He devised his famous globe projection and first used it in 1569; it had parallel lines of longitude to aid navigation by sea, as compass courses could be marked as straight lines.

The Mercator Museum in Sint-Niklaas, Belgium, features exhibits about Mercator's life and work. A simple, detailed description of his life and projection can be found in [mercator 05].

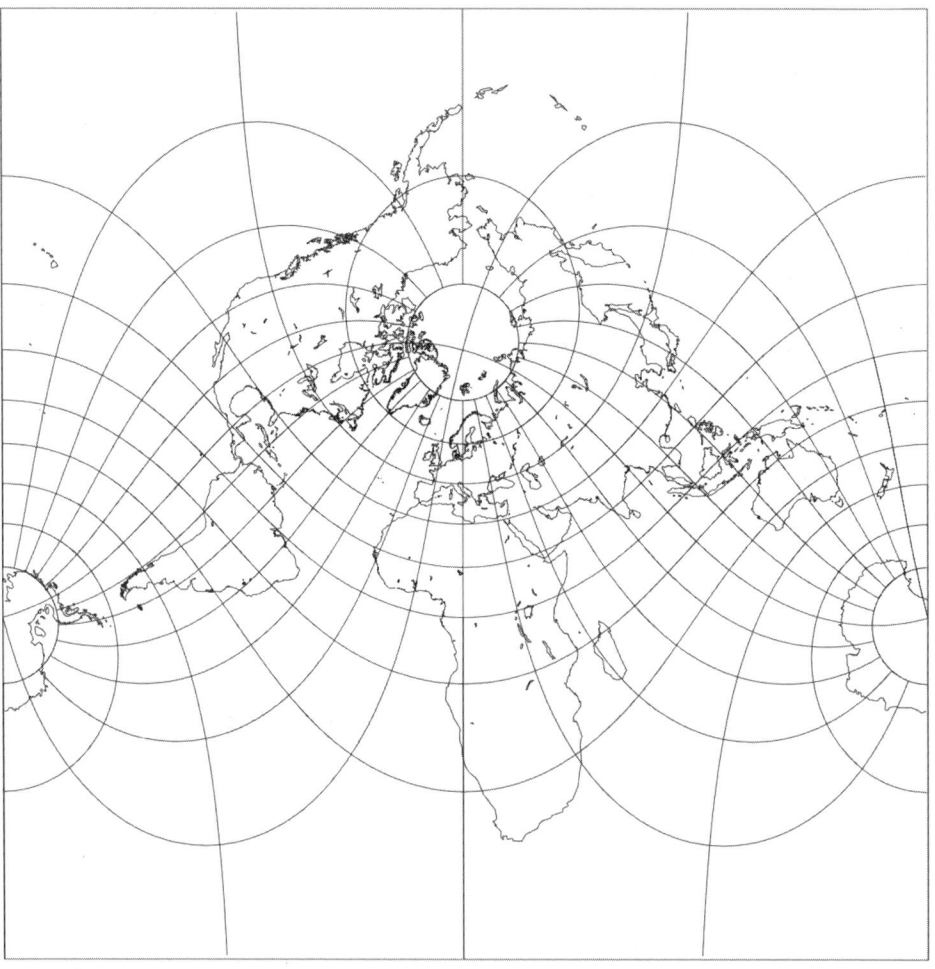

Figure 4.55: 45° Oblique Mercator Projection.

A distance measured along a meridian will have true scale because all the meridians have the same length. Given a projection with such meridians, how can we draw the latitudes so as to preserve scale along them too? There seems to be no solution to this problem because the latitudes get shorter as we approach the poles and the only way to fit shorter latitudes among the longitudes is to bend the longitudes. Thus, there is no cylindrical projection that preserves distances along both dimensions.

Pseudocylindrical projections. All the cylindrical projections discussed here (and also those not mentioned here) feature noticeable shape distortions at higher latitudes (where area is normally also greatly exaggerated). The poles are either infinitely stretched to lines or are impossible to include in the projection. Various pseudocylindrical projections have therefore been developed in attempts to correct these shortcomings. These projections feature (1) straight horizontal parallels, not necessarily equidistant, and (2) arbitrary curves for meridians, equidistant along every parallel.

The horizontal parallels help to compute and predict phenomena that depend on distance from the equator such as the lengths of day and night. The constant scale at any point of a parallel makes it easy to measure distances in the direction of a latitude.

Parallels and meridians do not always cross at right angles in a pseudocylindrical projection, which is why this type is nonconformal. Most pseudocylindrical projections are known to cause severe shape distortions at polar regions.

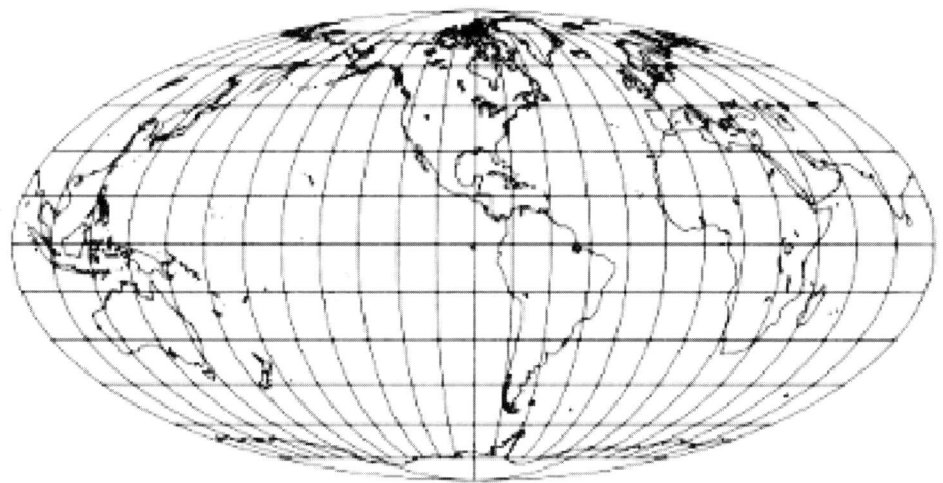

Figure 4.56: Mollweide Projection.

The following are examples of pseudocylindrical projections:

■   The Mollweide projection (Figure 4.56) was created in 1805 by Karl Mollweide and popularized by Jacques Babinet in 1857. This equal-area projection was designed to inscribe the world into a 2:1 ellipse, keeping the latitudes straight while still preserving areas. It was developed for educational purposes. All meridians except the central one are equally spaced semiellipses intersecting at the poles and concave toward the central meridian. Because of the aspect ratio chosen by Mollweide, the central meridian is half as long as the equator. The two meridians 90° east and west of the central meridian form a circle.

The mathematical expression of this projection starts with a point with longitude θ and latitude ϕ on the sphere. The point is mapped by this projection to the point

$$x = \frac{2\sqrt{2}(\theta - \theta_0)\cos\alpha}{\pi} \quad \text{and} \quad y = \sqrt{2}\sin\alpha$$

on the map, where θ_0 is the longitude at the center of the map and α is the solution to the equation $2\alpha + \sin(2\alpha) = \pi\sin\phi$.

This projection is also called homalographic, homolographic (from the Greek *homo*, meaning "same"), elliptical, or Babinet. There is also an interrupted version of the Mollweide projection. Mathematically, this projection is pseudocylindrical equal-area.

This projection is sometimes used in thematic world maps. It preserves scale up to latitude 40° (north and south). North and south of this latitude, distortions become more and more severe.

■   The sinusoidal projection (Figure 4.57), also known as the Sanson-Flamsteed projection and the Mercator equal-area projection, is the simplest pseudocylindrical equal-area projection.

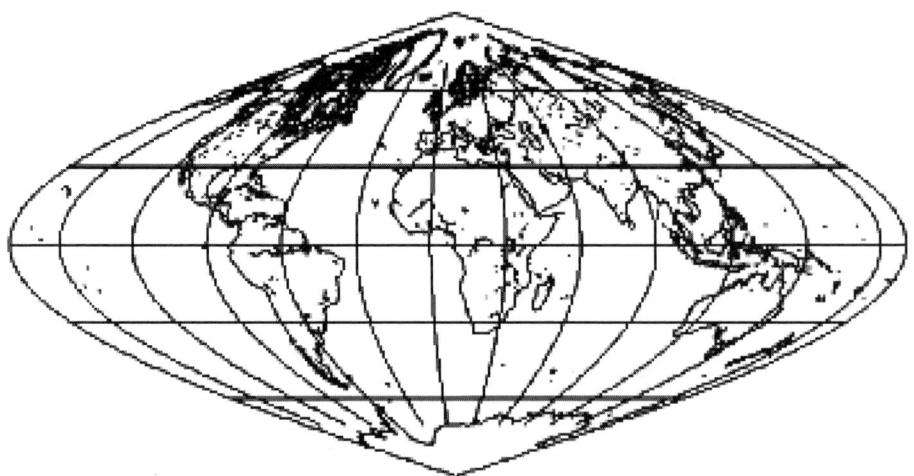

Figure 4.57: Sinusoidal Projection.

The width of a degree of longitude is proportional to the cosine of the latitude, and the lines of latitude become straight segments placed uniformly on the map. This combination preserves areas. Specifically, a point with longitude θ and latitude ϕ on the sphere will be mapped by this projection to the point $((\theta - \theta_0)\cos\phi, \phi)$ on the map (where θ_0 is the longitude at the center of the map).

This projection does not preserve shapes. Landmasses away from the central meridian are sheared, making them look extremely deformed or even unrecognizable.

An interrupted version of this projection reduces distortions considerably because (1) the scale on the equator is uniform, (2) the meridians cross it at right angles, and (3) the vertical scale of the projection does not vary along the equator for different longitudes.

It is worth mentioning that the sinusoidal and Mollweide projections handle polar regions in complementary ways; while the former crowds them together, the latter results in widely spaced meridians, which leads to more pronounced angular distortion. These two projections are combined in Goode's homolosine projection.

■ The Eckert IV equal-area world map projection (Figure 4.58) is the fourth in a set of six projections developed in the 1920s by Max Eckert as a pseudocylindrical compromise projection to obtain equal areas. The projection is in the form of a capsule, similar to an ellipse but larger, with curved lines of longitude (see also Figure 4.27). The outer meridians are semicircles, and the inner meridians are elliptical arcs. The central meridian is straight and its height is identical to the length of the equator.

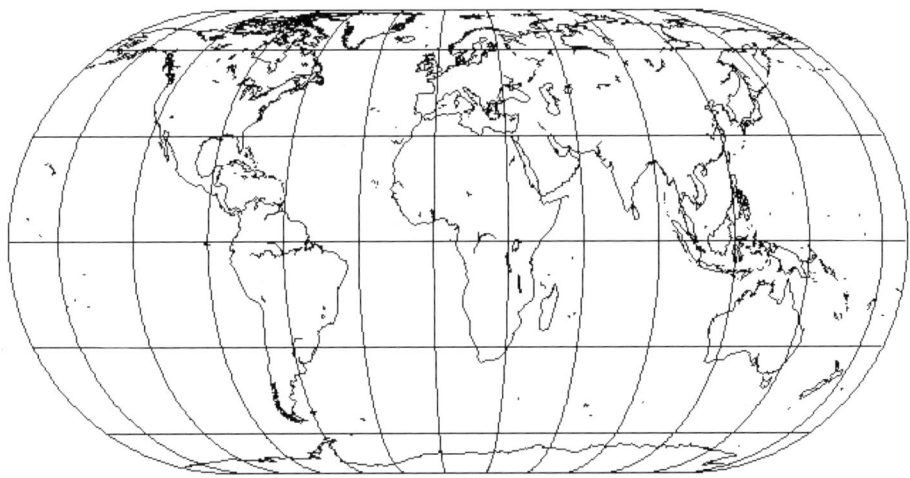

Figure 4.58: Eckert IV Projection.

The mathematical expression of this projection starts with a point with longitude θ and latitude ϕ on the sphere. The point is mapped by this projection to the point

$$x = \frac{2}{\sqrt{\pi(4+\pi)}}(\theta - \theta_0)(1 + \cos\alpha) \quad \text{and} \quad y = 2\sqrt{\frac{\pi}{4+\pi}}\sin\alpha$$

on the map, where θ_0 is the longitude at the center of the map and α is the solution to the equation $\alpha + \sin\alpha\cos\alpha + 2\sin\alpha = (2 + \pi/2)\sin\alpha$.

This projection is often the one favored by climatologists to display climate data. Sometimes it is used as a small inset inside another map (probably because of its pleasing shape), and the National Geographic Society in the United States used it for printing large wall maps of the world.

Conical projections. Projections that employ a cone as the developable surface have limited applications because they result in a noticeable distortion of shapes. Figure 4.59a portrays a cone of height h and radius R. We denote half its top angle by α and notice that α varies in the interval $[0, 90°)$. It is immediately obvious that $l^2 = h^2 + R^2$ and $\sin\alpha = R/l$. Part (b) of the figure shows the cone flattened, and the problem is to compute its top angle β. The bottom part of the flattened cone is a circular arc whose length equals the circumference $2\pi R$ of the original cone bottom. Thus $\beta l = 2\pi R$ or $\beta = 2\pi R/l = 2\pi \sin\alpha = 2\pi R/\sqrt{h^2 + R^2}$. For example, when $\alpha = 45°$, we get $\beta \approx 2\pi \cdot 0.7071 = 255°$.

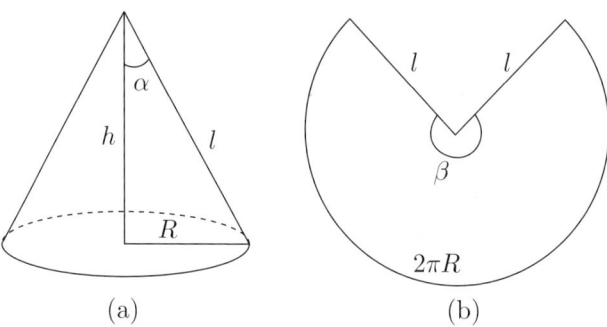

Figure 4.59: A Cone (a) Before and (b) After Flattening.

Clouds are not spheres, mountains are not cones, coastlines are not circles, and bark is not smooth, nor does lightning travel in a straight line.
—Benoît Mandelbrot, *The Fractal Geometry of Nature* (1982)

Figure 4.60 illustrates a simple equidistant conic projection of the Earth. This projection is appropriate for small regions regardless of their shape. It is also acceptable for large regions or even continents of predominant east–west extent. It illustrates the main features of a conic projection which are as follows:

1. Meridians are straight equidistant lines converging at the apex of the cone (normally a pole). The angular distance between meridians shrinks linearly as we move toward the apex, and the shrink factor is referred to as the cone constant.

2. Parallels are concentric circular arcs whose center is the point of convergence of the meridians. As a result, the parallels cross all the meridians at right angles and distortion is constant along each parallel.

3. In addition, the particular conical projection of Figure 4.60 is neither conformal nor equal-area, but such variations of the conical projection are possible.

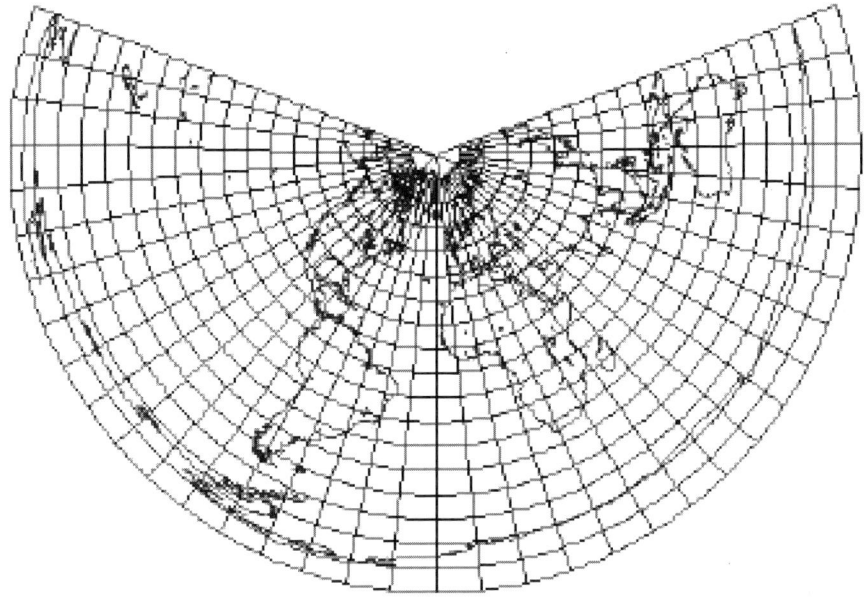

Figure 4.60: A Conical Projection.

Lambert's conic conformal projection. This type of projection was developed by Johann Lambert in 1772. After staying dormant for many years, it was revived during World War I and became the standard projection for intermediate and large-scale maps of regions in midlatitudes. Recall that "conformal" means shape preserving. A conformal mapping also preserves all angles between intersecting lines or curves. The principle of this projection is illustrated in Figure 4.61a. A cone is placed at a secant to the globe, intersecting the globe in two circles that become standard parallels. The distance between those parallels, which we denote by d, becomes $4/6$ of the vertical dimension of the projection. Thus, the projection covers a distance of $6d/4$ in the vertical direction. The projection extends $d/4$ above and $d/4$ below the standard parallels. The top and bottom of the cone are trimmed, and it is unrolled and takes a shape similar to that featured in Figure 4.61b. Notice the right angles between the (straight) meridians and the (curved) parallels. The scale along the two standard parallels is exact: The scale between them is less than 1 but its smallest value is only 0.99. The scale above and below them is greater than 1 but does not exceed 1.01.

Albers's conic equal-area projection. This projection, developed by Heinrich Albers in 1805, is very similar to Lambert's conical conformal projection. The cone is at a secant to the globe and intersects it at two latitudes. The difference between these two projections is that Albers shifts the parallels on the cone in order to preserve areas in a way similar to the cylindrical equal-area projection.

Given a point $\mathbf{P}$ on the globe, its projection $\mathbf{P}^*$ is determined by constructing a straight segment from $\mathbf{P}$ that is normal to the cone.

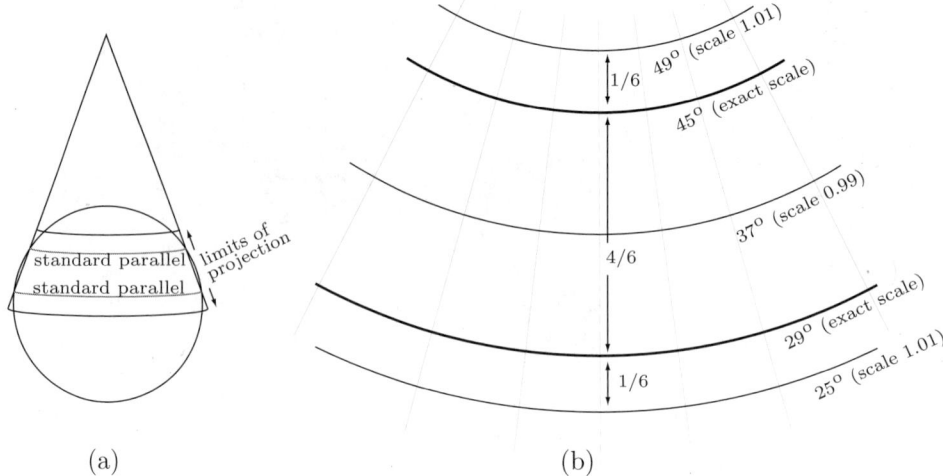

(a)   (b)

Figure 4.61: Lambert Conic Conformal Projection.

Perspective conic. Neither conformal nor equal-area, this projection maintains true scale at one standard latitude, while increasing distortion away from it. The principle is illustrated in Figure 4.62a. A cone covers part of the globe and is tangent to one latitude ϕ_0. Given a point **P** on the globe (inside the cone), we extend the straight segment from the center of the globe to **P** until it intersects the cone. The intersection point is the projection **P*** of **P**. A pole may be used instead of the center of the globe as the center of projection.

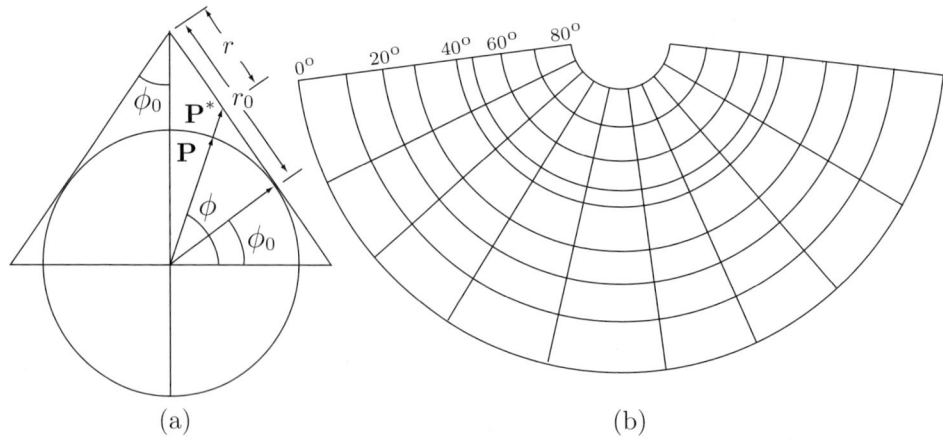

(a)   (b)

Figure 4.62: Conic Perspective Projection.

A simple application of similar triangles shows that ϕ_0, the latitude of tangency, is also half the apex angle of the cone. Thus, $r_0/R = \cot\phi_0$. The figure also shows that $r = r_0 - R\tan(\phi - \phi_0) = R\cot\phi_0 - R\tan(\phi - \phi_0)$. It is therefore natural to indicate the position of $\mathbf{P}^*$ on the flat projection by the polar coordinates (r, θ), where r is the distance from the top (the projection of the pole) and θ is simply the sine of the longitude of $\mathbf{P}$. Table 4.63 lists the ten latitudes from $0°$ to $90°$ for $R = 1$ and for $\phi_0 = 45°$ (where $R\cot\phi_0 = 1$). The differences between consecutive latitudes are also listed in this table, and it is clear that they increase as we move away (above or below) from ϕ_0.

ϕ	r	diff.	ϕ	r	diff.
0	1.999		50	0.913	0.175
10	1.700	0.300	60	0.732	0.180
20	1.466	0.234	70	0.534	0.198
30	1.268	0.198	80	0.300	0.234
40	1.087	0.180	90	0.001	0.300

Table 4.63: Ten Latitudes and Their Differences.

The most common example of this type of projection is the stereographic projection developed by Carl Braun in 1867. It wraps the globe in a cone aligned with the rotation axis. The cone is 1.5 times taller than the globe and is tangent to it at the $30°$ north parallel. The projection center is at the south pole, not at the center of the globe, and the resulting map is a perfect semicircle.

Pseudoconical projections. In this type of projection (Figure 4.64), the latitudes are still circular arcs with a common center (concentric), and the meridians still converge to this center but are no longer straight. Such projections have been known since the time of Ptolemy but are not commonly used today.

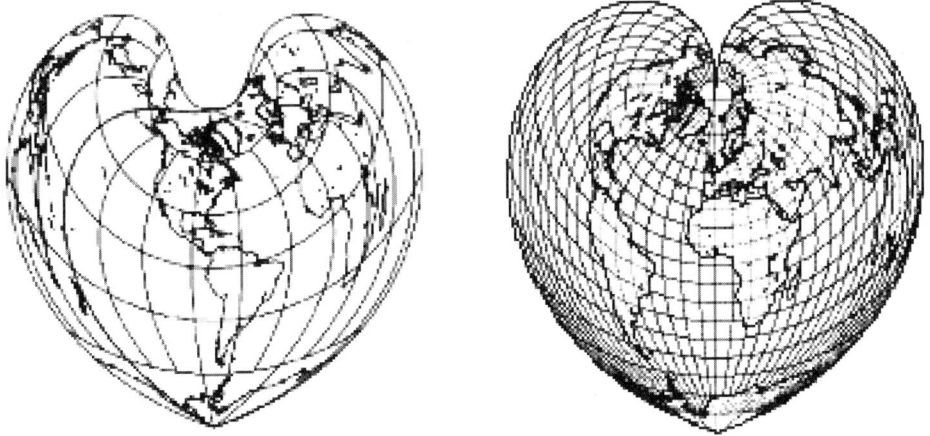

Figure 4.64: Pseudoconical Projections.

Figure 4.64 shows the Bonne (left) and the second Stabius-Werner (right) pseudo-conical projections. The first was developed by R. Bonne and the second is one of three pseudoconical designs by Johann Stabius and Johannes Werner.

Other Sphere Projections

Certain applications are best served by sphere projections that do not preserve any of the properties above but instead are a compromise where no feature is greatly distorted. The following are examples of such special projections. Perhaps the most original among them cut (or interrupt) the continuous map into slices or gores.

Projections that were especially developed to portray the entire world on one map often result in much distortion, mostly in regions located at the extremes of the projection. To improve the depiction of these distorted areas, "interrupted" forms splitting the projection into gores have been developed. In this approach, many landmasses (or oceans) can have their own central meridian, resulting in true shapes or conformality in each region of the projected map.

■   Goode's homolosine equal-area projection (Figure 4.65) is not a general sphere projection. It was developed in 1923 by J. Paul Goode specifically to project the entire Earth while trying to minimize the overall distortion of landmasses. Its main feature is discontinuity. It "interrupts" the map (splitting it into slices called "gores") in the oceans, with the result that the gores distort the shapes of the oceans while showing the continents in their true shapes. Mathematically, this projection is a combination of the homolographic and sinusoidal projections, hence the name homolosine.

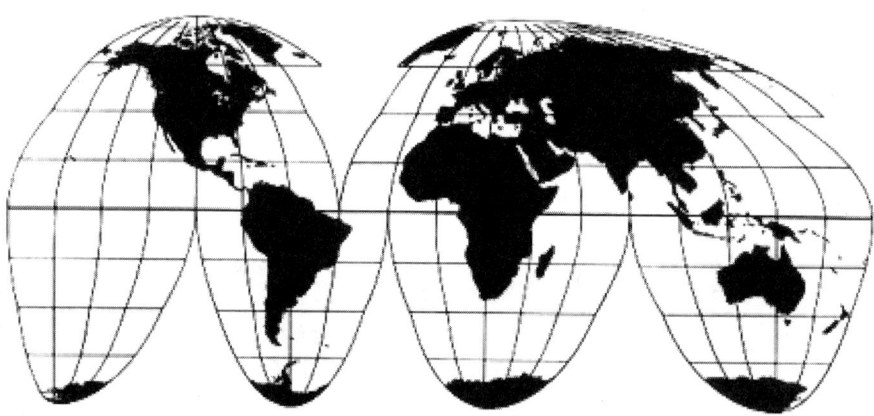

Figure 4.65: Goode's Homolosine Equal-Area Projection.

Gore is also:
1. A triangular point of land often found at road merges and diverges.
2. A triangular piece of cloth or metal used in three-dimensional fabrication.

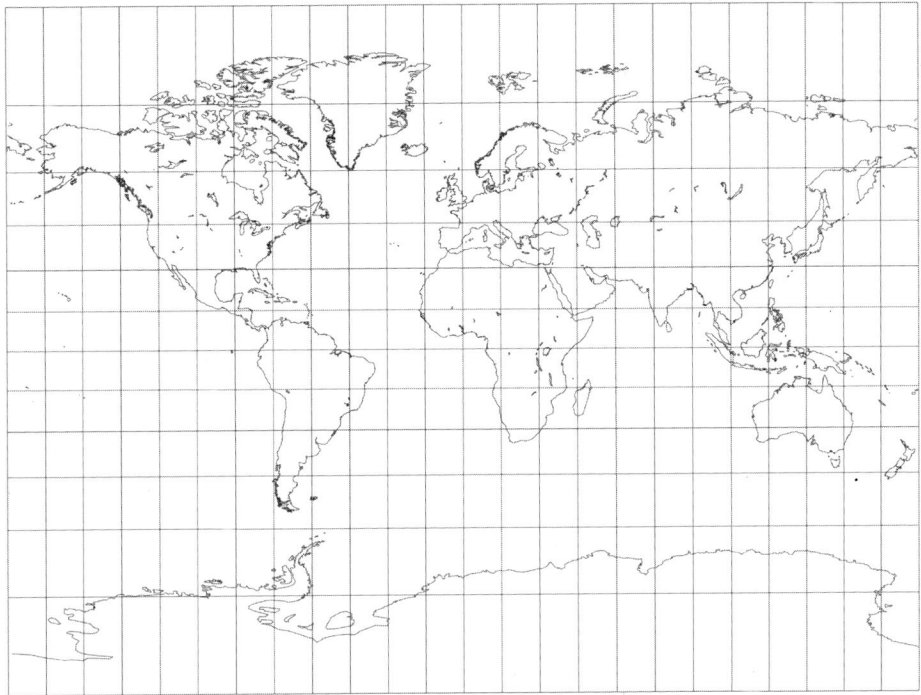

Figure 4.66: Miller Cylindrical Projection.

■   The Miller cylindrical projection (Figure 4.66) was developed by Osborn Miller in 1942 in an attempt to modify the Mercator projection to reduce the distortions around the poles and to make it possible to include the poles in the map. This projection is neither equal-area, equidistant, or conformal, nor is it perspective. Along the equator, scale is true, and near the equator there is no distortion (although the distortion increases away from the equator, becoming significant at the poles).

Miller started with the Mercator projection and moved the latitude lines closer to the equator. The distance L in the Mercator projection between each parallel and the equator was measured, and the parallel was moved to a distance of $0.8L$ from the equator. Thus, near the equator, this projection is virtually identical to Mercator. Another result of this shrinking of distances is that the height of the lines of longitude (the meridians) is 0.73 the length of the latitudes. Each pole, which on the Earth is a point, is displayed in this projection as a line of latitude, thereby causing maximum distortions at the poles.

The mathematical expression of this projection starts with a point with longitude θ and latitude ϕ on the sphere. The point is mapped by this projection to the point

$$x = \theta - \theta_0, \quad y = \frac{5}{4} \ln \left[\tan \left(\frac{\pi}{4} + \frac{2\phi}{5} \right) \right] = \frac{5}{4} \sinh^{-1} \left[\tan \left(\frac{4\phi}{5} \right) \right]$$

on the map (where θ_0 is the longitude at the center of the map).

The Miller cylindrical projection is often selected by cartographers for atlas maps of the world instead of the more popular Mercator projection. Evidently, some mapping experts feel that this variant is somewhat more appropriate or is simply more pleasing to the eye.

Nonlinear: Behaving in an erratic and unpredictable fashion; unstable. When used to describe the behavior of a machine or program, it suggests that said machine or program is being forced to run far outside of design specifications.

—Eric Raymond, *The Jargon File* (1997)

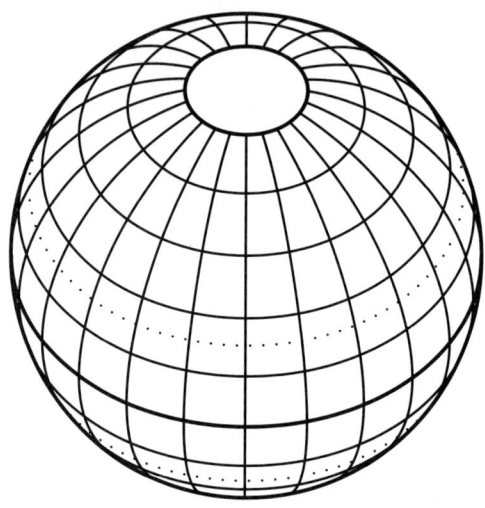

A
Vector Products

It is trivial to add and subtract vectors, but vectors can also be multiplied. This short appendix is a reminder of (or a refresher on) the two important operations of dot product and cross product.

The dot product (or inner product) of two vectors is denoted by $\mathbf{P} \bullet \mathbf{Q}$ and is defined as the scalar

$$(P_x, P_y, P_z)(Q_x, Q_y, Q_z)^T = \mathbf{P}\mathbf{Q}^T = P_x Q_x + P_y Q_y + P_z Q_z.$$

This simple definition implies that the dot product is commutative, $\mathbf{P} \bullet \mathbf{Q} = \mathbf{Q} \bullet \mathbf{P}$, and is also distributive with respect to vector addition or subtraction, $\mathbf{P} \bullet (\mathbf{Q} \pm \mathbf{T}) = \mathbf{P} \bullet \mathbf{Q} \pm \mathbf{P} \bullet \mathbf{T}$.

The dot product also has a simple and useful geometric interpretation; it equals $|\mathbf{P}| \, |\mathbf{Q}| \cos\theta$, where θ is the angle between the vectors. The dot product of perpendicular (or *orthogonal*) vectors is therefore zero. We use Figure A.1 to prove this interpretation. Part a shows a triangle with three sides a, b, and c and three angles A, B, and C opposite those sides. We draw a line from vertex B that is perpendicular to side b. This line divides the triangle into two right-angle triangles. The three sides of the triangle on the right are a, $a \sin C$, and $a \cos C$, while the sides of the triangle on the left are c, $a \sin C$, and $b - a \cos C$. Applying Pythagoras's theorem to the latter triangle yields the law of cosines

$$\begin{aligned}
c^2 &= (a \sin C)^2 + (b - a \cos C)^2 \\
&= a^2 \sin^2 C + b^2 - 2ab \cos C + a^2 \cos^2 C \\
&= a^2 (\sin^2 C + \cos^2 C) + b^2 - 2ab \cos C \\
&= a^2 + b^2 - 2ab \cos C.
\end{aligned}$$

This extends the Pythagorean theorem to arbitrary triangles.

Given two arbitrary vectors a and b separated by an angle θ, Figure A.1b shows how we can subtract them to obtain a third vector $c = a - b$, such that a, b and c form

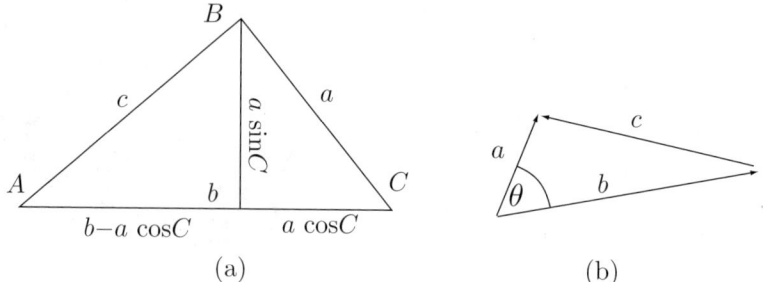

Figure A.1: Law of Cosines.

the three sides of a triangle. Applying the law of cosines to this triangle yields

$$c^2 = a^2 + b^2 - 2ab\cos\theta. \tag{A.1}$$

Applying the dot product to vector c yields

$$
\begin{aligned}
c^2 = c_x^2 + c_y^2 + c_z^2 = c \bullet c &= (a - b) \bullet (a - b) \\
&= a \bullet a + b \bullet b - 2(a \bullet b) \\
&= a^2 + b^2 - 2(a \bullet b).
\end{aligned} \tag{A.2}
$$

Equating Equations (A.1) and (A.2) yields $a \bullet b = ab\cos\theta$.

The triple product $(\mathbf{P} \bullet \mathbf{Q})\mathbf{R}$ is sometimes useful. It can be represented as

$$
\begin{aligned}
(\mathbf{P} \bullet \mathbf{Q})\mathbf{R} &= (P_x Q_x + P_y Q_y + P_z Q_z)(R_x, R_y, R_z) \\
&= ((P_x Q_x + P_y Q_y + P_z Q_z)R_x, (P_x Q_x + P_y Q_y + P_z Q_z)R_y, \\
&\quad (P_x Q_x + P_y Q_y + P_z Q_z))R_z \\
&= (Q_x, Q_y, Q_z)\begin{pmatrix} P_x R_x & P_y R_x & P_z R_x \\ P_x R_y & P_y R_y & P_z R_y \\ P_x R_z & P_y R_z & P_z R_z \end{pmatrix} \\
&= \mathbf{Q}(\mathbf{PR}),
\end{aligned} \tag{A.3}
$$

where the notation $(\mathbf{PR})$ stands for the 3×3 matrix above. (This material is used in Section 1.4.3.)

The cross product of two vectors (also called the *vector product*) is denoted by $\mathbf{P} \times \mathbf{Q}$ and is defined as the vector

$$(P_2 Q_3 - P_3 Q_2, -P_1 Q_3 + P_3 Q_1, P_1 Q_2 - P_2 Q_1). \tag{A.4}$$

It is easy to show that $\mathbf{P} \times \mathbf{Q}$ is perpendicular to both $\mathbf{P}$ and $\mathbf{Q}$.

⋄ **Exercise A.1:** Show it!

Perhaps the best proof is to construct the cross product from first principles. Given the two vectors $\mathbf{P}$ and $\mathbf{Q}$, we are looking for a vector $\mathbf{R}$ perpendicular to both $\mathbf{P}$ and $\mathbf{Q}$. This requirement does not fully define $\mathbf{R}$, since both $\mathbf{R}$ and $-\mathbf{R}$ satisfy it and since it says nothing about the magnitude of $\mathbf{R}$. We therefore extend our definition of the cross product by requiring that the triplet $(\mathbf{P}, \mathbf{Q}, \mathbf{R})$ be a right-handed triad of vectors and also that the magnitude of $\mathbf{R}$ be the product $|\mathbf{P}|\,|\mathbf{Q}|\sin\theta$, where θ is the angle between $\mathbf{P}$ and $\mathbf{Q}$.

The derivation exploits the orthogonality of the three coordinate axes $\mathbf{i} = (\mathbf{1}, \mathbf{0}, \mathbf{0})$, $\mathbf{j} = (\mathbf{0}, \mathbf{1}, \mathbf{0})$, and $\mathbf{k} = (\mathbf{0}, \mathbf{0}, \mathbf{1})$ and also uses our definition. The definition implies that $\mathbf{i}\times\mathbf{i} = \mathbf{0}$ because the angle between $\mathbf{i}$ and itself is zero, and the same holds for $\mathbf{j}$ and $\mathbf{k}$. It also implies that the cross product of any two of the three basis vectors is a unit vector because the basis vectors are unit vectors and because $\sin 90° = 1$. Once we realize that the triplet $(\mathbf{i}, \mathbf{j}, \mathbf{k})$ is a right-handed triad, we can deduce the following: $\mathbf{i}\times\mathbf{j} = \mathbf{k}$, $\mathbf{j}\times\mathbf{k} = \mathbf{i}$, $\mathbf{k}\times\mathbf{i} = \mathbf{j}$, $\mathbf{j}\times\mathbf{i} = -\mathbf{k}$, $\mathbf{k}\times\mathbf{j} = -\mathbf{i}$, and $\mathbf{i}\times\mathbf{k} = -\mathbf{j}$.

Armed with this information, we can easily derive the cross product $\mathbf{R}$:

$$
\begin{aligned}
\mathbf{R} = \mathbf{P}\times\mathbf{Q} &= (P_1\mathbf{i} + P_2\mathbf{j} + P_3\mathbf{k})\times(Q_1\mathbf{i} + Q_2\mathbf{j} + Q_3\mathbf{k}) \\
&= (P_1\mathbf{i} + P_2\mathbf{j} + P_3\mathbf{k})\times Q_1\mathbf{i} + (P_1\mathbf{i} + P_2\mathbf{j} + P_3\mathbf{k})\times Q_2\mathbf{j} + (P_1\mathbf{i} + P_2\mathbf{j} + P_3\mathbf{k})\times Q_3\mathbf{k} \\
&= (P_2Q_3 - P_3Q_2)\mathbf{i} + (-P_1Q_3 + P_3Q_1)\mathbf{j} + (P_1Q_2 - P_2Q_1)\mathbf{k} \\
&= (P_2Q_3 - P_3Q_2, -P_1Q_3 + P_3Q_1, P_1Q_2 - P_2Q_1).
\end{aligned}
$$

The magnitude of $\mathbf{R}$ can be calculated explicitly

$$
\begin{aligned}
|\mathbf{R}|^2 &= (P_2Q_3 - P_3Q_2)^2 + (-P_1Q_3 + P_3Q_1)^2 + (P_1Q_2 - P_2Q_1) \\
&= (P_1^2 + P_2^2 + P_3^2)(Q_1^2 + Q_2^2 + Q_3^2) - (P_1Q_1 + P_2Q_2 + P_3Q_3)^2 \\
&= |\mathbf{P}|^2|\mathbf{Q}|^2 - (\mathbf{P}\cdot\mathbf{Q})^2 = |\mathbf{P}|^2|\mathbf{Q}|^2 - (|\mathbf{P}||\mathbf{Q}|\cos\theta)^2 \\
&= |\mathbf{P}|^2|\mathbf{Q}|^2(1 - \cos^2\theta) = |\mathbf{P}|^2|\mathbf{Q}|^2\sin^2\theta.
\end{aligned}
$$

To illustrate the magnitude, we can draw the parallelogram defined by $\mathbf{P}$ and $\mathbf{Q}$ (with an angle θ between them) and show that vector $\mathbf{Q}\sin\theta$ is perpendicular to $\mathbf{P}$.

The following expressions show how $\mathbf{P}\times\mathbf{Q}$ can be expressed by means of a determinant,

$$
\mathbf{P}\times\mathbf{Q} = \begin{vmatrix} \mathbf{i} & \mathbf{j} & \mathbf{k} \\ P_1 & P_2 & P_3 \\ Q_1 & Q_2 & Q_3 \end{vmatrix} = \mathbf{i}\begin{vmatrix} P_2 & P_3 \\ Q_2 & Q_3 \end{vmatrix} - \mathbf{j}\begin{vmatrix} P_1 & P_3 \\ Q_1 & Q_3 \end{vmatrix} + \mathbf{k}\begin{vmatrix} P_1 & P_2 \\ Q_1 & Q_2 \end{vmatrix}
$$

$$
= (P_2Q_3 - P_3Q_2, -P_1Q_3 + P_3Q_1, P_1Q_2 - P_2Q_1),
$$

or, alternatively, by means of a matrix

$$
\mathbf{P}\times\mathbf{Q} = (Q_1, Q_2, Q_3)\begin{pmatrix} 0 & P_3 & -P_2 \\ -P_3 & 0 & P_1 \\ P_2 & -P_1 & 0 \end{pmatrix}. \tag{A.5}
$$

◇ **Exercise A.2:** The cross product $\mathbf{P} \times \mathbf{Q}$ is perpendicular to both $\mathbf{P}$ and $\mathbf{Q}$. In what direction does it point?

The cross product is not commutative and is not associative. It is, however, distributive with respect to addition or subtraction of vectors. Hence, $\mathbf{P} \times (\mathbf{Q} \pm \mathbf{T}) = \mathbf{P} \times \mathbf{Q} \pm \mathbf{P} \times \mathbf{T}$.

The magnitude of $\mathbf{P} \times \mathbf{Q}$ equals $|\mathbf{P}| |\mathbf{Q}| \sin \theta$, where θ is the angle between the two vectors. The cross product therefore has a simple geometric interpretation. Its magnitude equals the area of the parallelogram defined by the two vectors.

◇ **Exercise A.3:** Given that $\mathbf{P} \times \mathbf{Q} = 0$, what does it tell us about the vectors involved?

As an example, the vector equation of a straight line is shown below for the case where the direction of the line and one point on the line are known. Assume that $\mathbf{d}$ is a unit vector in the direction of the line and $\mathbf{P}_1$ is a given point on the line. The equation of the entire line is

$$\mathbf{P}(t) = \mathbf{P}_1 + t\mathbf{d}, \tag{A.6}$$

where t can take any real value.

◇ **Exercise A.4:** Derive the vector line equation for the straight segment between two given points $\mathbf{P}_1$ and $\mathbf{P}_2$.

Incidentally, there is a completely different way of looking at the cross product. It has to do with the following property of vectors. If we transform the coordinate system by reversing the direction of every coordinate axis, then any vector $\mathbf{v}$ will be transformed to $-\mathbf{v}$. However, the cross product of two vectors $\mathbf{P}$ and $\mathbf{Q}$ retains its sign when both $\mathbf{P}$ and $\mathbf{Q}$ are reversed (which can be seen directly from Equation (A.4)). The cross product, which otherwise behaves like a "normal" vector, differs from a vector in this respect and is therefore called a *pseudovector* or sometimes an *axial vector*. (A "normal" vector, incidentally, is called a *polar vector*.) It is also easy to show that the cross product of two pseudovectors is a pseudovector and that the cross product of a vector and a pseudovector is a vector.

The different way of looking at the cross product of two vectors is to interpret it as a rotation in the plane defined by the vectors. Imagine a flat plane spanned by two vectors. Select a point $\mathbf{P}$ in this plane and rotate it about another point $\mathbf{C}$. It is natural to visualize this rotation as if it takes place around a vector perpendicular to the plane (its normal vector) whose tail is at $\mathbf{C}$. The alternative interpretation of the cross product considers its three components not as the x, y, and z components of a vector but as numbers associated with the yz, zx, and xy planes defined by the three coordinate axes.

This interpretation makes more sense when we consider the cross product in higher dimensions. The four orthogonal coordinate axes of a four-dimensional space define six planes, so the cross product of two four-dimensional vectors should have six components. If it were a vector, it would have four components. In five dimensions, the five coordinate axes define ten planes, so the cross product of five-dimensional vectors should have ten components. In general, the cross product in n dimensions has $n(n-1)/2$ components, which is why the cross product in three dimensions happens to have three components.

Mathematicians long ago developed a notation to describe the components of the general cross product. It is called an *antisymmetric tensor* and is written as an $n \times n$ antisymmetric matrix whose diagonal elements are zero. There are $n(n-1)/2$ elements in the top half and the same number (with an opposite sign) in the bottom half of the matrix. Each element corresponds to a rotation in one of the $n(n-1)/2$ planes defined by the n coordinate axes.

The tensors for $n = 3$ and $n = 4$ are

$$(P_1, P_2, P_3) \times (Q_1, Q_2, Q_3) = \begin{bmatrix} 0 & P_1Q_2 - P_2Q_1 & P_1Q_3 - P_3Q_1 \\ P_2Q_1 - P_1Q_2 & 0 & P_2Q_3 - P_3Q_2 \\ P_3Q_1 - P_1Q_3 & P_3Q_2 - Q_2P_3 & 0 \end{bmatrix}.$$

$$(P_1, P_2, P_3, P_4) \times (Q_1, Q_2, Q_3, Q_4) =$$
$$\begin{bmatrix} 0 & Q_1P_2 - Q_2P_1 & Q_1P_3 - Q_3P_1 & Q_1P_4 - Q_4P_1 \\ P_1Q_2 - P_2Q_1 & 0 & Q_2P_3 - Q_3P_2 & Q_2P_4 - Q_4P_2 \\ P_1Q_3 - P_3Q_1 & P_2Q_3 - P_3Q_2 & 0 & Q_3P_4 - Q_4P_3 \\ P_1Q_4 - P_4Q_1 & P_2Q_4 - P_4Q_2 & P_3Q_4 - P_4Q_3 & 0 \end{bmatrix}.$$

We are here to study laws and vectors and constitutions, not to run in circles.

Mike Resnick, *Mwalimu in the Squared Circle* (1993)

Sir William R. Hamilton

B
Quaternions

Complex numbers can be interpreted as points in the xy plane. The complex number (a, b) can be interpreted as the point with coordinates (a, b). Is it possible to define *hypercomplex* numbers of the form (a, b, c) that could be interpreted as three-dimensional points? This question bothered the Irish mathematician William Rowan Hamilton for a long time. The problem was that multiplying complex numbers could be interpreted as a rotation in two dimensions (Section 1.4.5), so it made sense to require that multiplying the new hypercomplex numbers would be equivalent to a rotation in three dimensions. Readers of this book know (from Section 1.4.3) that a general rotation in three dimensions is fully defined by four numbers: one for the rotation angle and three for the rotation axis. Three numbers are not enough to fully specify such a rotation.

Hamilton could not come up with a reasonable rule for multiplying hypercomplex numbers that are triplets, and he eventually discovered, in October 1843, that he needed to add a fourth component to his triplets (i.e., turn them into 4-tuples) in order to multiply them in a way that made sense. He called these new entities *quaternions*. Using modern notation, a quaternion $\mathbf{q}$ can be represented as a 2×2 matrix of complex numbers

$$\mathbf{q} = \begin{pmatrix} z & w \\ -w^* & z^* \end{pmatrix} = \begin{pmatrix} a+ib & c+id \\ -c+id & a-ib \end{pmatrix},$$

where z and w are complex numbers and a, b, c, and d are real. This can also be written (by analogy with the complex numbers $a \cdot 1 + b \cdot i$) as $\mathbf{q} = a\mathbf{U} + b\mathbf{I} + c\mathbf{J} + d\mathbf{K}$, where

$$\mathbf{U} = \begin{pmatrix} 1 & 0 \\ 0 & 1 \end{pmatrix}, \quad \mathbf{I} = \begin{pmatrix} i & 0 \\ 0 & -i \end{pmatrix}, \quad \mathbf{J} = \begin{pmatrix} 0 & 1 \\ -1 & 0 \end{pmatrix}, \quad \text{and } \mathbf{K} = \begin{pmatrix} 0 & i \\ i & 0 \end{pmatrix}.$$

(Note that $\mathbf{U}$, not $\mathbf{I}$, is used here to denote the identity matrix. These matrices are closely related to the Pauli spin matrices used in particle physics.) From the definitions above, it follows that $\mathbf{I}^2 = -\mathbf{U}$, $\mathbf{J}^2 = -\mathbf{U}$, and $\mathbf{K}^2 = -\mathbf{U}$. We therefore conclude that

I, **J**, and **K** are three different solutions of the matrix equation $\mathbf{X}^2 = -\mathbf{U}$ and should be considered the square roots of minus the identity matrix.

Quaternions can also be viewed as elements of a four-dimensional *vector space*, one of whose bases are given by

$$i = \begin{pmatrix} 0 & 1 & 0 & 0 \\ -1 & 0 & 0 & 0 \\ 0 & 0 & 0 & 1 \\ 0 & 0 & -1 & 0 \end{pmatrix}, \quad j = \begin{pmatrix} 0 & 0 & 0 & -1 \\ 0 & 0 & -1 & 0 \\ 0 & 1 & 0 & 0 \\ 1 & 0 & 0 & 0 \end{pmatrix},$$

$$k = \begin{pmatrix} 0 & 0 & -1 & 0 \\ 0 & 0 & 0 & 1 \\ 1 & 0 & 0 & 0 \\ 0 & -1 & 0 & 0 \end{pmatrix}, \quad 1 = \begin{pmatrix} 1 & 0 & 0 & 0 \\ 0 & 1 & 0 & 0 \\ 0 & 0 & 1 & 0 \\ 0 & 0 & 0 & 1 \end{pmatrix}.$$

Quaternions satisfy the following identities, also known as Hamilton's Rules,

$$i^2 = j^2 = k^2 = -1, \quad ij = -ji = k, \quad jk = -kj = i, \quad ki = -ik = j.$$

They have the following multiplication table:

	1	i	j	k
1	1	i	j	k
i	i	−1	k	−j
j	j	−k	−1	i
k	k	j	−i	−1

The eight quaternions ± 1, $\pm i$, $\pm j$, and $\pm k$ form a group of order 8 with multiplication as the group operation.

Quaternions can also be interpreted as a combination of a scalar and a vector. They are consequently closely related to 4-vectors. Using this interpretation, a quaternion $\mathbf{q}$ can be represented as the sum $\mathbf{q} = w + x\mathbf{i} + y\mathbf{j} + z\mathbf{k}$, the 4-tuples (x, y, z, w) and (w, x, y, z), or the pair $[s, \mathbf{v}]$, where $s = w$ and $\mathbf{v} = (x, y, z)$.

The conjugate quaternion is given by $\mathbf{q}^* = w - x\mathbf{i} - y\mathbf{j} - z\mathbf{k}$. The sum or difference of two quaternions is the obvious

$$\mathbf{q}_1 \pm \mathbf{q}_2 = (w_1 + w_2) \pm (x_1 + x_2)\mathbf{i} \pm (y_1 + y_2)\mathbf{j} \pm (z_1 + z_2)\mathbf{k} = [s_1 \pm s_2, (\mathbf{v}_1 \pm \mathbf{v}_2)],$$

and the product is the nonobvious

$$\mathbf{q}_1 \cdot \mathbf{q}_2 = (w_1 w_2 - x_1 x_2 - y_1 y_2 - z_1 z_2) + (w_1 x_2 + x_1 w_2 + y_1 z_2 - z_1 y_2)\mathbf{i}$$
$$+ (w_1 y_2 - x_1 z_2 + y_1 w_2 + z_1 x_2)\mathbf{j} + (w_1 z_2 + x_1 y_2 - y_1 x_2 + z_1 w_2)\mathbf{k}$$
$$= [(s_1 s_2 - \mathbf{v}_1 \bullet \mathbf{v}_2), (s_1 \mathbf{v}_2 + s_2 \mathbf{v}_1 + \mathbf{v}_1 \times \mathbf{v}_2)].$$

A quaternion product is associative (i.e., $(\mathbf{q}_1 \mathbf{q}_2)\mathbf{q}_3 = \mathbf{q}_1(\mathbf{q}_2 \mathbf{q}_3)$) but not commutative.

An appropriate measure of the size of a quaternion is its *norm*, defined as

$$|\mathbf{q}| = \mathbf{q} \cdot \mathbf{q}^* = \mathbf{q}^* \cdot \mathbf{q} = \sqrt{w^2 + x^2 + y^2 + z^2} = \sqrt{s^2 + x^2 + y^2 + z^2}.$$

It is easy to verify that the norm is multiplicative, $|\mathbf{q}_1 \mathbf{q}_2| = |\mathbf{q}_1| \, |\mathbf{q}_2|$ (i.e., the norm of a product equals the product of the two individual norms). A *unit quaternion* is one for which $|\mathbf{q}| = 1$.

The inverse of a quaternion is given by

$$\mathbf{q}^{-1} = \frac{\mathbf{q}^*}{(\mathbf{q}\mathbf{q}^*)} = \frac{\mathbf{q}^*}{|\mathbf{q}|^2} = \frac{\mathbf{q}^*}{w^2 + x^2 + y^2 + z^2},$$

so quaternion division $\mathbf{q}_1/\mathbf{q}_2$ (except by zero) is performed by multiplying $\mathbf{q}_1$ by the inverse $\mathbf{q}_2^{-1}$. It's easy to verify that $\mathbf{q}\mathbf{q}^{-1} = [1, (0, 0, 0)] = [0, \mathbf{0}]$.

[Mathworld 05] and [WikiQuaternion 05] are basic references for quaternions.

$\diamond$ **Exercise B.1:** (If 4, why not more?) Quaternions are an extension of vectors. Are there extensions of quaternions?

> Every morning in the early part of the above-cited month [Oct. 1843]
> on my coming down to breakfast, your brother William Edwin
> and yourself used to ask me, "Well, Papa, can you multiply
> triplets?" Whereto I was always obliged to reply, with a sad
> shake of the head, "No, I can only add and subtract them."
>
> —William Rowan Hamilton

C
Color Figures

Most color figures in this book have been printed in place and in grayscale. All of them also appear here in color.

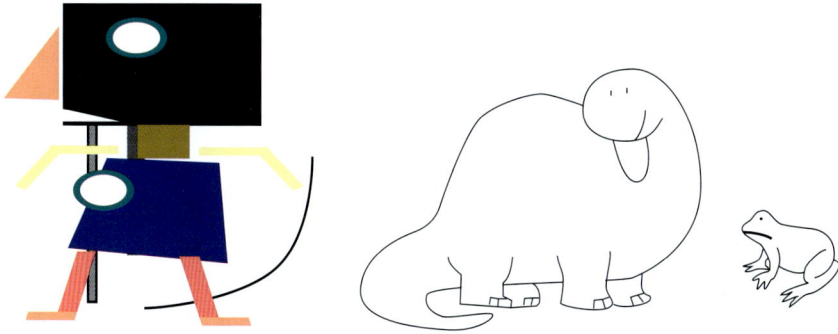

Figure 3.2. Modern Art (page 72).

I found I could say things with color and shapes that I couldn't say any other way— things I had no words for.

—Georgia O'Keeffe

Figure C.1: Many Vanishing Points (page 74).

Figure C.2: Masaccio's *Holy Trinity.* (page 82).

Figure C.3: Crude Perspective in Van Gogh's Yellow Chair (page 87).

"The most striking aspects of this work are the bright patches of contrasting color, the thickly applied paint and the odd perspective. The rear wall appears strangely angled. This is not a mistake: this corner of the Yellow House was, in fact, slightly skewed." (From www.vangoghmuseum.nl.)

Figure 3.49. A Stereo Pair, (photo taken by the author, page 134).

I don't paint things. I only paint the difference between things.
—Henri Matisse

(a)

(b)

(c)

Figure 3.50. Three Anaglyph Encodings (page 136).

Figure C.4: *A City of Cathay* (page 60).

When my daughter was about seven years old, she asked me one day what I did at work. I told her I worked at the college—that my job was to teach people how to draw. She stared at me, incredulous, and said, "You mean they forget?"

—Howard Ikemoto

Figure 4.10: Two Angular Fisheye Examples, page 155: (The one on the left is courtesy of Joseph Bly [joebly 06]. The one on the right is courtesy of Dick Termes.)

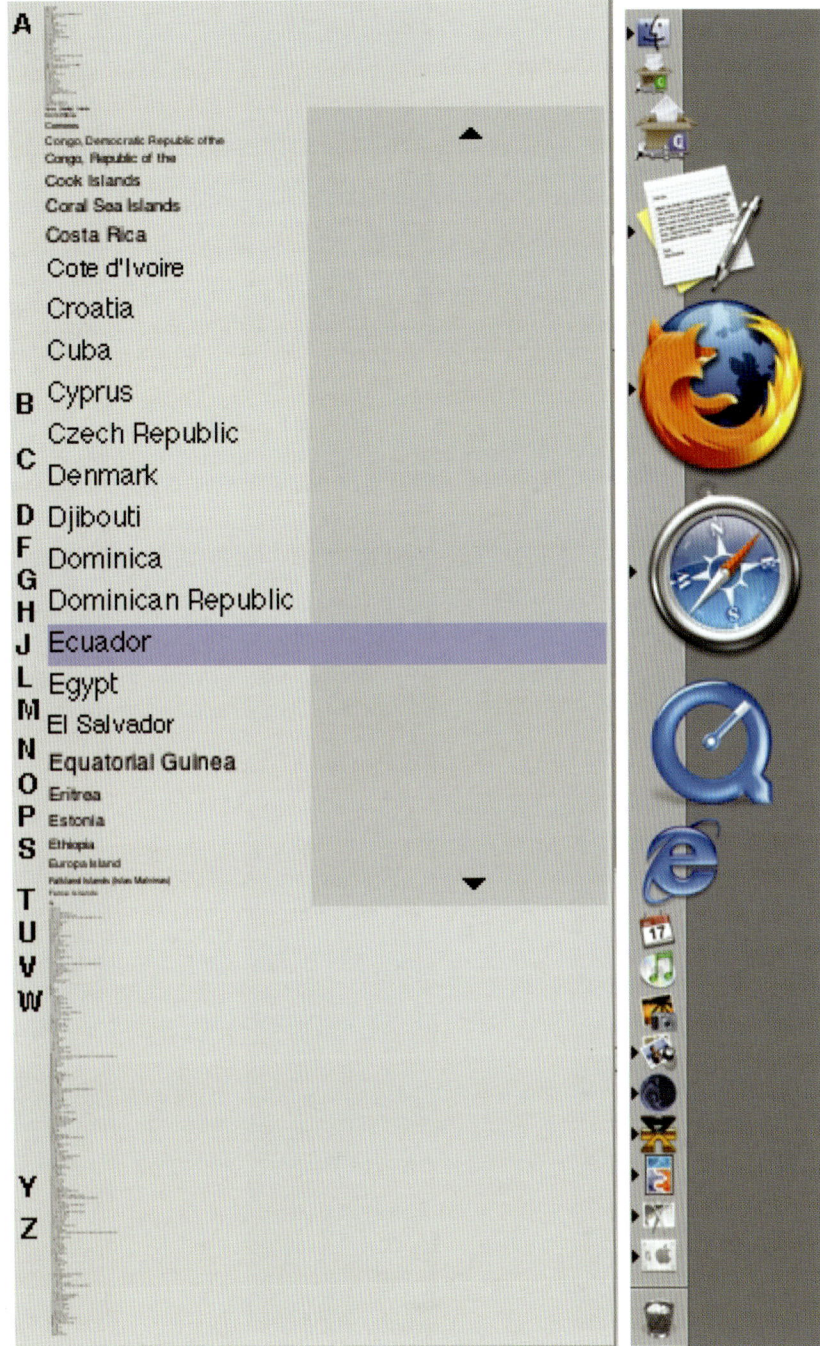

Figure 4.15. Fisheye Menus (page 161).

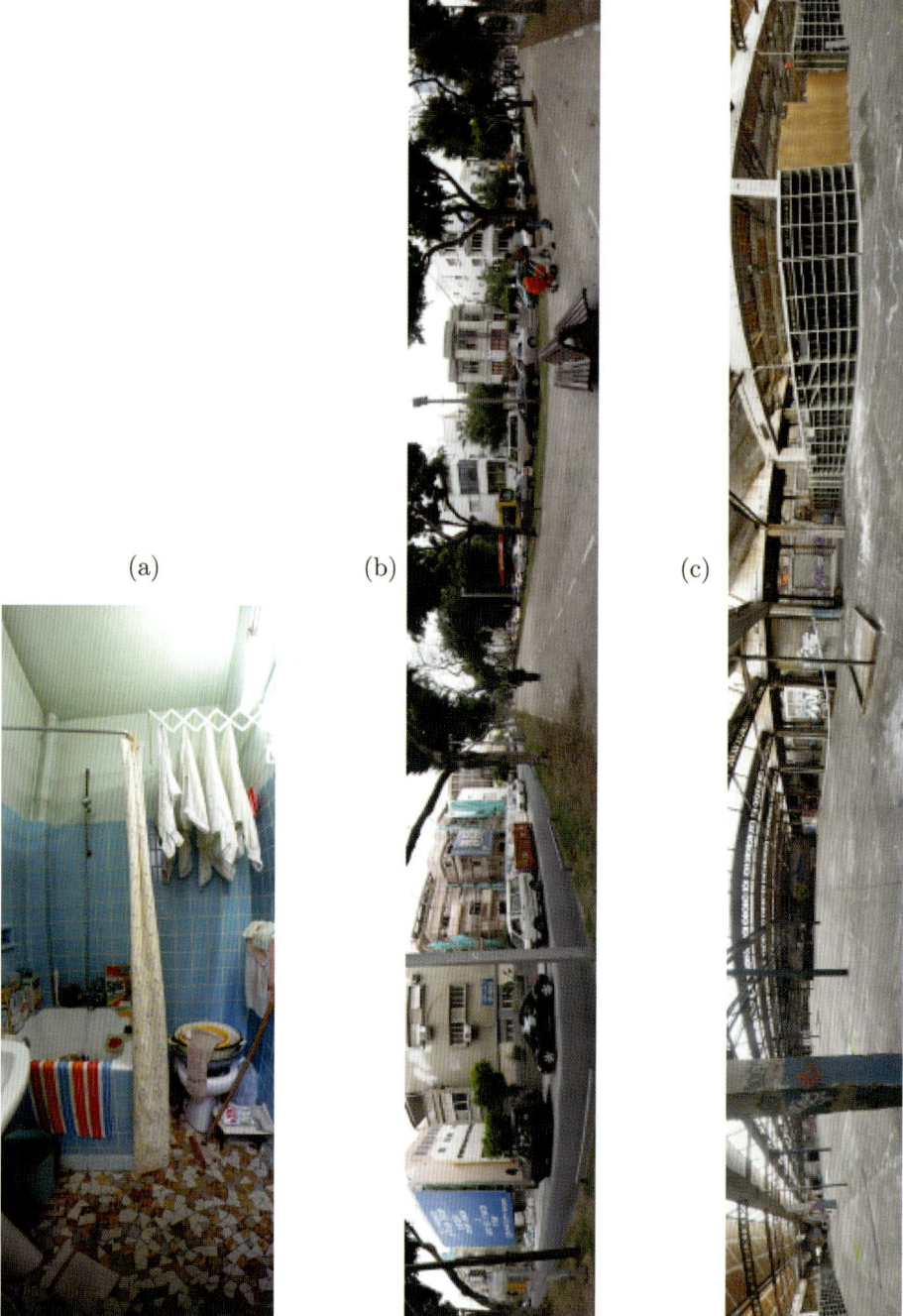

(a)   (b)   (c)

Figure 4.25. (a) Vertical and (b,c) Horizontal Cylindrical Projections (page 173, courtesy of Ari Salomon).

Figure 4.32. Cubic Panoramic Perspective.

(page 183, courtesy of Professor Shinji Araya, Fukuoka Institute of Technology.)

The world today doesn't make sense, so why should I paint pictures that do?

Every child is an artist. The problem is how to remain an artist once we grow up.

Painting is just another way of keeping a diary.

—Pablo Picasso

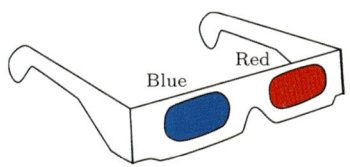

Glasses For Watching Anaglyphs (page 135).

Figure C.5: *Emptiness*, a 24" Termesphere (1986, courtesy of Dick Termes).

"This sphere shows rooms within rooms within rooms around you. Each room has one person which shows another type of emptiness" (Dick Termes).

Figure C.6: Four Variations of Anamorphosis.

Prepared by the author with *AnamorphMe* by Phillip Kent, free software for Windows [AnamorphMe 05]. Clockwise from top left: Conical mirror, cylindrical mirror, pyramid, and conical. The original (*Still Life*, by the author) is at the center.

As the sun colors flowers, so does art color life.
—John Lubbock

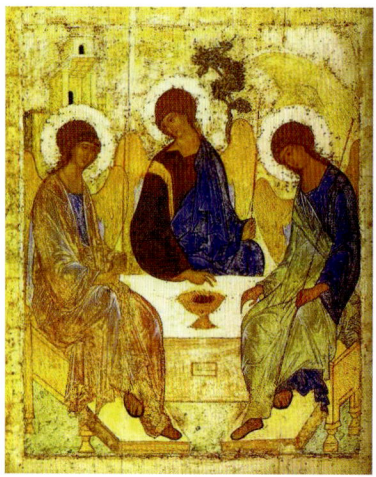

Figure C.7: Andrei Rublev, *The Old Testament Trinity*, c. 1410s.

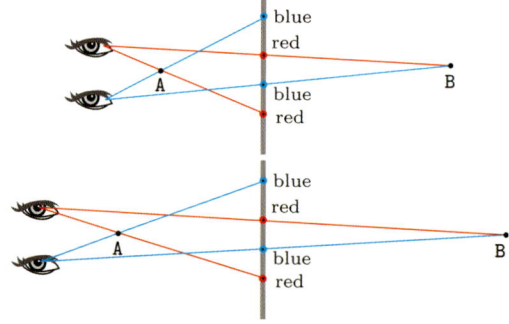

Figure 3.51. Relative Positions of the Red and Blue Parts (page 137).

Artists can color the sky red because they know it's
blue. Those of us who aren't artists must color things the
way they really are or people might think we're stupid.
—Jules Feiffer

Answers to Exercises

She smiled, but made no answer.
— Jane Austen, *Pride and Prejudice* (1813)

Pre.1: This is an impossible three-dimensional object. Such objects cannot be created in three dimensions but can be drawn in two dimensions because any projection causes loss of image detail. There are many examples of impossible objects, and this one, which is known as Schuster's conundrum or the Devil's fork, is especially simple. Notice that this impossible object cannot be colored.

The following original, succinct description of impossible objects is given by [Penrose and Penrose 58]. "Each individual part is acceptable as a representation of an object normally situated in three-dimensional space; and yet, owing to false [connections] of the parts, acceptance of the whole figure on this basis leads to the illusory effect of an impossible structure."

Intro.1: Map projections. Projecting a sphere on a flat surface always results in deformations, but such projections are important in cartography.

Intro.2: In one dimension the "universe" is a straight line. The only graphical elements on a line are points and line segments. They can be moved about on the line (translated), and segments can also be scaled. A rotation takes a line segment outside of the line, so rotation is not a one-dimensional transformation. A reflection is identical to moving a segment or a point, so it cannot be considered an independent transformation. Similarly, shearing a line segment either changes its length (which makes it identical to scaling) or takes it outside the line. Thus, the only basic independent transformations in one dimension are translation and scaling. The latter applies only to line segments.

1.1: Function f_1 is not onto since point $(-1, 0)$ is not the mapping of any real point. This function is also not one-to-one since the two different points (a, b) and $(-a, b)$ map to (a^2, b). Function f_2, however, is a valid geometric transformation.

1.2:   No. It is easy to come up with examples of two transformations f and g such that $f \circ g \neq g \circ f$. One example is a 90° counterclockwise rotation about the origin and a reflection about the x axis. When the point $(1,0)$ is first rotated 90° about the origin and then reflected about the x axis, it is first moved to $(0,1)$ and then ends up at $(0,-1)$. If the same point is first reflected and then rotated, it first moves to itself and then to $(0,1)$.

1.3:   This is a direct application of Equation (1.3). The result is

$$A(b_{11}x^* + b_{12}y^*)^2 + B(b_{11}x^* + b_{12}y^*)(b_{21}x^* + b_{22}y^*) + C(b_{21}x^* + b_{22}y^*)^2$$
$$+ D(b_{11}x^* + b_{12}y^*) + E(b_{21}x^* + b_{22}y^*) + F = 0,$$

which is a second-degree curve.

1.4:   A point (x, y) on a circle with radius R satisfies $x^2 + y^2 = R^2$ or $(x/R)^2 + (y/R)^2 = 1$. The transformed point (x^*, y^*) on an ellipse should satisfy $(x/a)^2 + (y/b)^2 = 1$. It is easy to guess that the transformation rule is $x^* = ax/R$, $y^* = by/R$, but this can also be proved as follows. The general scaling transformation is $x^* = k_1 x$, $y^* = k_2 y$. For the transformed point to be on an ellipse, it should satisfy $(k_1 x/a)^2 + (k_2 y/b)^2 = 1$, which can be simplified to $k_1^2 b^2 x^2 + k_2^2 a^2 y^2 = a^2 b^2$. Substituting $y^2 = R^2 - x^2$ yields

$$(k_1^2 b^2 - k_2^2 a^2)x^2 = a^2 b^2 - k_2^2 a^2 R^2.$$

This equation must hold for every value of x, which is possible only if $k_1^2 b^2 - k_2^2 a^2 = 0$ and $a^2 b^2 - k_2^2 a^2 R^2 = 0$. Solving these equations yields $k_1 = a/R$ and $k_2 = b/R$.

1.5:   The transformation can be written $(x, y) \rightarrow (x, -x + y)$, so $(1,0) \rightarrow (1,-1)$, $(3,0) \rightarrow (3,-3)$, $(1,1) \rightarrow (1,0)$, and $(3,1) \rightarrow (3,-2)$. The original rectangle is therefore transformed into a parallelogram.

1.6:   From $\cos 45° = 0.7071$ and $\tan 45° = 1$, we get the 45° rotation matrix as the product:

$$\begin{pmatrix} 0.7071 & 0 \\ 0 & 0.7071 \end{pmatrix} \begin{pmatrix} 1 & -1 \\ 1 & 1 \end{pmatrix}.$$

Figure Ans.1 shows how a 2×2 square centered on the origin (Figure Ans.1a) is first shrunk to about 70% of its original size (Figure Ans.1b), then sheared by the second matrix according to $(x^*, y^*) = (x + y, -x + y)$, and then becomes the rotated diamond shape of Figure Ans.1c. Direct calculations show that the two original corners $(-1,1)$ and $(1,1)$ are transformed to $(0, 1.4142)$ and $(1.4142, 0)$, respectively.

1.7:   Figure 1.3 gives the polar coordinates $\mathbf{P} = (r, \alpha)$ and $\mathbf{P}^* = (r, \phi) = (r, \alpha - \theta)$. There is no 2×2 matrix for this rotation because our transformation matrices are linear, while the transformation between polar and Cartesian coordinates is nonlinear. This is true because, for example, $(\alpha x, \alpha y) = (\alpha r, \theta)$. Multiplying both the x and y components by a constant multiplies r by that constant but does not change θ.

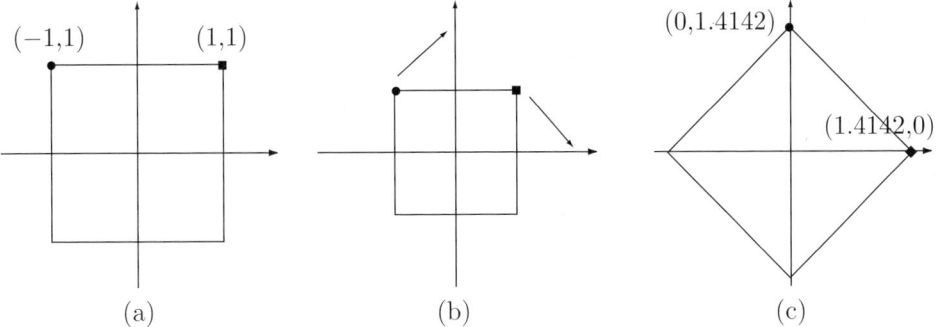

Figure Ans.1: A 45° Rotation As Scaling and Shearing.

We can artificially construct a matrix $\mathbf{T} = \begin{pmatrix} a\,b \\ c\,d \end{pmatrix}$ such that $\mathbf{P}^*$ will equal the product $\mathbf{PT}$. It does not take much to figure out that what we are looking for is

$$\mathbf{T} = \begin{pmatrix} 1 & -\theta/r \\ 0 & 1 \end{pmatrix}.$$

However, this matrix, which should be independent of the coordinates of the rotated point, depends on r. Also, it is not orthonormal (and not even orthogonal).

1.8: A reflection about the x axis transforms a point (x, y) to a point $(x, -y)$. A reflection about $y = -x$ similarly transforms a point (x, y) to a point $(-y, -x)$. [This is matrix $\mathbf{T}_3$ of Equation (1.5).] Thus, the combination of these two transformations transforms (x, y) to $(y, -x)$, which is another form of the negate and exchange rule, corresponding to a 90° clockwise rotation about the origin. This rotation can also be expressed by the matrix [compare with Equation (1.6)]

$$\begin{pmatrix} \cos 90° & \sin 90° \\ -\sin 90° & \cos 90° \end{pmatrix} = \begin{pmatrix} 0 & 1 \\ -1 & 0 \end{pmatrix}.$$

1.9: The determinant of this matrix equals

$$\left(\frac{1-t^2}{1+t^2}\right)^2 - \frac{-4t^2}{(1+t^2)^2} = \frac{(1-t^2)^2 + 4t^2}{(1+t^2)^2} = +1,$$

which shows that it generates pure rotation. Also, if we denote this matrix by

$$\begin{pmatrix} a_{11} & a_{12} \\ a_{21} & a_{22} \end{pmatrix},$$

it is easy to see that $a_{11} = a_{22}$, $a_{12} = -a_{21}$, $a_{11}^2 + a_{12}^2 = 1$, and $a_{21}^2 + a_{22}^2 = 1$. These properties are all satisfied by a rotation matrix.

1.10: The determinant of this matrix is

$$\left(\frac{a}{A}\right)^2 - \frac{b}{A}\left(-\frac{b}{A}\right) = \frac{a^2 + b^2}{A^2}.$$

It equals 1 for $A = \pm\sqrt{a^2 + b^2}$ but cannot equal -1 since it is the ratio of the two nonnegative numbers $a^2 + b^2$ and A^2. We consequently conclude that this matrix can represent pure rotation but never pure reflection. An example of pure rotation is $a = b = 1$, which produces $A = \pm\sqrt{2} \approx \pm 1.414$. The rotation matrices for this case are

$$\left(\begin{array}{cc} 1/\sqrt{2} & 1/\sqrt{2} \\ -1/\sqrt{2} & 1/\sqrt{2} \end{array}\right) = \left(\begin{array}{cc} 0.7071 & 0.7071 \\ -0.7071 & 0.7071 \end{array}\right),$$

$$\left(\begin{array}{cc} -1/\sqrt{2} & -1/\sqrt{2} \\ 1/\sqrt{2} & -1/\sqrt{2} \end{array}\right) = \left(\begin{array}{cc} -0.7071 & -0.7071 \\ 0.7071 & -0.7071 \end{array}\right),$$

and they correspond to $45°$ rotations about the origin.

1.11: The combined transformation matrix is the product

$$\left(\begin{array}{ccc} 1 & 0 & 0 \\ 0 & -1 & 0 \\ 0 & 0 & 1 \end{array}\right) \cdot \left(\begin{array}{ccc} 1 & 0 & 0 \\ 0 & 1 & 0 \\ -1 & -1 & 1 \end{array}\right) \left(\begin{array}{ccc} \cos 180° & -\sin 180° & 0 \\ \sin 180° & \cos 180° & 0 \\ 0 & 0 & 1 \end{array}\right) = \left(\begin{array}{ccc} -1 & 0 & 0 \\ 0 & 1 & 0 \\ 1 & 1 & 1 \end{array}\right).$$

This matrix combines a reflection of the x coordinates with a one-unit translation in the x and y directions. Applying it to the four points yields $(0,2)$, $(0,0)$, $(2,2)$, and $(2,0)$. This is the same square but is now located in the first quadrant (Figure Ans.2).

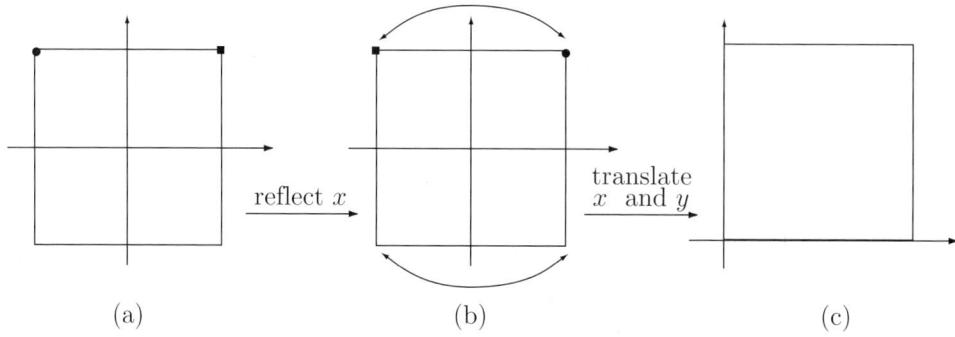

(a) (b) (c)

Figure Ans.2: An x Reflection and Translation.

1.12: Using angles ϕ and θ from Figure 1.3 but assuming that the rotation is counterclockwise about (x_0, y_0), we get

$$x^* = x_0 + (x - x_0)\cos\theta - (y - y_0)\sin\theta,$$
$$y^* = y_0 + (x - x_0)\sin\theta + (y - y_0)\cos\theta.$$

We are looking for a matrix $\mathbf{T}$ that satisfies

$$(x^*, y^*, 1) = (x, y, 1) \begin{pmatrix} a & b & 0 \\ c & d & 0 \\ m & n & 1 \end{pmatrix}.$$

The simple solution is

$$\mathbf{T} = \begin{pmatrix} \cos\theta & \sin\theta & 0 \\ -\sin\theta & \cos\theta & 0 \\ x_0(1-\cos\theta) + y_0\sin\theta & y_0(1-\cos\theta) - x_0\sin\theta & 1 \end{pmatrix}.$$

In a similar way, it can be shown that a clockwise rotation about (x_0, y_0) is produced by

$$\mathbf{T} = \begin{pmatrix} \cos\theta & -\sin\theta & 0 \\ \sin\theta & \cos\theta & 0 \\ x_0(1-\cos\theta) - y_0\sin\theta & y_0(1-\cos\theta) + x_0\sin\theta & 1 \end{pmatrix}.$$

1.13: If a point $\mathbf{P} = (x, y, 1)$ is reflected to a point $\mathbf{P}^* = (x^*, y^*, 1) = (y-1, x+1, 1)$ about the line $y = x+1$, then their midpoint [which is $(\mathbf{P}+\mathbf{P}^*)/2 = (x+y-1, y+x+1)/2$] should be on the line. It's easy to see that the midpoint is on the line because its y coordinate equals 1 more than its x coordinate.

1.14: This is easily done with the help of appropriate mathematical software, and the result is

$$\begin{pmatrix} 0.5 & 0.866 & 0 \\ 0.866 & -0.5 & 0 \\ -0.866 & 1.5 & 1 \end{pmatrix}.$$

1.15: Such a thing is possible but would not improve the algorithm. Transforming a point from octant 1 to octant 2 is done by reflecting it about the $45°$ line $y = x$. A point (x, y) is therefore transformed to the point (y, x). The similar transformation between half-octants amounts to reflection about the $22.5°$ line $y = ax$ (where $a = \tan 22.5° \approx 0.414$). This transforms point (x, y) to $(0.7071x + 0.7071y, 0.7071x - 0.7071y)$ (see the following proof) and would slow down the algorithm since it involves real-number arithmetic.

　　Proof. Let's denote $\alpha = \sin 22.5°$, $\beta = \cos 22.5°$. To reflect about the $22.5°$ line, we rotate clockwise by $22.5°$, reflect about the x axis, and rotate back. The combined transformation matrix is

$$\begin{pmatrix} \beta & -\alpha & 0 \\ \alpha & \beta & 0 \\ 0 & 0 & 1 \end{pmatrix} \begin{pmatrix} 1 & 0 & 0 \\ 0 & -1 & 0 \\ 0 & 0 & 1 \end{pmatrix} \begin{pmatrix} \beta & \alpha & 0 \\ -\alpha & \beta & 0 \\ 0 & 0 & 1 \end{pmatrix}$$

$$= \begin{pmatrix} \beta^2 - \alpha^2 & 2\alpha\beta & 0 \\ 2\alpha\beta & \alpha^2 - \beta^2 & 0 \\ 0 & 0 & 1 \end{pmatrix} \approx \begin{pmatrix} .7071 & .7071 & 0 \\ .7071 & -.7071 & 0 \\ 0 & 0 & 1 \end{pmatrix}.$$

The last equality is true because

$$0.7071 \approx \sin 45° = \sin 22.5° \cos 22.5° + \cos 22.5° \sin 22.5° = 2\alpha\beta,$$
$$0.7071 \approx \cos 45° = \cos 22.5° \cos 22.5° - \sin 22.5° \sin 22.5° = \beta^2 - \alpha^2.$$

1.16:   In order for the general line $ax + by + c = 0$ to pass through the origin, it must satisfy $c = 0$. This implies $y = -(a/b)x$, so $-a/b$ is the slope (i.e., $\tan\theta$) and a and b equal $\sin\theta$ and $\cos\theta$, respectively, up to a sign. This also implies $a^2 + b^2 = 1$ and $ab = \sin\theta\cos\theta$. When this is substituted in Equation (1.12), it reduces to

$$
\begin{aligned}
x^* &= x - 2a(ax + by) = x(1 - 2a^2) - 2aby \\
&= x\cos(2\theta) + y\sin(2\theta), \\
y^* &= y - 2b(ax + by) = -2abx + y(1 - 2b^2) \\
&= x\sin(2\theta) - y\cos(2\theta).
\end{aligned}
\tag{Ans.1}
$$

1.17:   Reflecting a point (x, y) about the line $y = c$ moves it to $(x, 2c - y)$. Reflecting this about line $y = 0$ simply reverses the y coordinate. Thus, the two reflections move (x, y) to $(x, y - 2c)$. This is a translation of $-2c$ units in the y direction.

1.18:   Starting with $\sin 90° = 1$, $\cos 90° = 0$, we multiply the matrices to obtain

$$
\begin{pmatrix} 0 & 1 & 0 \\ 2 & 0 & 0 \\ 0 & 0 & 1 \end{pmatrix}
\begin{pmatrix} 0 & -1 & 0 \\ 1 & 0 & 0 \\ 0 & 0 & 1 \end{pmatrix}
=
\begin{pmatrix} 1 & 0 & 0 \\ 0 & -2 & 0 \\ 0 & 0 & 1 \end{pmatrix},
$$

which is a reflection and scaling in the y dimension.

1.19:   Direct multiplication yields

$$
\begin{pmatrix}
\cos\theta_1\cos\theta_2 - \sin\theta_1\sin\theta_2 & -\cos\theta_1\sin\theta_2 - \cos\theta_2\sin\theta_1 & 0 \\
\sin\theta_1\cos\theta_2 + \cos\theta_1\sin\theta_2 & -\sin\theta_1\sin\theta_2 + \cos\theta_1\cos\theta_2 & 0 \\
0 & 0 & 1
\end{pmatrix}
$$
$$
=
\begin{pmatrix}
\cos(\theta_1 + \theta_2) & -\sin(\theta_1 + \theta_2) & 0 \\
\sin(\theta_1 + \theta_2) & \cos(\theta_1 + \theta_2) & 0 \\
0 & 0 & 1
\end{pmatrix},
$$

thereby proving that two-dimensional rotations are additive.

1.20:   Direct multiplication yields

$$
\mathbf{T}_1\mathbf{T}_2 =
\begin{pmatrix}
1 + bc & b & 0 \\
c & 1 & 0 \\
0 & 0 & 1
\end{pmatrix}.
$$

This is a combination of shearing and scaling in the x direction. It is pure shearing only if $bc = 0$. This shows that shearing is not an additive transformation.

1.21: The product of the three shears is

$$\begin{pmatrix} 1 & a \\ 0 & 1 \end{pmatrix} \begin{pmatrix} 1 & 0 \\ b & 1 \end{pmatrix} \begin{pmatrix} 1 & c \\ 0 & 1 \end{pmatrix} = \begin{pmatrix} ab+1 & a+abc+c \\ b & bc+1 \end{pmatrix}.$$

When we equate this to the standard rotation matrix

$$\begin{pmatrix} \cos\theta & -\sin\theta \\ \sin\theta & \cos\theta \end{pmatrix},$$

we end up with

$$a = c = \frac{\cos\theta - 1}{\sin\theta} = -\tan\frac{\theta}{2}, \quad \text{and} \quad b = \sin\theta,$$

which shows how to calculate a, b, and c from θ. Notice that both $(\cos\theta - 1)$ and $\sin\theta$ approach zero for small angles. The ratio of two small numbers is hard to calculate with any precision, which is why it is preferable to use $\tan(\theta/2)$ instead. This particular combination of transformations does not save any time because we still have to calculate $\sin\theta$ and $\cos\theta$ in order to obtain a, b, and c. Still, it is an interesting, unexpected result that's illustrated in Figure Ans.3 for $\theta = 45°$.

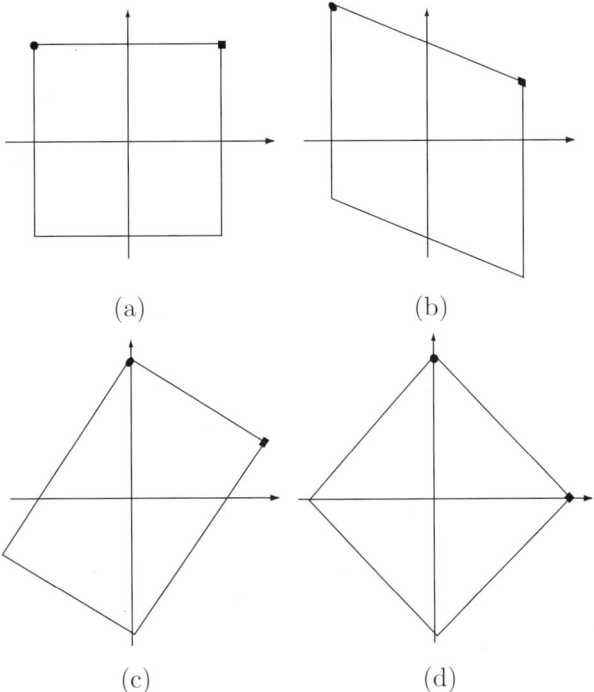

(a)　　　　　　　　(b)

(c)　　　　　　　　(d)

Figure Ans.3: A 45° Rotation As Three Successive Shearings.

1.22:   The transformation matrices are

$$\begin{pmatrix} \cos\theta & -\sin\theta & 0 \\ \sin\theta & \cos\theta & 0 \\ 0 & 0 & 1 \end{pmatrix} \begin{pmatrix} a & 0 & 0 \\ 0 & d & 0 \\ 0 & 0 & 1 \end{pmatrix} \begin{pmatrix} \cos\theta & \sin\theta & 0 \\ -\sin\theta & \cos\theta & 0 \\ 0 & 0 & 1 \end{pmatrix}$$

$$= \begin{pmatrix} a\cos^2\theta + d\sin^2\theta & (d-a)\cos\theta\sin\theta & 0 \\ (d-a)\cos\theta\sin\theta & a\sin^2\theta + d\cos^2\theta & 0 \\ 0 & 0 & 1 \end{pmatrix}.$$

When $a = d$, this reduces to

$$\begin{pmatrix} a & 0 & 0 \\ 0 & a & 0 \\ 0 & 0 & 1 \end{pmatrix},$$

which does not depend on θ! This proves that uniform scaling produces identical results regardless of the particular axes used.

1.23:   We simply multiply

$$\begin{pmatrix} \cos\theta & -\sin\theta & 0 \\ \sin\theta & \cos\theta & 0 \\ 0 & 0 & 1 \end{pmatrix} \begin{pmatrix} 1 & b & 0 \\ c & 1 & 0 \\ 0 & 0 & 1 \end{pmatrix} \begin{pmatrix} \cos\theta & \sin\theta & 0 \\ -\sin\theta & \cos\theta & 0 \\ 0 & 0 & 1 \end{pmatrix}$$

$$= \begin{pmatrix} \cos^2\theta - c\sin\theta\cos\theta - & \sin\theta\cos\theta - c\sin^2\theta + & 0 \\ -b\sin\theta\cos\theta + \sin^2\theta & +b\cos^2\theta - \sin\theta\cos\theta & \\ \sin\theta\cos\theta + c\cos^2\theta - & \sin^2\theta + c\sin\theta\cos\theta + & 0 \\ -b\sin^2\theta - \sin\theta\cos\theta & +b\sin\theta\cos\theta + \cos^2\theta & \\ 0 & 0 & 1 \end{pmatrix}$$

$$= \begin{pmatrix} 1 - (b+c)\sin\theta\cos\theta & b\cos^2\theta - c\sin^2\theta & 0 \\ c\cos^2\theta - b\sin^2\theta & 1 + (b+c)\sin\theta\cos\theta & 0 \\ 0 & 0 & 1 \end{pmatrix}.$$

This expression does depend on θ! When $b = c = 0$, the expression reduces to the identity matrix. However, when $b = c \neq 0$, this does not reduce to anything simple or elegant.

1.24:   A direct scaling of point $\mathbf{P} = (x, y)$ relative to (x_0, y_0) is done by

$$x^* = x_0 + (x - x_0)s_x = x \cdot s_x + x_0(1 - s_x),$$
$$y^* = y_0 + (y - y_0)s_y = y \cdot s_y + y_0(1 - s_y).$$

Using matrix notation, this is written as

$$(x^*, y^*, 1) = (x, y, 1) \begin{pmatrix} s_x & 0 & 0 \\ 0 & s_y & 0 \\ x_0(1 - s_x) & y_0(1 - s_y) & 1 \end{pmatrix}. \qquad \text{(Ans.2)}$$

Performing the same transformation by means of translation, scaling, and reverse translation is done by the matrix product

$$\begin{pmatrix} 1 & 0 & 0 \\ 0 & 1 & 0 \\ -x_0 & -y_0 & 1 \end{pmatrix} \begin{pmatrix} s_x & 0 & 0 \\ 0 & s_y & 0 \\ 0 & 0 & 1 \end{pmatrix} \begin{pmatrix} 1 & 0 & 0 \\ 0 & 1 & 0 \\ x_0 & y_0 & 1 \end{pmatrix},$$

which produces the same result.

1.25: Substituting $k_1 = k_2 = k$ in Equation (1.16) yields

$$\begin{pmatrix} k^2 & 0 & 0 \\ 0 & k^2 & 0 \\ k(1-k)x_1 + (1-k)x_2 & k(1-k)y_1 + (1-k)y_2 & 1 \end{pmatrix}.$$

This is equivalent to a single scaling by a factor k^2 about point

$$\mathbf{P}_c = \frac{k(1-k)}{1-k^2}\mathbf{P}_1 + \frac{1-k}{1-k^2}\mathbf{P}_2 = \frac{k}{1+k}\mathbf{P}_1 + \frac{1}{1+k}\mathbf{P}_2.$$

1.26: Using homogeneous coordinates, we transform

$$(t^2, t, 1)\begin{pmatrix} -1 & 0 & 1 \\ 0 & 2 & 0 \\ 1 & 0 & 1 \end{pmatrix} = (1 - t^2, 2t, 1 + t^2),$$

which, after dividing by the third component, becomes the point

$$\left(\frac{1 - t^2}{1 + t^2}, \frac{2t}{1 + t^2}\right).$$

This point satisfies the relation $x^2 + y^2 = 1$, so it is located on the unit circle.

1.27: The *Mathematica* code

```
t14=2^14;
Print["(x*=",(8192-(2 14189.))/t14,",y*=",(14189.+(2 8192))/t14,")"]
Print["(x*=",Cos[60 Degree]-2. Sin[60 Degree],
  ",y*=",Sin[60 Degree]+2. Cos[60 Degree], ")"]
```

calculates the rotated point twice, first using integers and then using *Mathematica*'s built-in sine and cosine functions. The results are identical: $(x^* = -1.23206, y^* = 1.86603)$.

For an 80° rotation, the code

```
t14=2^14;
```

```
Print["(x*=",(2845.-(2 16135.))/t14,",y*=",(16135.+(2 2845.))/t14,
  ")"]
Print["(x*=",Cos[80 Degree]-2. Sin[80 Degree],
  ",y*=",Sin[80 Degree]+2. Cos[80 Degree], ")"]
```

produces $(x^* = -1.79596, y^* = 1.33209)$ and $(x^* = -1.79597, y^* = 1.3321)$ (a slightly different result).

1.28: From the definition of θ_i, we know that the ratio $\tan\theta_{i+1}/\tan\theta_i$ is $1/2$. Small angles satisfy $\tan\theta \approx \theta$, so we conclude that the ratio θ_{i+1}/θ_i equals approximately $1/2$, except for the first few θ_i's. This can also be confirmed by manually checking the ratios from Table 1.11. Given an infinite sequence of numbers t, $t/2$, $t/4$,..., $t/2^i$, we can express every number from 0 (which is obtained by subtracting all the numbers in the sequence from the first one) to $2t$ (which is obtained by adding all the numbers in the sequence). Our sequence of θ_i is finite and the ratio of consecutive elements isn't always precisely $1/2$, but [Walther 71] proves that every number in the range $[0, 90°)$ can be reached, up to a certain precision, by adding and subtracting a number of consecutive θ_i's.

1.29: The method proposed here is based on the fact that the magnitude of the rotated vector (x^*, y^*) should be identical to that of the original vector (x, y). This can be achieved by first normalizing (x^*, y^*) and then multiplying it by the magnitude of (x, y),

$$(x^*, y^*) \leftarrow (x^*, y^*)\frac{\sqrt{x^2 + y^2}}{\sqrt{x^{*2} + y^{*2}}} = (x^*, y^*)\sqrt{\frac{x^2 + y^2}{x^{*2} + y^{*2}}},$$

a calculation involving four exponentiations, one division, one multiplication, and one square root.

1.30: The traditional way of calculating a sine function is by its power series

$$\sin(\theta) = \frac{\theta}{1!} - \frac{\theta^3}{3!} + \frac{\theta^5}{5!} - \frac{\theta^7}{7!} + \cdots,$$

and similarly for cosine. These series, however, converge very slowly, requiring many multiplications and divisions. If a graphics application needs just rotations, the method of Section 1.2.3 may be simpler and faster than CORDIC. The advantage of CORDIC is that it can be adapted to the calculation of many different functions. A general software package that is concerned not just with rotations may benefit from the application of CORDIC.

1.31: From the definition $k = \sqrt{a^2 + c^2}$, it follows that $k = 0$ implies $a = c = 0$. In this case, the similarity becomes $x^* = m$, $y^* = n$, and this is not a transformation because it is not one-to-one.

1.32:   Transforming point $(x - 2P_x + 2Q_x, y - 2P_y + 2Q_y)$ through another halfturn yields

$$(x - 2P_x + 2Q_x, y - 2P_y + 2Q_y, 1) \begin{pmatrix} -1 & 0 & 0 \\ 0 & -1 & 0 \\ 2R_x & 2R_y & 1 \end{pmatrix}$$

$$= (-x + 2P_x - 2Q_x + 2R_x, -y + 2P_y - 2Q_y + 2R_y, 1).$$

Comparing this with Equation (1.19) shows that the result of three halfturns is a halfturn about the point $\mathbf{S} = \mathbf{P} - \mathbf{Q} + \mathbf{R}$. Writing this as $\mathbf{S} - \mathbf{P} = \mathbf{R} - \mathbf{Q}$ shows that $\mathbf{PQRS}$ is a parallelogram (Figure 1.14c). Thus, point $\mathbf{S}$ completes the original three points to a parallelogram.

1.33:   The first part results in

$$(x^*, y^*) = (x, y) \begin{pmatrix} 3 & 4 & 0 \\ -2 & 5 & 0 \\ 1 & -6 & 1 \end{pmatrix}.$$

The decomposition is simple because $A = \sqrt{9 + 16} = 5$:

$$\begin{pmatrix} 1 & 0 & 0 \\ 14/25 & 1 & 0 \\ 0 & 0 & 1 \end{pmatrix} \begin{pmatrix} 5 & 0 & 0 \\ 0 & 23/5 & 0 \\ 0 & 0 & 1 \end{pmatrix} \begin{pmatrix} 3/5 & 4/5 & 0 \\ -4/5 & 3/5 & 0 \\ 0 & 0 & 1 \end{pmatrix} \begin{pmatrix} 1 & 0 & 0 \\ 0 & 1 & 0 \\ 1 & -6 & 1 \end{pmatrix}.$$

1.34:   From Equation (1.22), we get the following.

1. For scaling, the inverse of

$$\mathbf{T} = \begin{pmatrix} a & 0 & 0 \\ 0 & d & 0 \\ 0 & 0 & 1 \end{pmatrix} \quad \text{is} \quad \mathbf{T}^{-1} = \frac{1}{ad} \begin{pmatrix} d & 0 & 0 \\ 0 & a & 0 \\ 0 & 0 & 1 \end{pmatrix},$$

so

$$(x, y)\mathbf{T}^{-1} = \left(\frac{x}{a}, \frac{y}{d}, \frac{1}{ad} \right) \rightarrow (x^*, y^*) = (dx, ay),$$

which is also scaling by factors d and a.

2. For shearing, the inverse of

$$\mathbf{T} = \begin{pmatrix} 1 & b & 0 \\ c & 1 & 0 \\ 0 & 0 & 1 \end{pmatrix} \quad \text{is} \quad \mathbf{T}^{-1} = \frac{1}{-bc} \begin{pmatrix} 1 & -b & 0 \\ -c & 1 & 0 \\ 0 & 0 & 1 \end{pmatrix},$$

so

$$(x, y, 1)\mathbf{T}^{-1} = \left(\frac{x - yc}{-bc}, \frac{-xb + y}{-bc}, \frac{1}{-bc} \right) \rightarrow (x^*, y^*) = (x - yc, -xb + y),$$

which is a combination of shearing and scaling.

3. For rotation, the inverse of

$$\mathbf{T} = \begin{pmatrix} \cos\theta & -\sin\theta & 0 \\ \sin\theta & \cos\theta & 0 \\ 0 & 0 & 1 \end{pmatrix}$$

is

$$\mathbf{T}^{-1} = \frac{1}{\cos^2\theta + \sin^2\theta} \begin{pmatrix} \cos\theta & \sin\theta & 0 \\ -\sin\theta & \cos\theta & 0 \\ 0 & 0 & 1 \end{pmatrix} = \begin{pmatrix} \cos\theta & \sin\theta & 0 \\ -\sin\theta & \cos\theta & 0 \\ 0 & 0 & 1 \end{pmatrix}.$$

This is a rotation in the opposite direction.

4. For translation, the inverse of

$$\begin{pmatrix} 1 & 0 & 0 \\ 0 & 1 & 0 \\ m & n & 1 \end{pmatrix} \quad \text{is} \quad \begin{pmatrix} 1 & 0 & 0 \\ 0 & 1 & 0 \\ -m & -n & 1 \end{pmatrix}.$$

This is a reverse of the original translation.

1.35: We denote the transformation matrix by $\begin{pmatrix} a & b \\ c & d \end{pmatrix}$ and write the four equations

$$\mathbf{P}_i \begin{pmatrix} a & b \\ c & d \end{pmatrix} = \mathbf{P}_i^* \quad \text{for } 1 \le i \le 4.$$

These are easy to solve and yield $a = 6$, $b = 1$, $c = 2$, and $d = 3$.

1.36: The plane should pass through the three points $(0,0,0)$, $(0,0,1)$, and $(1,1,0)$. Equation (1.24) gives

$$A = \begin{vmatrix} 0 & 0 & 1 \\ 0 & 1 & 1 \\ 1 & 0 & 1 \end{vmatrix} = -1, \quad B = -\begin{vmatrix} 0 & 0 & 1 \\ 0 & 1 & 1 \\ 1 & 0 & 1 \end{vmatrix} = 1,$$

$$C = \begin{vmatrix} 0 & 0 & 1 \\ 0 & 0 & 1 \\ 1 & 1 & 1 \end{vmatrix} = 0, \quad D = -\begin{vmatrix} 0 & 0 & 0 \\ 0 & 0 & 1 \\ 1 & 1 & 0 \end{vmatrix} = 0.$$

The expression of the plane is therefore $-x + y = 0$.

1.37: They are the points where the plane $x/a + y/b + z/c = 1$ intercepts the three coordinate axes.

1.38: $s = \mathbf{N} \bullet \mathbf{P}_1 = (1,1,1) \bullet (1,1,1) = 3$, so the plane is given by $x + y + z - 3 = 0$. It intercepts the three coordinate axes at points $(3,0,0)$, $(0,3,0)$, and $(0,0,3)$ (Figure 1.20a).

1.39: The expression is

$$\mathbf{P}(u, w) = \mathbf{P}_1 + u(\mathbf{P}_2 - \mathbf{P}_1) + w(\mathbf{P}_3 - \mathbf{P}_1) = (3, 0, 0) + u(-3, 3, 0) + w(-3, 0, 3).$$

1.40: This is trivial. The origin is point $(0, 0, 0)$, and Equation (1.27) shows that the distance between it and the plane $Ax + By + Cz + D = 0$ is

$$\frac{D}{\sqrt{A^2 + B^2 + C^2}}.$$

1.41: Because d is the signed distance. If the normal points from the plane in the direction of $\mathbf{P}$, then d is positive, but we have to travel in the direction of $-\mathbf{N}$. If the normal points in a direction opposite that of $\mathbf{P}$, then we travel from $\mathbf{P}$ to $\mathbf{P}^*$ in the direction of $\mathbf{N}$ but d is negative.

1.42: The product $\mathbf{T}_r \mathbf{R}_x \mathbf{T}_{rr}$ yields

$$\begin{pmatrix} 1 & 0 & 0 & 0 \\ 0 & \cos\theta & \sin\theta & 0 \\ 0 & -\sin\theta & \cos\theta & 0 \\ 0 & m(\cos\theta - 1) - n\sin\theta & n(\cos\theta - 1) + m\sin\theta & 1 \end{pmatrix}.$$

Substituting $\theta = 30°$ produces the matrix

$$\begin{pmatrix} 1 & 0 & 0 & 0 \\ 0 & 0.866 & 0.5 & 0 \\ 0 & -0.5 & 0.866 & 0 \\ 0 & 0.634 & -0.366 & 1 \end{pmatrix},$$

which transforms point $(1, 2, 3, 1)$ to $(1, 0.866, 3.232, 1)$.

1.43: Using the rule for quaternion multiplication and the three trigonometric identities

$$\cos\theta = \cos^2\tfrac{\theta}{2} - \sin^2\tfrac{\theta}{2}, \quad \sin\theta = 2\sin\tfrac{\theta}{2}\cos\tfrac{\theta}{2}, \quad \text{and} \quad \cos\theta = 1 - 2\sin^2\tfrac{\theta}{2},$$

we can write

$$\begin{aligned}
\mathbf{q} \cdot [0, \mathbf{P}] \cdot \mathbf{q}^{-1} &= \left[\cos\tfrac{\theta}{2}, \mathbf{u}\sin\tfrac{\theta}{2}\right] \cdot [0, \mathbf{P}] \cdot \left[\cos\tfrac{\theta}{2}, -\mathbf{u}\sin\tfrac{\theta}{2}\right] \\
&= \left\{\left[\cos\tfrac{\theta}{2}, \mathbf{u}\sin\tfrac{\theta}{2}\right] \cdot [0, \mathbf{P}]\right\} \cdot \left[\cos\tfrac{\theta}{2}, -\mathbf{u}\sin\tfrac{\theta}{2}\right] \\
&= \left[-\sin\tfrac{\theta}{2}(\mathbf{u} \bullet \mathbf{P}), \cos\tfrac{\theta}{2}\mathbf{P} + \sin\tfrac{\theta}{2}(\mathbf{u} \times \mathbf{P})\right] \cdot \left[\cos\tfrac{\theta}{2}, -\mathbf{u}\sin\tfrac{\theta}{2}\right] \\
&= \left[-\sin\tfrac{\theta}{2}\cos\tfrac{\theta}{2}(\mathbf{u} \bullet \mathbf{P}) + \sin\tfrac{\theta}{2}\cos\tfrac{\theta}{2}(\mathbf{P} \bullet \mathbf{u}) - \sin^2\tfrac{\theta}{2}(\mathbf{u} \times \mathbf{P}) \bullet \mathbf{u}, \right. \\
&\quad \left. \sin^2\tfrac{\theta}{2}(\mathbf{u} \bullet \mathbf{P})\mathbf{u} + \cos^2\tfrac{\theta}{2}\mathbf{P} + \sin\tfrac{\theta}{2}\cos\tfrac{\theta}{2}(\mathbf{u} \times \mathbf{P})\right.
\end{aligned}$$

$$-\sin\tfrac{\theta}{2}\cos\tfrac{\theta}{2}(\mathbf{P}\times\mathbf{u})-\sin^2\tfrac{\theta}{2}(\mathbf{u}\times\mathbf{P})\times\mathbf{u}]$$
$$=[0,\sin^2\tfrac{\theta}{2}(\mathbf{u}\bullet\mathbf{P})\mathbf{u}+\cos^2\tfrac{\theta}{2}\mathbf{P}+2\sin\tfrac{\theta}{2}\cos\tfrac{\theta}{2}(\mathbf{u}\times\mathbf{P})$$
$$\qquad-\sin^2\tfrac{\theta}{2}(\mathbf{P}-(\mathbf{u}\bullet\mathbf{P})\mathbf{u})]$$
$$=[0,2\sin^2\tfrac{\theta}{2}(\mathbf{u}\bullet\mathbf{P})\mathbf{u}+(\cos^2\tfrac{\theta}{2}-\sin^2\tfrac{\theta}{2})\mathbf{P}+2\sin\tfrac{\theta}{2}\cos\tfrac{\theta}{2}(\mathbf{u}\times\mathbf{P})]$$
$$=[0,(1-\cos\theta)(\mathbf{u}\bullet\mathbf{P})\mathbf{u}+\cos\theta\mathbf{P}+\sin\theta(\mathbf{u}\times\mathbf{P})]$$
$$=[0,(\mathbf{u}\bullet\mathbf{P})\mathbf{u}+\cos\theta(\mathbf{P}-(\mathbf{u}\bullet\mathbf{P})\mathbf{u})+\sin\theta(\mathbf{u}\times\mathbf{P})],$$

that is Equation (1.31).

2.1:   They could be (a) a cube, (b) the same cube seen edge on, and (c) the same cube seen rotated through 30° with one front edge and one back edge.

2.2:   Given $s_z = 0.625$, we calculate θ and ϕ

$$\theta = \sin^{-1}\left(\pm\frac{0.625}{\sqrt{2}}\right) = \sin^{-1}(\pm0.44194) = \pm26.23°,$$

$$\phi = \sin^{-1}\left(\pm\frac{0.625}{\sqrt{2-0.625^2}}\right) = \sin^{-1}(\pm0.49266) = \pm29.52°.$$

2.3:   Equation (2.4) shows that $s_x^2 = s_z^2$ is equivalent to

$$\cos^2\phi + \sin^2\phi\sin^2\theta = \sin^2\phi + \cos^2\phi\sin^2\theta.$$

This can be simplified to $(\sin^2\phi - \cos^2\phi)\cos^2\theta = 0$, with the two solutions $\cos^2\theta = 0 \to \theta = \pm90°$ and $\sin^2\phi - \cos^2\phi = 0$, which implies $\sin\phi = \pm\cos\phi$ and results in $\phi = 90° \pm 45°$ and $270° \pm 45°$.

3.1:   Such examples abound, mostly in modern art, which is one reason why many consider modern art trivial or false. Figure C.7 on page 242 [*The Old Testament Trinity*, c. 1410s, by the Russian painter Andrei Rublev is an example of reversed perspective. A well-known example of diverging lines is *Woman in Mirror* (1937) by Picasso.

3.2:   A rolodex [eldonoffice 05] features many vanishing points because each of its index cards is oriented differently, causing its sides to seem to converge to a different point. A striped shirt may feature several vanishing points because the groups of parallel stripes on a sleeve, on the shirt itself, and on the flat parts of the collar may point in different directions. Long, meandering railway tracks may feature straight segments that go in different directions and create different vanishing points. Many scenes feature multiple vanishing points, as illustrated by the flat rectangles of Figure Ans.4. The well-known drawing *High and Low* by Escher [Ernst 76] features five vanishing points, four near the four corners of the figure and the fifth one at the center.

Figure Ans.4: Many Vanishing Points.

3.3: Yes, by viewing it through a telescope. This device "telescopes" a scene and brings objects closer to the observer rather than magnifying them, but it does not affect the perspective. See Section 4.11 for the telescopic projection.

3.4: Figure Ans.5 illustrates the construction. First, lines "a" and "b" are constructed, followed by the two lines labeled "c." This is followed by the eight "d" lines, four of which are equally spaced on the left-hand side of "b" and the other four equally spaced on the right-hand side of "b." The last step is to construct the eight "e" vertical line segments.

> We shall therefore borrow all our rules for the finishing of our proportions, from the musicians, who are the greatest masters of this sort of numbers, and from those things wherein nature shows herself most excellent and compleat.
>
> —Leon Battista Alberti

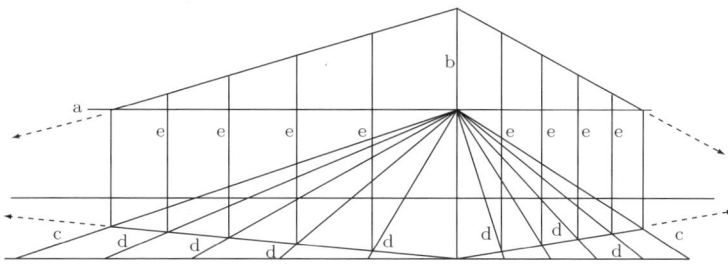

Figure Ans.5: Two-Point Perspective With Equally-Spaced Lines.

3.5: Because the seven horizontal lines of the grid of part (b) are no longer equally spaced. Instead, they converge toward the top of the grid.

3.6: We start with a rectangle in one-point perspective and determine its single vanishing point (Figure Ans.6). We then copy the bottom line of the original rectangle (with the five numbered key points) and move it between the converging lines to form line "a." This makes it easy to construct the three lines "b." Next, the left-hand side of the original rectangle (with the seven points labeled "A" through "G") is placed to the right of the perspective rectangle and point "G" is connected with point "x." This segment is continued until it intercepts line "h" to determine point "f." Connecting points

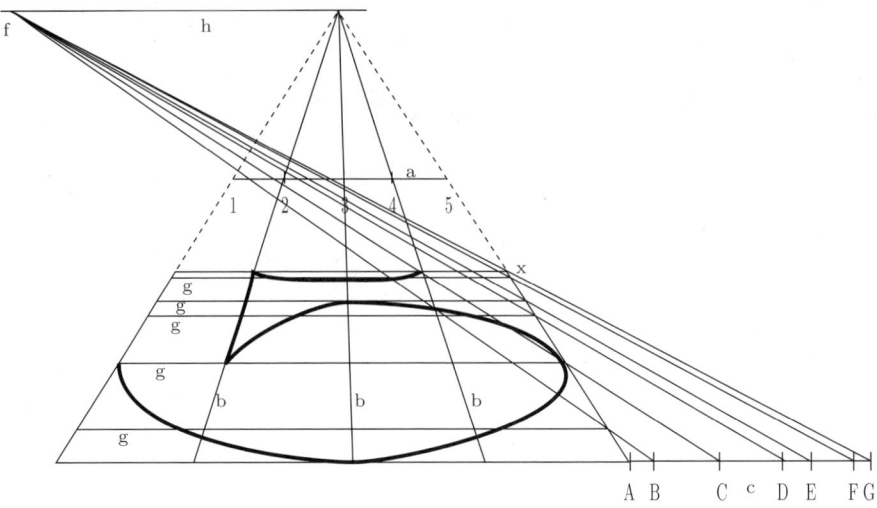

Figure Ans.6: A Large Digit "5" in One-Point Perspective.

"B" through "F" to point "f" determines the locations of the five horizontal guidelines "g," which completes the construction of the perspective grid. It is now obvious how to move the various key points of the large digit to their new locations.

3.7:   In the standard position, the line of sight of the viewer is the z axis. In order for a line segment to be perpendicular to this direction, all its points must have the same z coordinate (i.e., the segment must be contained in a plane parallel to the xy plane). We therefore select two endpoints with $z = 1$ and two other endpoints with $z = 3$. The first two points are selected, somewhat arbitrarily, as $\mathbf{P}_1 = (2, 3, 1)$ and $\mathbf{P}_2 = (3, -1, 1)$. The third point is chosen as $\mathbf{P}_3 = (0, 2, 3)$ and the last point is determined from $\mathbf{P}_4 = \mathbf{P}_2 - \mathbf{P}_1 + \mathbf{P}_3 = (1, -2, 3)$. The four points are now projected to $\mathbf{P}_1^* = (1, 3/2)$, $\mathbf{P}_2^* = (3/2, -1/2)$, $\mathbf{P}_3^* = (0, 1/2)$, and $\mathbf{P}_4^* = (1/4, -1/2)$.

It is easy to show that the two straight segments defined by the four projected points are parallel by computing the differences $\mathbf{v}_1 = \mathbf{P}_2^* - \mathbf{P}_1^* = (1/2, -2)$ and $\mathbf{v}_2 = \mathbf{P}_4^* - \mathbf{P}_3^* = (1/4, -1)$. The difference of two points is a vector, and the two vectors $\mathbf{v}_1$ and $\mathbf{v}_2$ point in the same direction.

3.8:   We are looking for a t value for which $\mathbf{P}^*(t) = (0, 1/4)$. This can be written as the vector equation

$$(1 - t)^2(-1/2, 0) + 2t(1 - t)(0, 1/3) + t^2(1/4, 1/4) = (0, 1/4)$$

or as the two separate scalar equations $(1 - t)^2(-1/2) + 2t(1 - t)(0) + t^2(1/4) = 0$ and $(1 - t)^2(0) + 2t(1 - t)(1/3) + t^2(1/4) = (1/4)$. The first equation yields the solutions $t \approx 0.5858$ and $t \approx 3.414$, while the second equation has the solutions $t = 0$ and $t = 1.6$. The two equations are therefore contradictory.

3.9: Appropriate mathematical software produces the result $(0, 2, 4, 1)$. The rotation transforms $(0, 1, -4, 1)$ to $(0, 4, 1, 1)$, the translation transforms this to $(0, 4, 4, 1)$, and the scaling produces $(0, 2, 4, 1)$.

3.10: When T_1 or T_2 gets large, the object is magnified. However, when T_3 gets large, the object is scaled in the z direction *relative to the origin*. All the z coordinates become large, effectively moving the object away from the observer. When all three scale factors get large, the magnification in the x and y directions is canceled out by the effect of moving away in the z direction, so the object does not seem to change in size.

3.11: Equation (3.5) gives us

$$\mathbf{T} = \begin{bmatrix} 1 & 0 & 0 & 0 \\ 0 & 0 & 0 & 1 \\ 0 & -1/2 & 0 & 0 \\ 0 & 0 & 0 & 4 \end{bmatrix},$$

and we know that $(0, 1, -4, 1)\mathbf{T} = (0, 2, 0, 5)$. We are looking for a point $\mathbf{P} = (x, y, z)$ such that $(x, y, z, 1)\mathbf{T} = (0, 0, 0, w)$ for any $w \neq 0$. The explicit form of this set of equations is $(x, -z/2, 0, y + 4) = (0, 0, 0, w)$, and this is satisfied by all the points of the form $(0, y, 0)$, where $y \neq -4$. The interpretation of this result is simple. The rotation brings the points on the y axis to the z axis, where they are translated by three units and remain on the z axis. The scaling doesn't move these points any farther. Point $(0, -4, 0)$ is rotated to $(0, 0, -4)$ and translated to $(0, 0, -1)$, which is the viewer's position. All the points on the z axis are projected to the origin except the viewer's location. The projection of the viewer is undefined because the case $z = -k$ results in Equation (3.1) having a zero denominator. The next example sheds more light on the perspective projection of points with negative z coordinates.

3.12: The terms *clockwise* and *counterclockwise* fully describe rotations in two dimensions. Our example, however, is in three dimensions, where rotations are more complex and can have more directions. The rotation produced by matrix (3.6) is from the positive z to the positive x direction (or, alternatively, from the negative x direction to the negative z direction).

3.13: Because of the special orientation of the projection plane. This equation says that any point (x, y, z) satisfying $\alpha x = -\beta z$ lies on the projection plane, regardless of its y coordinate.

3.14: The case $\theta = 0$ means $\alpha = 0$ and $\beta = 1$. Matrix (3.7) reduces to

$$\begin{pmatrix} k & 0 & 0 & 0 \\ 0 & k & 0 & 0 \\ 0 & 0 & 0 & 1 \\ 0 & 0 & 0 & k \end{pmatrix} = k \begin{pmatrix} 1 & 0 & 0 & 0 \\ 0 & 1 & 0 & 0 \\ 0 & 0 & 0 & r \\ 0 & 0 & 0 & 1 \end{pmatrix}.$$

The case $\theta = 45°$ implies $\alpha = \beta = 1/\sqrt{2}$. Matrix (3.7) is reduced to

$$
\begin{pmatrix}
k/2 & 0 & -k/2 & 1/\sqrt{2} \\
0 & k & 0 & 0 \\
-k/2 & 0 & k/2 & 1/\sqrt{2} \\
0 & 0 & 0 & k
\end{pmatrix}.
$$

The case $\theta = 90°$ means $\alpha = 1$ and $\beta = 0$. Matrix (3.7) reduces to

$$
\begin{pmatrix}
0 & 0 & 0 & 1 \\
0 & k & 0 & 0 \\
0 & 0 & k & 0 \\
0 & 0 & 0 & k
\end{pmatrix}
= k
\begin{pmatrix}
0 & 0 & 0 & r \\
0 & 1 & 0 & 0 \\
0 & 0 & 1 & 0 \\
0 & 0 & 0 & 1
\end{pmatrix}.
$$

3.15: Direct multiplication yields

$$
(\beta l, m, -\alpha l, 1)
\begin{pmatrix}
k\beta^2 & 0 & -k\alpha\beta & \alpha \\
0 & k & 0 & 0 \\
-k\alpha\beta & 0 & k\alpha^2 & \beta \\
0 & 0 & 0 & k
\end{pmatrix}
$$

$$
= (kl\beta^3 + kl\alpha^2\beta, mk, -kl\alpha\beta^2 - kl\alpha^3, l\alpha\beta - l\alpha\beta + k)
$$
$$
= (kl\beta, km, -kl\alpha, k).
$$

The transformed point is $\mathbf{P}^* = (l\beta, m, -l\alpha) = \mathbf{P}$. Point $\mathbf{P}$ is thus transformed to itself! This happens because $\mathbf{P}$ resides on the projection plane. The equation of the plane is $\alpha x = -\beta z$, and a simple check verifies that the coordinates of point $\mathbf{P}$ satisfy this relation.

3.16: The steps are similar to the ones used to derive matrix (3.7):

■   Use the relation $(-k\alpha, -k\beta\gamma, -k\beta\delta) \bullet (x, y, z) = 0$ to derive the equation of the projection plane. This is trivial, and the equation is $-xk\alpha - yk\beta\gamma - zk\beta\delta = 0$.

■   Compute the straight segment from the viewer to a general point $\mathbf{P} = (l, m, n)$:

$$
(l + k\alpha, m + k\beta\gamma, n + k\beta\delta)t + (-k\alpha, -k\beta\gamma, -k\beta\delta).
$$

■   Calculate the value of t_0 at the intersection point of the segment and the plane. From

$$
\big[(l + k\alpha)t_0 - k\alpha\big]k\alpha + \big[(m + k\beta\gamma)t_0 - k\beta\gamma\big]k\beta\gamma + \big[(n + k\beta\delta)t_0 - k\beta\delta\big]k\beta\delta = 0,
$$

we get

$$
t_0 = \frac{k(\alpha^2 + \beta^2\gamma^2 + \beta^2\delta^2)}{(l + k\alpha)\alpha + (m + k\beta\gamma)\beta\gamma + (n + k\beta\delta)\beta\delta}
$$
$$
= \frac{k(\alpha^2 + \beta^2\gamma^2 + \beta^2\delta^2)}{l\alpha + m\beta\gamma + n\beta\delta + k(\alpha^2 + \beta^2\gamma^2 + \beta^2\delta^2)}.
$$

■ The coordinates of the projected point can now be determined. The x^* coordinate is

$$
x^* = (l + k\alpha)t_0 - k\alpha = (l + k\alpha)\frac{k(\alpha^2 + \beta^2\gamma^2 + \beta^2\delta^2)}{l\alpha + m\beta\gamma + n\beta\delta + k(\alpha^2 + \beta^2\gamma^2 + \beta^2\delta^2)} - k\alpha
$$
$$
= \frac{lk\beta^2(\gamma^2 + \delta^2) - mk\alpha\beta\gamma - nk\alpha\beta\delta}{l\alpha + m\beta\gamma + n\beta\delta + k(\alpha^2 + \beta^2\gamma^2 + \beta^2\delta^2)}.
$$

■ The y^* coordinate is

$$
y^* = (m + k\beta\gamma)t_0 - k\beta\gamma
$$
$$
= (m + k\beta\gamma)\frac{k(\alpha^2 + \beta^2\gamma^2 + \beta^2\delta^2)}{l\alpha + m\beta\gamma + n\beta\delta + k(\alpha^2 + \beta^2\gamma^2 + \beta^2\delta^2)} - k\beta\gamma
$$
$$
= \frac{-lk\alpha\beta\gamma + mk(\alpha^2 + \beta^2\delta^2) - nk\beta^2\gamma\delta}{l\alpha + m\beta\gamma + n\beta\delta + k(\alpha^2 + \beta^2\gamma^2 + \beta^2\delta^2)}.
$$

■ The z^* coordinate is

$$
z^* = (n + k\beta\delta)t_0 - k\beta\delta
$$
$$
= (n + k\beta\delta)\frac{k(\alpha^2 + \beta^2\gamma^2 + \beta^2\delta^2)}{l\alpha + m\beta\gamma + n\beta\delta + k(\alpha^2 + \beta^2\gamma^2 + \beta^2\delta^2)} - k\beta\delta
$$
$$
= \frac{-lk\alpha\beta\delta - mk\beta^2\gamma\delta + nk(\alpha^2 + \beta^2\gamma^2)}{l\alpha + m\beta\gamma + n\beta\delta + k(\alpha^2 + \beta^2\gamma^2 + \beta^2\delta^2)}.
$$

■ The projection matrix is now easy to calculate. It is

$$
\begin{pmatrix}
k\beta^2(\gamma^2 + \delta^2) & -k\alpha\beta\gamma & -k\alpha\beta\delta & \alpha \\
-k\alpha\beta\gamma & k(\alpha^2 + \beta^2\delta^2) & -k\beta^2\gamma\delta & \beta\gamma \\
-k\alpha\beta\delta & -k\beta^2\gamma\delta & k(\alpha^2 + \beta^2\gamma^2) & \beta\delta \\
0 & 0 & 0 & k(\alpha^2 + \beta^2\gamma^2 + \beta^2\delta^2)
\end{pmatrix}. \quad \text{(Ans.3)}
$$

To check our result, we consider the special case of no rotation about the x axis. In this case, $\phi = 0$, $\gamma = 0$, and $\delta = 1$. It is easy to see that this reduces matrix (Ans.3) to matrix (3.7).

3.17: After the two rotations, the viewer may end up at any point in space, but the projection plane still passes through the origin. This is why our case is not completely general.

3.18: These two translation matrices can easily be written, and it is obvious that their product is a translation from the origin to **B**.

$$
\mathbf{T}_3 = \begin{bmatrix} 1 & 0 & 0 & 0 \\ 0 & 1 & 0 & 0 \\ 0 & 0 & 1 & 0 \\ 0 & 0 & -k & 1 \end{bmatrix}, \quad
\mathbf{T}_4 = \begin{bmatrix} 1 & 0 & 0 & 0 \\ 0 & 1 & 0 & 0 \\ 0 & 0 & 1 & 0 \\ a & b & c+k & 1 \end{bmatrix}, \quad
\mathbf{T}_3 \cdot \mathbf{T}_4 = \begin{bmatrix} 1 & 0 & 0 & 0 \\ 0 & 1 & 0 & 0 \\ 0 & 0 & 1 & 0 \\ a & b & c & 1 \end{bmatrix}.
$$

3.19: Recall that the basic rule of perspective projection is to connect an image point to the viewer with a line that intercepts the projection plane. The viewer and the image points should therefore be on different sides of the projection plane. In our case, point $(0,0,0)$ is behind the viewer, so it is on the same side of the projection plane as the viewer and, consequently, it does not make sense to project it.

3.20: Direct multiplication yields

$$(\beta l, m, -\alpha l, 1) \begin{pmatrix} \beta & 0 & 0 & \alpha r \\ 0 & 1 & 0 & 0 \\ -\alpha & 0 & 0 & \beta r \\ 0 & 0 & 0 & 1 \end{pmatrix} = (l\beta^2 + l\alpha^2, m, 0, lr\alpha\beta - lr\alpha\beta + 1) = (l, m, 0, 1),$$

so the transformed point is $\mathbf{P}^* = (l, m, 0)$. Figure Ans.7 shows that point $\mathbf{P} = (\beta l, m, -\alpha l)$ resides on the projection plane. After the transformations, it is still located on the projection plane, only now this is the xy plane.

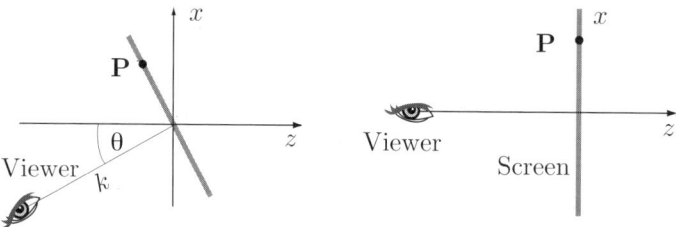

Figure Ans.7: Transforming and Projecting.

3.21: Because they project on different projection planes. Matrix (3.7) projects on plane $\alpha x = -\beta z$, where the z coordinate is proportional to the x coordinate, whereas matrix (3.9) projects on the xy plane, where the z coordinate is zero.

3.22: Figure Ans.8a shows the geometry of the problem. Notice that the viewer looks in the direction of *negative z* and also down. The *Mathematica* code

```
<< LinearAlgebra'Orthogonalization'
k = 3.; r = 1/k;
{a, b, c} = {0, 2k, -k}; {d, e, f} = Normalize[{0, -1, -1}]
T = {{(e^2 + f + f^2)/(1 + f), -d e/(1 + f), 0, d r},
 {-d e/(1 + f), (d^2 + f + f^2)/(1 + f), 0, e r},
 {-d, -e, 0, f r},
 {(c d + b d e - a e^2 - a f + c d f - a f^2)/(1 + f),
 (-b d^2 + c e + a d e - b f + c e f - b f^2)/(1 + f),
 0, -(a d + b e + c f)  r}};
{0,0,-4k,1}.T
```

computes the normalized components of $\mathbf{D}$ as $(0, -0, 7071, -0.7071)$ and the projected point as the 4-tuple $(0, -2.12132, 0, 3.53553)$ (i.e., point $(0, -0.6)$ on the xy plane, shown

in the diagram). This is the first example where the viewer is not looking in the positive
z direction or anywhere near that direction, and this fact raises the issue of the top of
the screen. If the viewer is looking at or near the positive z direction, the rotation that
aligns the screen with the xy plane is about a small angle. In such a case, the screen
does not change its orientation much, so there is no problem with the direction of the
top of the screen. If we assume that the top of the screen was in the positive y direction
(or close to it) before the rotation, then the rotation aligns the top of the screen with
the positive y axis.

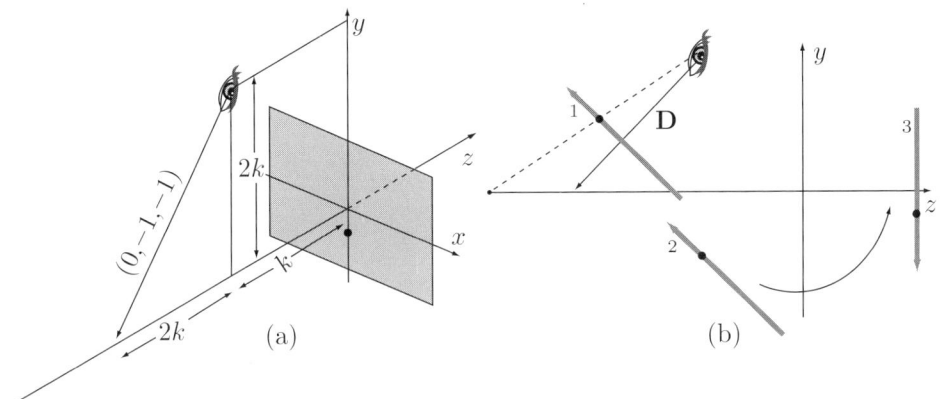

Figure Ans.8: A Viewer Looking "Backward."

In this example, however, the rotation is about an angle that is close to $180°$,
so the direction of the top of the screen becomes important. Figure Ans.8b suggests
that the top of the screen is a vector in the direction $(0, 1, -1)$ because this direction is
perpendicular to the line of sight of the viewer and isn't very different from the direction
of positive y. If this is so, then after the large rotation this top becomes the bottom of
the screen in the xy plane. Figure Ans.8b shows how the top of the screen retains its
orientation when the screen is translated from 1 to 2 but becomes the bottom when the
screen is rotated about the x axis from 2 to 3. Thus, a complete treatment of general
perspective should include an optional rotation of the screen about the z axis. (See the
discussion of the top vector in Section 3.7.)

3.23:   This is easily done with the help of appropriate software. The results for the
two cases are

$$\begin{pmatrix} 1 & 0 & 0 & 0 \\ 0 & 1 & 0 & 0 \\ 0 & 0 & 0 & r \\ -a & -b & 0 & -cr \end{pmatrix} \quad \text{and} \quad \begin{pmatrix} 1 & 0 & 0 & 0 \\ 0 & 1 & 0 & 0 \\ 0 & 0 & 0 & r \\ 0 & 0 & 0 & 1 \end{pmatrix}. \tag{Ans.4}$$

Notice how the second matrix of (Ans.4) is the standard perspective projection matrix
$\mathbf{T}_p$ of Equation (3.4).

3.24: We substitute $(a, b, c) = (0, 1, 0)$ and $(d, e, f) = (0, 1/\sqrt{2}, 1/\sqrt{2})$ in matrix (3.13). The transformation is therefore

$$(0, 1, 10, 1)\begin{pmatrix} 1 & 0 & 0 & 0 \\ 0 & \frac{1}{\sqrt{2}} & 0 & \frac{r}{\sqrt{2}} \\ 0 & \frac{-1}{\sqrt{2}} & 0 & \frac{r}{\sqrt{2}} \\ 0 & \frac{-1}{\sqrt{2}} & 0 & \frac{-r}{\sqrt{2}} \end{pmatrix} = \left(0, \frac{1 - 10 - 1}{\sqrt{2}}, 0, \frac{r + 10r - r}{\sqrt{2}}\right),$$

so $\mathbf{P}^* = (0, -1/r, 0) = (0, -k, 0)$. The following *Mathematica* code may be helpful for further experimentation:

```
<<LinearAlgebra'Orthogonalization'
{a,b,c}={0,1.,0}; {d,e,f}=Normalize[{0,1,1}]
T = {{(e^2 + f + f^2)/(1 + f), -d e/(1 + f), 0, d r},
 {-d e/(1 + f), (d^2 + f + f^2)/(1 + f), 0, e r},
 {-d, -e, 0, f r},
 {(c d + b d e - a e^2 - a f + c d f - a f^2)/(1 + f),
 (-b d^2 + c e + a d e - b f + c e f - b f^2)/(1 + f),
 0, -(a d + b e + c f)  r}};
{0,1,10,1}.T
```

3.25:   1. The magnitude of vector $\mathbf{D}$ is $\sqrt{1^2 + 1^2} = \sqrt{2} = k$, so the choice $k = \sqrt{2}$ implies that $\mathbf{D}$ has the correct length. It goes from point $\mathbf{B}$ to the center of the projection plane. The coordinates of the center are therefore $\mathbf{B} + \mathbf{D} = (0, 2k - 1, -2k - 1)$. To find the equation of the projection plane, we consider an arbitrary point $\mathbf{P} = (x, y, z)$ on this plane. The vector from the center point to $\mathbf{P}$ is the difference $\mathbf{P} - (\mathbf{B} + \mathbf{D}) = (x, y - 2k + 1, z + 2k + 1)$. This vector is perpendicular to $\mathbf{D}$, so their dot product $(x, y - 2k + 1, z + 2k + 1) \bullet (0, -1, -1)$ must be zero. This produces the equation $y + z = -2$, and Figure Ans.9 illustrates how this plane is parallel to the x axis, which is why its equation does not depend on x. Thus, a general point on the projection plane has coordinates $(x, y, -y - 2)$

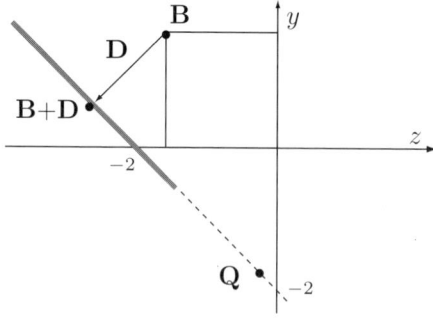

Figure Ans.9: Projection Plane $y + z = -2$.

2. Appropriate mathematical software produces the 4×4 transformation matrix

$$\mathbf{T}_{123} = \tag{Ans.5}$$

$$\begin{bmatrix} \frac{e^2+f-e^2f-f^3}{1-f^2} & -\frac{de}{1+f} & 0 & dr \\ -\frac{de}{1+f} & \frac{d^2+f-d^2f-f^3}{1-f^2} & 0 & er \\ -d & -e & 0 & fr \\ cd + \frac{bde}{1+f} - \frac{a(e^2+f-e^2f-f^3)}{1-f^2} & ce + \frac{ade}{1+f} - \frac{b(d^2+f-d^2f-f^3)}{1-f^2} & 0 & 1+(-ad-be-cf)r \end{bmatrix}.$$

Notice the denominators in this matrix. They imply that values $f = \pm 1$ require special treatment. In our first case, vector $\mathbf{D} = (0, -1, -1)$ has $f = -1$, so we change it to $(0, -1, -0.99)$. We pick up the point $\mathbf{Q} = (0, 2k-3, -2k+1) \approx (0, -0.17, -1.83)$ that's on the projection plane, located "below" the center of the plane. The product $\mathbf{Q} \cdot \mathbf{T}_{123}$ yields the 4-tuple $(0, 3.97, 0, 2.42)$, which is point $(0, 1.64, 0)$, located on the xy plane but above the origin.

3.26: We first determine α

$$\alpha = \frac{|\mathbf{a}|^2}{\mathbf{a} \bullet (\mathbf{p} - \mathbf{b})} = \frac{8}{(0, 2, 2) \bullet (x - 0, y - 1, z - 0)} = \frac{4}{y + z - 1}.$$

[Note that $\mathbf{P} = (0, 1, 10)$, implying $\alpha = 4/(1 + 10 - 1) = 2/5$.]
  Next, we compute vector $\mathbf{d}$

$$\mathbf{d} = \mathbf{b} + \alpha(\mathbf{p} - \mathbf{b})$$
$$= (0, 1, 0) + \frac{4}{y + z - 1}(x, y - 1, z)$$
$$= \frac{4}{y + z - 1}\left(x, (5y - z - 5)/4, z\right).$$

[A check verifies that $\mathbf{P} = (0, 1, 10) \Rightarrow \mathbf{d} = (0, 1, 4)$.]
  Vector $\mathbf{c}$ can now be calculated

$$\mathbf{c} = \alpha(\mathbf{p} - \mathbf{b}) - \mathbf{a}$$
$$= \frac{4}{y + z - 1}(x, y - 1, z) - (0, 2, 2)$$
$$= \frac{4}{y + z - 1}\left(x, (y - z - 1)/2, -(y - z - 1)/2\right).$$

Thus, the screen coordinates are

$$\mathbf{u} \bullet \mathbf{c} = (1, 0, 0) \bullet \frac{4}{y + z - 1}\left(x, (y - z - 1)/2, -(y - z - 1)/2\right)$$
$$= \frac{4x}{y + z - 1},$$

$$\mathbf{w} \bullet \mathbf{c} = (0, 1/\sqrt{2}, -1/\sqrt{2}) \bullet \frac{4}{y + z - 1}\left(x, (y - z - 1)/2, -(y - z - 1)/2\right)$$

$$= \frac{4(y - z - 1)}{\sqrt{2}(y + z - 1)}.$$

Again, a direct check verifies that $\mathbf{P} = (0, 1, 10)$ results in

$$\mathbf{u} \bullet \mathbf{c} = 0 \quad \text{and} \quad \mathbf{w} \bullet \mathbf{c} = \frac{4(1 - 10 - 1)}{\sqrt{2}(1 + 10 - 1)} = \frac{-4}{\sqrt{2}} = -\sqrt{8}.$$

Also, the screen coordinates of point $\mathbf{P} = (0, 5, 4)$ are

$$\mathbf{u} \bullet \mathbf{c} = 0, \quad \text{and} \quad \mathbf{w} \bullet \mathbf{c} = \frac{4(5 - 4 - 1)}{\sqrt{2}(5 + 4 - 1)} = 0,$$

as should be expected (why?).

3.27:   Figure Ans.10 shows that in a right-handed coordinate system, the positive y axis is in the direction of vector $\mathbf{w}$ and vector $\mathbf{u}$ is in the direction of *negative* x.

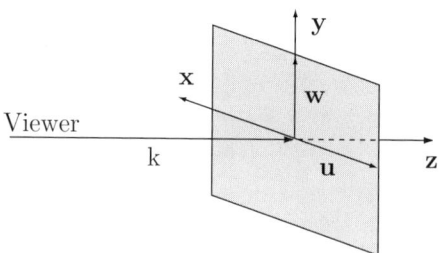

Figure Ans.10: A Right-Handed Coordinate System.

3.28:   The proof is straightforward but a little messy. We start with two three-dimensional points, $\mathbf{P}_1 = (x_1, y_1, z_1)$ and $\mathbf{P}_2 = (x_2, y_2, z_2)$. Their projections are

$$\mathbf{P}_1^* = \left(\frac{x_1 k}{k + z_1}, \frac{y_1 k}{k + z_1}, \frac{z_1}{k + z_1}\right) \quad \text{and} \quad \mathbf{P}_2^* = \left(\frac{x_2 k}{k + z_2}, \frac{y_2 k}{k + z_2}, \frac{z_2}{k + z_2}\right).$$

Now consider the two lines $\mathbf{P}(t) = \mathbf{P}_1 + (\mathbf{P}_2 - \mathbf{P}_1)t$ and $\mathbf{P}^*(u) = \mathbf{P}_1^* + (\mathbf{P}_2^* - \mathbf{P}_1^*)u$. We need to prove that every point on $\mathbf{P}(t)$ is transformed to a point on $\mathbf{P}^*(u)$, where u depends on t, k, $\mathbf{P}_1$, and $\mathbf{P}_2$ only.

The coordinates of a general point on $\mathbf{P}(t)$ are

$$\left(\frac{(x_1 + (x_2 - x_1)t)k}{k + z_1 + (z_2 - z_1)t}, \frac{(y_1 + (y_2 - y_1)t)k}{k + z_1 + (z_2 - z_1)t}, \frac{(z_1 + (z_2 - z_1)t)k}{k + z_1 + (z_2 - z_1)t}\right).$$

The coordinates of a general point on $\mathbf{P}^*(u)$ are

$$
\left(\frac{x_1 k}{k+z_1} + \left(\frac{x_2 k}{k+z_2} - \frac{x_1 k}{k+z_1}\right) u, \; \frac{y_1 k}{k+z_1} + \left(\frac{y_2 k}{k+z_2} - \frac{y_1 k}{k+z_1}\right) u, \right.
$$

$$
\left. \frac{z_1}{k+z_1} + \left(\frac{z_2}{k+z_2} - \frac{z_1}{k+z_1}\right) u\right).
$$

In order for the points to be equal, the following two equations have to hold:

$$
\frac{(x_1 + (x_2 - x_1)t)k}{k+z_1+(z_2-z_1)t} = \frac{x_1 k}{k+z_1} + \left(\frac{x_2 k}{k+z_2} - \frac{x_1 k}{k+z_1}\right) u,
$$

$$
\frac{(z_1 + (z_2 - z_1)t)k}{k+z_1+(z_2-z_1)t} = \frac{z_1}{k+z_1} + \left(\frac{z_2}{k+z_2} - \frac{z_1}{k+z_1}\right) u.
$$

(There are actually three equations, but the second one, for y, is equivalent to the first one, so it is not included here.) Because of the way the depth transformation is defined, both equations are satisfied if u is defined by

$$
u = t\frac{k+z_2}{k+z_1+(z_2-z_1)t}.
$$

Note that $t = 0 \Rightarrow u = 0$ and $t = 1 \Rightarrow u = 1$.

3.29: The tangent of half the angle is $(W/2)/k = 1/2$. Therefore, half the angle equals 26.5° and the entire field of view is twice that, or 53° wide. (See also the discussion of Brunelleschi's peepshow experiment on page 82.)

3.30: Since k is scaled by the same factor of 5, we should scale e by this factor, bringing it down from 2.5 to 0.5.

3.31: The difference between the two pictures of a stereo pair is a horizontal shift, but various parts of the pictures are shifted by different amounts. Parts close to the camera are shifted more than parts that are far away. Thus, the two pictures are not simply shifted versions of each other. However, a person looking at a picture can often tell the approximate distance of each picture element from the camera. This makes it possible, at least in principle, to create a copy of a picture and have the user specify the amount of shift of every picture element in the copy. In practice, such a method is slow and cumbersome, and the result depends on the depth estimates of the user, so it should be used only as a last resort, when only one picture is available and it is important to see it in three dimensions.

4.1: The *Mathematica* code

```
(* exercise for hemispherical fisheye projection *)
```

```
k=1;
scal[q_]:=(k Tan[ArcTan[q/k]/2])/q;
{scal[1.],scal[10.], scal[100.], scal[1000.], scal[10000.]}
```

produces the values 0.414214, 0.0904988, 0.0099005, 0.000999, and 0.00009999.

4.2: Figure Ans.11 illustrates this effect. We see a few points on a vertical line. In the fisheye projection, each point is moved toward the origin, but points that are close to the origin are moved less than points that are further away. The result is a curve. Applying this argument to straight lines that pass through the center of the circle shows that they are not bent.

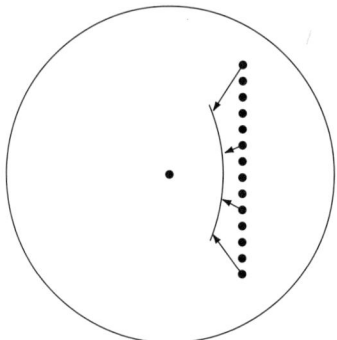

Figure Ans.11: Vertical Distortion in Fisheye Projection.

4.3: The only differences are that (1) w varies in the intervals $[0, 90°]$ (for the top half of space) and $[270°, 360°]$ (for the bottom half) and (2) the entire radius-k circle, not just half of it, is now devoted to the hemisphere of space in front of the viewer. As a result, the new table (Ans.12) has just two rows.

w	r	r interval	u	$\sin w$
$0 \to 90$	$k \sin w$	$[0, k]$	top	$0 \to 1$
$270 \to 360$	$-k \sin w$	$[k, 0]$	bottom	$-1 \to 0$

Table Ans.12: Two Cases of w, r, and u.

4.4: This point corresponds to $w = 0°$ (implying $r = 0$), and the pair of polar coordinates $(0, u)$ corresponds to the center of the radius-k circle regardless of u. This special point is therefore mapped to the center of the circle.

4.5: Imagine a straight segment parallel to the x axis (Figure Ans.13a). The angle w is the same for all the points of this segment, so a point in direction (u, w) on the segment is projected on the circle into a point with polar coordinates $(\frac{k}{2}\sin w, u)$. The result is a set of points with polar coordinates (r, u), where r is constant (i.e., a circular arc). When this straight segment is slightly perturbed, as in Figure Ans.13b, its projection does not vary much and remains a curve. On the other hand, when a straight segment passes through the viewer's line of sight, as in Figure Ans.13c, all its points have the same angle u. The projection of such a segment is a set of points (r, u), where r normally varies but u is constant, a straight segment.

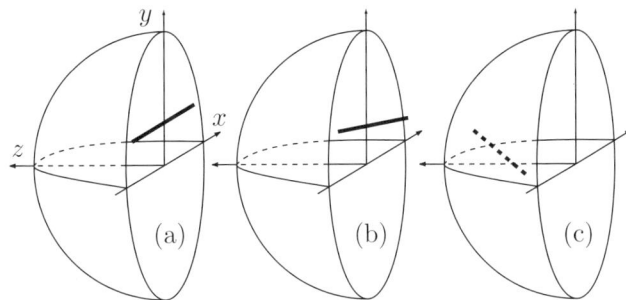

Figure Ans.13: Straight Segments in the Angular Fisheye Projection.

4.6: The Cassinian oval is another anallagmatic curve. Recall that an ellipse is the locus of all the points the sum of whose distances from two foci is constant. The Cassinian oval is similarly defined as the locus of the points the product of whose distances from two foci is constant.

See [xahlee 05] for a detailed discussion and figures.

4.7: Figure Ans.14 shows a circle C that does not pass through the origin. (Notice that the circle of inversion itself is not shown.) We construct the line L from O through the center of C and examine the intersection points $\mathbf{P}$ and $\mathbf{Q}$. Their projections $\mathbf{P}^*$ and $\mathbf{Q}^*$ must be on L. We select an arbitrary point $\mathbf{R}$ on C and denote its projection $\mathbf{R}^*$. From $OP \cdot OP^* = OQ \cdot OQ^* = OR \cdot OR^*$, we get $OR^*/OP = OP^*/OR$, indicating that the two triangles ORP and OR^*P^* are similar. This implies that angles OP^*R^* and ORP are equal and also that angles OQ^*R^* and ORQ are equal. We subtract angles and find that $OP^*R^* - OQ^*R^* = ORP - ORQ = 90°$, which implies that angle $P^*R^*Q^* = 90°$. Since this is true for a general point $\mathbf{R}$, we conclude that all the points R^* (which together constitute the projection of C) are located on the circle C^* centered on L with diameter P^*Q^*.

4.8: The rule $\mathbf{P}^* = 1/\mathbf{P}$ is generalized to $\mathbf{P}^* = R^2/\mathbf{P}$. This projection retains all the features mentioned in the text with regard to the unit circle.

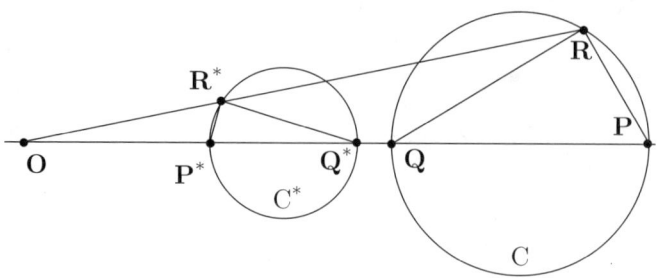

Figure Ans.14: Circular Inversion of a Circle.

4.9: The z^* coordinate depends on Z in the sense that point **P** should be projected on the cylinder only if $|z^*| \leq Z$.

4.10: Figure Ans.15a shows a cylinder of radius R centered on the origin, with its axis in the z direction. We start with a circle in the xy plane. The circle's equation is $\big(R\cos(2\pi t), R\sin(2\pi t), 0\big)$. The circle is now rotated θ degrees about the y axis, as shown in Figure Ans.15b. The new circle is given by

$$
\big(R\cos(2\pi t), R\sin(2\pi t), 0\big)
\begin{pmatrix}
\cos\theta & 0 & \sin\theta \\
0 & 1 & 0 \\
-\sin\theta & 0 & \cos\theta
\end{pmatrix}
$$
$$
= \big(R\cos(2\pi t)\cos\theta, R\sin(2\pi t), R\cos(2\pi t)\sin\theta\big).
$$

Figure Ans.15c shows that in order to convert this tilted circle into an ellipse, its x and z coordinates should be scaled by a factor of $1/\cos\theta$. The equation of this ellipse is thus

$$
\big(R\cos(2\pi t), R\sin(2\pi t), R\cos(2\pi t)\tan\theta\big). \tag{Ans.6}
$$

In order to prove that this is an ellipse, we can rotate it back to the xy plane. The result is

$$
\big(R\cos(2\pi t), R\sin(2\pi t), R\cos(2\pi t)\tan\theta\big)
\begin{pmatrix}
\cos\theta & 0 & -\sin\theta \\
0 & 1 & 0 \\
\sin\theta & 0 & \cos\theta
\end{pmatrix}
$$
$$
= \big(R(\cos\theta + \sin^2\theta)\cos(2\pi t), R\sin(2\pi t), 0\big),
$$

an expression that satisfies $x^2/a^2 + y^2/b^2 = 1$ for $a = R(\cos\theta + \sin^2\theta)$ and $b = R$. Figure Ans.15d shows the unrolled cylinder, cut along the $y = 0$ line, with the origin at its center.

The behavior of the resulting flat curve can be figured out when we notice that the x and y coordinates of the ellipse of Equation (Ans.6) form a circle, which is a curve with constant speed. This means that when the curve is flattened, it has constant speed in the horizontal direction (i.e., incrementing t in equal steps moves us equal horizontal

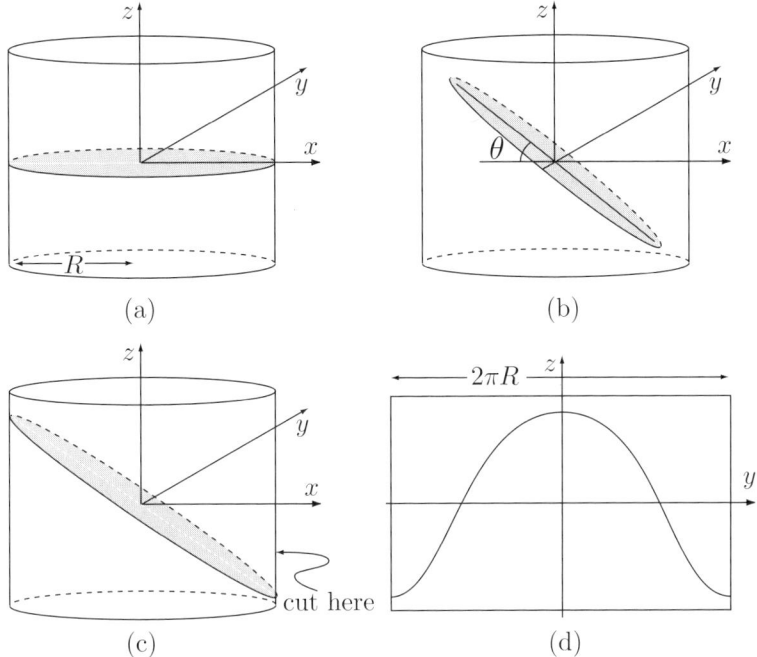

Figure Ans.15: Ellipse and Sinusoid.

increments on the unrolled cylinder). The vertical behavior of the flattened curve is determined by the z coordinate of Equation (Ans.6), and this coordinate behaves like a sine curve with an amplitude $R \tan \theta$. The result is the parametric curve

$$\Big((2t-1)\pi R, R \tan \theta \cos\big[(2t-1)\pi\big] \Big), \qquad 0 \le t \le 1.$$

As t varies from zero to one, the horizontal coordinate varies from $-\pi R$ to $+\pi R$ and the vertical coordinate varies as a sine curve from -1 to 0 to $+1$, back to 0, and ends up at -1.

It is also interesting to consider the curvature of this sine curve. The curvature is essentially given by the second derivative, which, in the case of $\sin(t)$, equals $-\sin(t)$. We are interested only in the absolute magnitude of the curvature, so we can disregard the minus sign. The result is that for $t = 0$ and $t = \pi$ the curvature is zero, while for $t = \pi/2$ it is maximum. The conclusion is that when a straight line is projected by curved perspective into a sinusoid, those parts of the line that are close to the observer become highly curved, while the distant parts remain straight or close to straight. Figure 4.24 is a typical example of this behavior.

4.11: Figure Ans.16 illustrates the geometry of the problem. Parts a and b show that the distance between the two points on the hemisphere is $R(\sin \theta)\phi$, and part c shows that the distance between them on the circle is $R\theta\phi$. The ratio of the distances is $\theta/\sin\theta$ and this number, which is undefined for $\theta = 0$, starts at 1 for small angles, reaches 1.01 for $30°$, and becomes $\pi/2 \approx 1.57$ for $90°$.

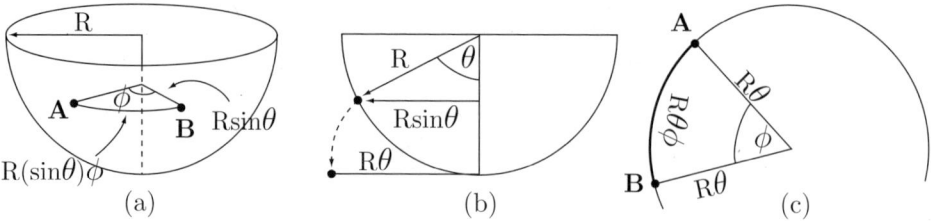

Figure Ans.16: Distance Between Points in Curvilinear Perspective.

4.12:   Imagine a straight horizontal line of the form $(x(t), y(t), 0)$. All its z coordinates are zero, which makes it easy to show (and also to visualize) that the projected segments of this line (there can be up to three segments) are all horizontal and therefore have identical slopes. Figure Ans.17 illustrates an example. Given the two points $\mathbf{P}_1 = (3k, -3k/2, 0)$ and $\mathbf{P}_2 = (k, 5k/4, 0)$, it is trivial to determine that $\mathbf{P}_1^* = (k, -k/2, 0)$ and $\mathbf{P}_2^* = (4k/5, k, 0)$. Since both $\mathbf{P}_1$ and $\mathbf{P}_2$ have z coordinates of zero, the entire line segment connecting them has $z = 0$. Thus, even though we don't know the precise location of point $\mathbf{P}_0$, we know that its z coordinate is zero. The coordinates of its projection $\mathbf{P}_0^*$ are between those of $\mathbf{P}_0$ and the origin, implying that the z coordinate of $\mathbf{P}_0^*$ is also zero. Thus, the two segments $\mathbf{P}_1^*\mathbf{P}_0^*$ and $\mathbf{P}_0^*\mathbf{P}_2^*$ have $z = 0$ and therefore have the same slope.

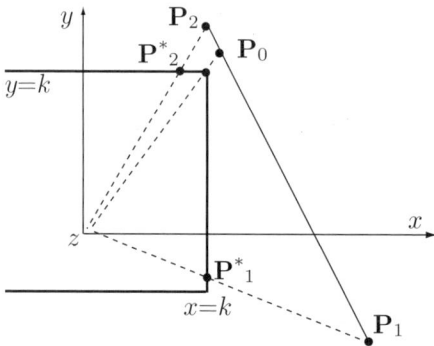

Figure Ans.17: Cubic Projection of a Horizontal Straight Segment.

On the other hand, a straight horizontal line of the form $(x(t), y(t), a)$ for $a \neq 0$ features an interpanel slope discontinuity that's proportional to a. Here is an illustrative example. Given the two points $\mathbf{P}_1 = (2k, 0, a)$ and $\mathbf{P}_2 = (2k, 100k, a)$, the line segment connecting them is $\mathbf{L}(t) = (2k, 100tk, a)$. Point $\mathbf{P}_1$ is projected to $\mathbf{P}_1^* = (k, 0, a/2)$, and point $\mathbf{P}_2$ is projected to $\mathbf{P}_2^* = (k/50, k, a/100)$. Point $\mathbf{P}_0$ is a point $\mathbf{L}(t_0)$ on this segment with the property that its x and y coordinates are equal. This produces the equation $2k = 100t_0 k$, which yields $t_0 = 1/50$. Thus, $\mathbf{P}_0$ is the point $(2k, 2k, a)$ and is projected to $\mathbf{P}_0^* = (k, k, a/2)$. The result is two projected segments. The one on panel $x = k$ goes from $\mathbf{P}_1^* = (k, 0, a/2)$ to $\mathbf{P}_0^* = (k, k, a/2)$, so its slope is zero. The projected segment on the $y = k$ segment goes from $\mathbf{P}_0^* = (k, k, a/2)$ to $\mathbf{P}_2^* = (k/50, k, a/100)$, so its slope is

$$\left(\frac{a}{2} - \frac{a}{100} \right) \bigg/ \left(k - \frac{k}{50} \right).$$

Assuming that the y coordinate of $\mathbf{P}_2$ is very large (more than 100), we obtain the approximate slope $a/(2k)$. The slope discontinuity is proportional to the z coordinate a. It is zero for $a = 0$ and becomes large (positive or negative) with a.

4.13:   The image is circular because the main mirror is circular. It has a hole in the middle because the main mirror has a hole in it (more accurately, because light hitting the top of the main mirror, around its hole, cannot reach the secondary mirror).

> I've finally figured out what's wrong with photography. It's a one-eyed man looking through a little 'ole. Now, how much reality can there be in that?
> —David Hockney

4.14:   It is easy to see from Equation (4.4) that $z = k \rightarrow z^* = k/2$.

4.15:   These concepts are defined for the Earth or for any rotating sphere. The rotation naturally defines two special points, the poles. These, in turn, define the equator (the great circle at equal distances from the poles and parallel to the axis of rotation). Now imagine a point $\mathbf{P}$ on the surface of the sphere. Draw a vertical great circle arc from $\mathbf{P}$ to the equator. (The term "vertical" means part of a great circle that passes through the poles.) This arc meets the equator at a point $\mathbf{Q}$. The angle $\mathbf{POQ}$ (where $\mathbf{O}$ is the center of the sphere) is the latitude of $\mathbf{P}$. It varies in the interval $[0, 90°]$ for each hemisphere. Thus, latitude is a natural coordinate on the rotating sphere. Its definition does not require any arbitrary choices.

> She wanted to live in Canada, he wanted to live in Mexico, so they parted. Years later, when asked the reason she replied simply "I just didn't like his latitude!"
> —Charles Schultz, *Peanuts*

The definition of longitude, on the other hand, is arbitrary and depends on a special direction that must be chosen by general agreement. This direction, which is referred to as longitude zero (or meridian zero), is perpendicular to the rotation axis. The longitude of a point $\mathbf{P}$ is the angle between its direction (the segment connecting it to the axis) and longitude zero. Thus, longitude varies in the interval $[0, 360°]$, although many maps show it in the interval $[0, 180°]$ and add the designation "east" or "west."

The antipode of point $\mathbf{P}$ is the point on the surface of the sphere at maximum distance from $\mathbf{P}$.

A graticule is a spherical grid of coordinate lines, latitudes and longitudes, over the surface of the sphere. The latitudes are circles perpendicular to the axis, which is why they are also called parallels. Each longitude is a semicircular arc (a meridian) with the axis as its chord. All the meridians meet at each pole, and every parallel crosses every meridian at a right angle.

4.16:   Yes, there are infinitely many developable surfaces, one of which is shown in Figure Ans.18. Notice that at every point on a developable surface it is possible to draw a straight line that lies completely on the surface.

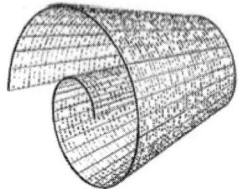

Figure Ans.18: A Developable Surface.

A.1:   This is easily proved by showing that both dot products $(\mathbf{P}\times\mathbf{Q})\bullet\mathbf{P}$ and $(\mathbf{P}\times\mathbf{Q})\bullet\mathbf{Q}$ equal zero,

$$(\mathbf{P}\times\mathbf{Q})\bullet\mathbf{P} = P_1(P_2Q_3 - P_3Q_2) + P_2(-P_1Q_3 + P_3Q_1) + P_3(P_1Q_2 - P_2Q_1) = 0,$$

and similarly for $(\mathbf{P}\times\mathbf{Q})\bullet\mathbf{Q}$.

A.2:   In the special case where $\mathbf{i} = (1,0,0)$ and $\mathbf{j} = (0,1,0)$, it is easy to verify that the product $\mathbf{i}\times\mathbf{j}$ equals $(0,0,1) = \mathbf{k}$. Thus, the triplet $(\mathbf{i},\mathbf{j},\mathbf{i}\times\mathbf{j} = \mathbf{k})$ has the handedness of the coordinate system. (It is either right-handed or left-handed, depending on the coordinate system.) In a right-handed coordinate system, the right-hand rule says: If your thumb points in the direction of $\mathbf{P}$ and your second finger in the direction of $\mathbf{Q}$, then your middle finger will point in the direction of $\mathbf{P}\times\mathbf{Q}$. In a left-handed coordinate system, a similar left-hand rule applies.

A.3:   They either point in the same direction or in opposite directions.

A.4:   We are looking for a vector $\mathbf{P}(t)$ that's linear in t and that satisfies $\mathbf{P}(0) = \mathbf{P}_1$ and $\mathbf{P}(1) = \mathbf{P}_2$. It is easy to guess that

$$\mathbf{P}(t) = (1 - t)\mathbf{P}_1 + t\mathbf{P}_2 = t(\mathbf{P}_2 - \mathbf{P}_1) + \mathbf{P}_1 \qquad \text{(Ans.7)}$$

satisfies both conditions. This equation can also be considered a weighted sum of $\mathbf{P}_1$ and $\mathbf{P}_2$ with the weights $1 - t$ and t. It is an important and useful relation often employed in graphics.

B.1: Yes, there are two and only two such extensions, octonions and sedenions.

Octonions are a nonassociative extension of the quaternions. They form an 8-dimensional normed division algebra over the real numbers. Octonions were defined and developed in 1843 by John T. Graves, an associate of William Hamilton, who referred to them as octaves.

Sedenions form a 16-dimensional algebra over the real numbers obtained by applying the Cayley-Dickson construction to the octonions. Like octonions, multiplication of sedenions is neither commutative nor associative.

Two references to these exotic mathematical objects are [Carmody 88] and [Carmody 97].

He always answered at extreme speed, as though the questions
were reflected instantaneously off the front of his head.

—C. P. Snow, *The Light and the Dark (1947)*

Bibliography

[Abelson and DiSessa 82] Abelson, H., and A. A. DiSessa (1982) *Turtle Geometry*, Cambridge, MA, MIT Press.

[anabuilder 05] `http://anabuilder.free.fr/Concurrents.html`.

[AnamorphMe 05] `http://myweb.tiscali.co.uk/artofanamorphosis/software.html`.

[anamorphosis 05] `http://www.anamorphosis.com/`.

[BeHere 05] `http://www.behere.com/`.

[berezin 06] `http://www.berezin.com/3d/slide_bar.htm`.

[Bourke 05] `http://astronomy.swin.edu.au/~pbourke/projection/fisheye/`.

[cameraInproduction 05] `http://www.panoramicassociation.org/IAPP_cameraInproduction.html`.

[cameraTimeline 05] `http://www.panoramicassociation.org/IAPP_cameraTimeline.html`.

[Carmody 88] Carmody, Kevin, (1988) "Circular and Hyperbolic Quaternions, Octonions and Sedenions," *Applied Mathematics and Computation*, **28**:47–72.

[Carmody 97] Carmody, Kevin, (1997) "Circular and Hyperbolic Quaternions, Octonions and Sedenions—Further results," *Applied Mathematics and Computation*, **84**:27–47.

[Coxeter 69] Coxeter, H. S. M. (1969) "Inversion in a Circle" and "Inversion of Lines and Circles" Sections 6.1 and 6.3 (pages 77–83) in *Introduction to Geometry*, 2nd ed., New York, John Wiley and Sons.

[davidlebovitz 05] `http://www.davidlebovitz.com/archives/2005/10/`.

[deeplight 06] `http://www.deeplight.com/`.

[eclipsechaser 05] http://www.eclipsechaser.com/eclink/astrotec/allsky.htm.

[Edmund Scientific 05] http://www.edsci.com/.

[eldonoffice 05] http://www.eldonoffice.com/rolodex/.

[Ernst 76] Ernst, Bruno (1976) *The Magic Mirror of M. C. Escher*, New York, Random House.

[fisheyemenu 05] http://www.cs.umd.edu/hcil/fisheyemenu/.

[flatearthsociety 05] http://www.theflatearthsociety.org/.

[Flocon and Barre 68] Flocon, Albert and André Barre (1968), *La Perspective Curviligne*, Flammarion.   English translation by Robert Hansen, 1987 *Curvilinear Perspective*, Berkeley, CA, University of California Press.

[funsci 05] http://www.funsci.com/fun3_en/panoram2/pan2_en.htm.

[Furuti 97] http://www.progonos.com/furuti/MapProj/.

[Gardner 84] Gardner, M. (1984) *The Sixth Book of Mathematical Games from Scientific American*, Chicago, IL, University of Chicago Press.

[Givens 58] Givens, Wallace (1958) "Computation of Plane Unitary Rotations Transforming a General Matrix to Triangular Form," *Journal of the Society for Industrial and Applied Mathematics*, **6**(1):26–50, March.

[Globuscope 05] http://everent.com/.

[handprint 06] http://www.handprint.com/HP/WCL/tech10.html.

[helloari 05] http://www.helloari.com/gallery/.

[hocg 06] http://hem.passagen.se/des/hocg/hocg_1960.htm.

[Holbein 05] http://www.nationalgallery.org.uk/cgi-bin/WebObjects.dll/CollectionPublisher.woa/wa/work?workNumber=NG1314.

[hulchercamera 05] http://www.hulchercamera.com/.

[Huntley 70] Huntley, H. E. (1970) *The Divine Proportion: A Study in Mathematical Beauty*, New York, Dover Publications.

> There is also a subset of RFCs called FYIs (For Your Information). They are written in a language much more informal than that used in the other, standard RFCs. Topics range from answers to common questions for new and experienced users to a suggested bibliography.
> —Brendan P. Kehoe, *Zen and the Art of the Internet* (1992)

[IAPP 05] http://www.panoramicassociation.org/.

[IPIX 05] http://www.ipix.com/products_realestate.html.

[Jacobs 05] Jacobs, Corinna (2005) *Interactive Panoramas: Techniques for Digital Panoramic Photography*, translated by J. Parrish, New York, Springer-Verlag.

[Jarvis 90] Jarvis, P. (1990) "Implementing CORDIC Algorithms," *Dr. Dobb's Journal*, 152–158, October.

[joebly 06] `http://www.joebly.com/7-Fisheye-Lawn.jpg`.

[Kaidan 05] `www.Kaidan.com`.

[Keith 01] Keith, Sandra (2001) "Dick Termes and His Spheres," *Math Horizons*, September.

[King 00] King, Ross (2000) *Brunelleschi's Dome: How a Renaissance Genius Reinvented Architecture*, New York, Walker and Company; London, Chatto and Windus.

[Krikke 00] Krikke, Jan (2000) "Axonometry: A Matter of Perspective," *IEEE Computer Graphics and Applications*, **20**(4):7–11, July/August.

[kspark 05] kspark (2005) is `http://kspark.kaist.ac.kr/Escher/Escher.htm`.

[lampshade 05] `http://www.philohome.com/lampshade/lampshade.htm`.

[Lindenmayer 68] Lindenmayer, A. (1968) "Mathematical Models for Cellular Interaction in Development," *Journal of Theoretical Biology* **18**:280–315.

[Manetti 88] Manetti, Antonio (1488) *The Life of Brunelleschi*, translated by Catherine Enggass, University Park, PA, Pennsylvania State University Press, 1970.

[Mathworld 05] `http://mathworld.wolfram.com/Quaternion.html`.

[mercator 05] `http://mathsforeurope.digibel.be/mercator.htm`.

[Mesdag Documentation Society 98] `http://www.mesdag.com/index.html`.

[Newbold 05] `http://dogfeathers.com/java/pulfrich.html`.

[New Perspective 98] Termes, Dick (1998) *New Perspective Systems*, published privately, `http://www.termespheres.com/perspective.html`.

[Noblex 05] `http://www.kamera-werk-dresden.de/`.

[Orosz 05] `http://www.geocities.com/SoHo/Museum/8716/images.html`.

[Pearson 90] Pearson, F. (1990) *Map Projections: Theory and Applications*, Boca Raton, FL., CRC Press.

[Penrose and Penrose 58] Penrose, L. S., and Roger Penrose (1958) "Impossible objects: A Special Type of Visual Illusion," *British Journal of Psychology*, **49**(1):31–33.

[Petersik 05] `http://www.stereofoto.de/index.html`.

[philohome 05] `http://www.philohome.com/panorama.htm`.

[photo3d 06] `http://www.photo3-d.com/#`.

[Prusinkiewicz 89] Prusinkiewicz, Przemyslaw (1989) *Lindenmayer Systems, Fractals, and Plants*, New York, Springer-Verlag.

[remotereality 05] `http://www.remotereality.com/`.

[roundshot 05] `http://www.roundshot.ch/`.

[Salomon 99] Salomon, David (1999) *Computer Graphics and Geometric Modeling*, New York, Springer-Verlag.

[Salomon 05] Salomon, David (2005) *Curves and Surfaces for Computer Graphics*, New York, Springer-Verlag.

[Sgrilli 33] Sgrilli, Bernardo Sansone (1733) *Descrizione e studi dell'insigne fabbrica di Santa Maria del Fiore metropolitana fiorentina*, Florence, Bernardo Paperini.

[shortcourses 05] `http://www.shortcourses.com/`.

[Snyder 87] Snyder, J. P. (1987) *Map Projections: A Working Manual*, US Geological Survey Professional Paper 1395, Washington, DC, US Government Printing Office.

[Snyder 93] Snyder, J. P. (1993) *Flattening the Earth*, Chicago, Ill, The University of Chicago Press.

[Steinhaus 83] Steinhaus, Hugo (1983) "Platonic Solids, Crystals, Bees' Heads, and Soap," Chapter 8 in *Mathematical Snapshots*, 3rd ed. New York, Dover, pp. 199–201 and 252–256.

[StereoGraphics 05] `http://www.stereographics.com/`.

[Stothers 05] `http://www.maths.gla.ac.uk/~wws/cabripages/inversive/` file `inversive0.html`.

[Swartzlander 90] Swartzlander, Earl E. (1990) *Computer Arithmetic*, Silver Spring, MD, IEEE Computer Society Press.

[Termes 80] United States patent 4,214,821, available at `http://patft.uspto.gov/netacgi/nph-Parser?Sect1=PTO2&Sect2=HITOFF&p=1&u=` `/netahtml/search-bool.html&r=10&f=G&l=50&co1=AND&d=ptxt&` `s1=4214821&OS=4214821&RS=4214821`.

[termespheres 05] `http://www.termespheres.com/perspective.html`.

[Thimbleby et al. 94] Thimbleby, Harold, Stuart Inglis, and Ian Witten (1994) "Displaying 3D Images: Algorithms for Single-Image Random-Dot Stereograms," *IEEE Computer*, **27**(10):38–48.

[Travis 90] Travis, A. R. L., (1990) "Autostereoscopic 3-D Display," *Applied Optics*, **29**(29):4341–4342.

[Travis 92] "Three-Dimensional Display Device," United States patent 5,132,839 issued July 21, 1992 to Adrian R. L. Travis.

[Vachss 87] Vachss, Raymond (1987) "The CORDIC Magnification Function," *IEEE Micro*, **7**(5):83–84, October.

[Vitruvius 06] Vitruvius (2006) *De Architectura*. The Latin text and English translation is at `http://penelope.uchicago.edu/Thayer/E/Roman/Texts/Vitruvius/home.html`.

Also see D. Rowland and T.N. Howe, *Vitruvius: Ten Books on Architecture*, Cambridge, Cambridge University Press, 1999.

[Volder 59] Volder, Jack E. (1959) "The CORDIC Trigonometric Computing Technique," *IRE Transactions on Electronic Computers*, **EC-8**:330–334.

[Walker 02] Walker, Paul Robert (2002) *The Feud That Sparked the Renaissance: How Brunelleschi and Ghiberti Changed the Art World*, New York, HarperCollins.

[Walther 71] Walther, John S. (1971) "A Unified Algorithm for Elementary Functions," *Proceedings of the Spring Joint Computer Conference*, **38**:379–385.

[wikipedia 05] http://en.wikipedia.org/wiki/Flat_earth.

[WikiQuaternion 05] http://en.wikipedia.org/wiki/Quaternion.

[Wolfram 06] http://mathworld.wolfram.com/topics/MapProjections.html.

[xahlee 05] http://www.xahlee.org/SpecialPlaneCurves_dir/CassinianOval_dir/cassinianOval.html.

"So few?" asked Frost. "Many of them contained bibliographies of books I have not yet scanned." "Then those books no longer exist," said Mordel. "It is only by accident that my master succeeded in preserving as many as there are."

—Roger Zelazny, *For a Breath I Tarry* (1966)

Index

This index reflects my belief that a detailed index is invaluable in a scientific/technical book. A special effort was made to include full names (first and middle names instead of initials) and dates of persons mentioned in the book. Any mistakes, inaccuracies, and omissions found in the index and brought to my attention will be included in the errata list and corrected in any future editions of the book.

How index-learning turns no student pale,
Yet holds the eel of science by the tail!

—Alexander Pope, *The Dunciad* (1743)

Colophon

The idea of writing this book originated during 2003–2004, when I noticed a number of new computer programs for creating panoramas from several individual overlapping images and for processing images in various ways. It became obvious that more and more computer users and digital camera owners were interested in manipulating images, which could justify a book of this kind. Most of the material was written during the second half of 2005. This material is based on my long experience with computer graphics methods, on text from my 1999 book on computer graphics, and on material found on the Internet. Chapter 3 has material from the history of art, but otherwise this book is mathematically oriented. The many inserts with quotations have been included to liven up the book and also to push the text up or down in order to improve the page breaks.

The book was designed and typeset by me in plain TeX (plus about 150 macros). The figures and diagrams were generated in *Adobe Illustrator*. The following numbers convey an idea of the amount of work that went into the book:

- The book contains about 101,500 words, consisting of about 600,000 characters.

- The text is typeset mainly in font cmr10, but about 30 other fonts were used.

- The raw index file has about 810 items.

- There are about 330 cross references in the book.

The difference between a mountain and a molehill is your perspective.
—Anonymous